D1085303

LIFE IN THE UNIVERSE

LIFE IN THE UNIVERSE

John Billingham, Editor

The MIT Press
Cambridge, Massachusetts
London, England

Second printing, 1982
First MIT Press edition, 1981

This book was printed and bound in the United States of America.

Library of Congress Cataloging in Publication Data

Main entry under title:
Life in the universe.

  Proceedings of the Conference on Life in the Universe, held at NASA Ames
Research Center, June 19-20, 1979.
  Includes bibliographies and index.
  1. Life on other planets—Congresses.
I. Billingham, John.   II. Conference on Life in
the Universe (1979: Ames Research Center)
III. Ames Research Center.
QB54.L483   574.999   81-15626
ISBN 0-262-52062-1 (paper)   AACR2
     0-262-02155-2 (hard)

Contents

IV DETECTABILITY OF TECHNOLOGICAL CIVILIZATIONS

V REFLECTIONS   419

Preface

Over the past twenty years, there has emerged a new direction in science, that of the study of life outside the Earth, or exobiology. Stimulated by the advent of space programs, this fledgling science has now evolved to a stage of reasonable maturity and respectability. Its central tenet, based on a wide range of studies in a score or more of scientific disciplines, is that it is probable that life has emerged and evolved not only on Earth, but in many other places in the Universe. We do not yet have any evidence of the existence of extraterrestrial life, but its existence is predicted from the following arguments.

Modern astrophysical theory predicts that planets are the rule rather than the exception. Planets are therefore likely to number in the hundreds of billions in our Galaxy alone. Given a suitable location and environment for any single planet, current theories of chemical evolution and the origin of life predict that life will begin. And given a period of billions of years of comparative stability on the planetary surface, life will sometimes evolve to the stage of intelligence. The next step may be the emergence of a technological civilization, and it is possible that civilizations may be in communication with each other. It is relevant to ask whether we are now in a position to detect them.

As exobiology developed, it was only reasonable that the early work should concentrate on our Solar System. Studies of chemical evolution and the origin of life were complemented by the search for life and life-related chemistry on the planets of the Solar System. These investigations, of course, continue and are the essence of exobiology today. They draw heavily and inevitably on the story of the origin and evolution of life on Earth.

More recently, it has become apparent that our horizons should properly be enlarged to begin to look beyond the Solar System, to the Galaxy, and to the Universe. What is the nature and distribution of Life in the Universe? — a simple question, even bold, and clearly a question whose answers will be most difficult to achieve. But it is a question that surely should be asked. And it is the question that stimulated us to convene a Conference on Life in the Universe.

This volume records the proceedings of that conference. The meeting was held at NASA Ames Research Center, June 19-20, 1979. It was designed to address the broad questions of life in a roughly chronological sequence, dealing in turn with present theories of the physical and chemical events in the story of cosmic evolution, with the environments in which life can originate and evolve, to the evolutionary patterns which allow the emergence of complex biological systems, including intelligent beings, and finally to the possibility that we may now be able to detect the existence of extraterrestrial intelligent life.

The Conference on Life in the Universe was made possible only through the efforts of many people. It is not possible to acknowledge everyone, but special recognition should go to Bob Frosch, Administrator of NASA, who gave the Introduction at the meeting and who has encouraged us in this venture; to Sy Syvertson and Tom Young, Director and Deputy Director at Ames Research Center, who have equally lent their support; to our Chairmen, Harold Klein, George Herbig, Mark Stull, and Frank Drake, whose remarks appear at the front of the four sections of this volume; to the many Ames scientists who have helped me with editing this volume; to Larry Cohen and Charles Seeger for editing the manuscripts and preparing the Glossary; to the Ames people, particularly Mark Stull, who organized and arranged the meeting; to Paul Bennett and his colleagues in the Technical Information Division at Ames — Beulah Gossett for editing the manuscript and Ken Atchley and Marianne Rudolf for their work on the illustrations; to Charlotte Barton, Mary Jessup Young, and Al McCahon of Management Systems Associates for preparing the camera-ready copy; to Vera Buescher and Lorraine Bendik for pulling everything together at every stage; and last but not least to the authors of the wide range of presentations given at this meeting, whose papers are presented here and whose interest in life in the Universe was responsible, sine qua non, for the success of the Conference.

John Billingham
Chief, Extraterrestrial Research Division
Ames Research Center, NASA

Conference Overview

LIFE IN THE UNIVERSE as a title for a conference is both suggestive and debatable. We know of only one existence of life in the Universe — that being ourselves on planet Earth. We know that our Earth is an enormously small part of our Universe. A perplexing question evolves as to whether life abounds or are we unique. Attacking this question is not easy, and we are left with two distinctly different, though complementary, approaches.

The first is indirect and deals with developing an understanding of the origin of life and environments necessary to support life. Armed with some understanding in these areas, one looks to the Universe to study chemical characteristics and physical conditions. Combining our knowledge concerning the origin of life, life supporting environments, and the chemical character and physical nature of our Universe, we can develop theories concerning the evolution of complex life in our Universe. The first three sessions of this Conference focused on the indirect approach. This approach has strengths which were effectively illustrated by the Conference. There is a constant feedback as we stretch ourselves to consider life in the Universe, to further develop our views on the origin of life and the characteristics of life supporting environments. Additionally, we are continually striving to better understand our Universe. While we recognize the indirect approach will never uniquely determine the existence of life in the Universe beyond Earth, the pursuit will result in enormous growth in our understanding of life, its origin, and the necessary conditions for its existence.

The fourth session reviewed the direct search approach, referred to as the Search for Extraterrestrial Intelligence (SETI). It involves listening for potential radio transmissions from intelligent life somewhere in the Universe. The approach includes both a survey and the examination of suitable stars as targets. A research program has been established to begin a modest, direct search for life in the Universe.

The Conference very effectively dealt with the indirect and direct approaches to studying life in the Universe. Some people ask why such studies should be performed. The answers are numerous, varying from the scientific contributions that will result from the pursuit to the more intangible benefits derived from pondering our place in the vastness that surrounds us. The idea of a conference on life in the Universe is a tribute to the future. It demonstrates a faith, confidence, and concern about the future that is critically needed as we face the demands of the present.

A. Thomas Young
Deputy Director
Ames Research Center, NASA
(now Director of Goddard
Space Flight Center, NASA)

Introduction

When I was invited to attend this meeting, I decided to come for two reasons, one being to demonstrate by coming to talk to you that I think this subject is of central importance to the NASA job. My second reason, which is always a problem to the formal system, is that I wanted to come to the meeting to hear the papers because of my interest in the subject. As it happens, the system will not let me escape completely, so I will get to hear papers today, but not tomorrow.

We are in a unique time and in a unique situation. All generations feel that is the case. I think we can make the statement that for the first time we can see what the sketch of the connection between the origin of the universe and ourselves may be. As I understand the situation, it is only a sketch. There are lots of gaps in the chain of connection between the first instants of the universe, the formation of matter, the whole evolution of the universe, the galaxies, the stars, the solar system, the planets, the organic molecules and so on to ourselves. But at least we can say we have a sketch. We have pieces that we think we understand and there are pieces that we do not understand.

There are perhaps two gigantic missing pieces in this chain where I think it is fair to say we have hardly a clue. One is at the end nearest ourselves when we contemplate the nature of consciousness; the other is at the very other end where we contemplate the first instants of the universe. At best we may simply have reduced the whole problem to a lack of understanding of the beginning, which I suppose could be described as reducing it to a previously unsolved problem.

At least we have the satisfaction of knowing that a quest for understanding of ourselves has now become more than a blind groping for ideas. It has become a matter of connecting ideas from a distant past through a set of processes of which we have a glimmering and some knowledge to a present, and looking on beyond that into some distant future that we can only imagine. Our job now is to begin to test this chain of logic by looking at the universe and to ask seriously: "If the chain of logic is correct, is it only us?" If it is not only us — and that is what the chain of logic implies — then how can we not seek our siblings? How can we not find out who else is part of the same chain of logic? If it should turn out that it is only us, then a very much greater problem is posed for us — how come? And what does that mean for the tasks that we have to set ourselves next? If the task is to find an understanding of ourselves, and if our understanding of origins implies others, surely we must find the others; but if there are no others, then we have the problem of generating our own next tasks.

All of this understanding comes at a very difficult time for attempts to understand. It comes at a time in which we seem to have a faltering in global and national interest in knowledge for its own sake. We have become hyperpractical and are expected to explain the use of things we do not understand, before we understand them.

It is a time of the "golden fleece" for SETI, and I presume it will be a time of golden fleeces for other things we try to do. The "golden fleece" idea, the idea that searches, gropings for knowledge whose purpose we do not understand are silly and some kind of a ripoff, results from sheer lack of understanding, lack of imagination, and lack of perception of the meaning of the history of the human race.

It is a lack of understanding of the nature of the increase of knowledge, whether immediately practical knowledge or not, and a lack of understanding of the intellectual risk-taking which is an essential part of any groping for knowledge. In any noisy system the only way in which we can be sure of making no false alarm errors is to turn the system off. There can be no detections if there is no possibility of false alarm. To grope we must have the possibility of error, the whole game is making errors, finding them, and disposing of them.

The second strain of trouble with which we battle in doing this work is the idea that the universe has been around for a long time, so let us postpone the expensive effort to understand it until we have saved the human race. The question is again backward — a misinterpretation of the nature of our history as a race. The question is not "how can we pursue knowledge until we have saved the race?" The historical question is more "how and why should we bother to save the race unless we seek to understand where we came from, who we are, and what the universe around us is about?"

The very process of saving ourselves is imbedded in and indissolubly tangled up with the process of understanding ourselves and what is around us. The cathedral of knowledge of self cannot be built only from the inside. It must also be built by testing what is built inside against what is outside. Our inner perception must somehow be congruent with the outer fact of the universe in which we are imbedded. What we are doing is not seeking, as is sometimes suggested, only power, technological capability, and what is sometimes described as bloodless scientific knowledge. Scientific knowledge is tied up intimately with our internal attempts to search for ourselves and to understand ourselves because it provides the true context for what we are trying to understand.

I look forward to participating in this meeting because I think it is of profound importance scientifically, socially, and philosophically to pursue this subject. I think we are at the point where we must seriously talk about how we do experiments and seek success; we are beyond the point where we can only talk about how we grope. I look forward to the next steps.

Dr. Robert A. Frosch
NASA Administrator
(now President, American Association
of Engineering Societies)

Contributors

Duwayne M. Anderson
State University of New York
Buffalo, NY

Gustaf Arrhenius
Scripps Institution of Oceanography
La Jolla, CA

John Billingham
Extraterrestrial Research Division
NASA Ames Research Center
Moffett Field, CA

David C. Black
NASA Ames Research Center
Moffett Field, CA

Ronald N. Bracewell
Stanford University
Stanford, CA

Bernard Campbell
L. S. B. Leakey Foundation
Pasadena, CA

Eric J. Chaisson
Harvard University
Cambridge, MA

Sherwood Chang
NASA Ames Research Center
Moffett Field, CA

Jung Chen
University of Hawaii
Honolulu, HA

Martin Cohen
University of California
Berkeley, CA

R. Bruce Crow
Jet Propulsion Laboratory
Pasadena, CA

Thomas M. Donahue
University of Michigan
Ann Arbor, MI

Frank D. Drake
Cornell University
Ithaca, NY

Robert E. Edelson
GTE Laboratories
Waltham, MA

Samuel Gulkis
Jet Propulsion Laboratory
Pasadena, CA

Robert S. Harrington
U.S. Naval Observatory
Washington, DC

George H. Herbig
University of California
Santa Cruz, CA

Donald M. Hunten
University of Arizona
Tucson, AR

Kenneth Janes
Boston University
Boston, MA

James F. Kasting
University of Michigan
Ann Arbor, MI

Harold P. Klein
NASA Ames Research Center
Moffett Field, CA

C. Owen Lovejoy
Kent State University
Kent State, OH

James E. Lovelock
Reading University
Reading, England

Lynn Margulis
Boston University
Boston, MA

Philip Morrison
Massachusetts Institute of
  Technology
Cambridge, MA

Bernard M. Oliver
Hewlett-Packard Corporation
Palo Alto, CA

Edward T. Olsen
Jet Propulsion Laboratory
Pasadena, CA

Allen M. Peterson
Stanford University
Stanford, CA

Alexander Rich
Massachusetts Institute of
  Technology
Cambridge, MA

Dale A. Russell
National Museum of Natural
  Sciences
Ottawa, Canada

Stephen H. Schneider
National Center for Atmospheric
  Research
Boulder, CO

Charles L. Seeger
San Francisco State University
San Francisco, CA

B. Z. Siegel
University of Hawaii
Honolulu, HA

S. M. Siegel
University of Hawaii
Honolulu, HA

Mark A. Stull
Extraterrestrial Research Division
NASA Ames Research Center
Moffett Field, CA

Woodruff T. Sullivan III
University of Washington
Seattle, WA

Jill C. Tarter
University of California
Berkeley, CA

Starley L. Thompson
National Center for Atmospheric
  Research
Boulder, CO

Kenneth M. Towe
Smithsonian Institution
Washington, DC

Wallace H. Tucker
Harvard-Smithsonian Observatory
  and U.S. International University
Bonsall, CA

Karl K. Turekian
Yale University
New Haven, CT

James W. Valentine
University of California
Santa Barbara, CA

John H. Wolfe
NASA Ames Research Center
Moffett Field, CA

Simon P. Worden
Space Division
Los Angeles Air Force Station
Los Angeles, CA

LIFE IN THE UNIVERSE

Three Eras of Cosmic Evolution

ERIC J. CHAISSON

This conference-opening paper sketches the global hypothesis that permeated the discussions at this interdisciplinary meeting, the purpose of which was to explore prospects for research into the nature and distribution of life in the Universe.

A friend who is a high-energy physicist once suggested that everything of importance happened within the first few minutes of the Universe. All subsequent events, he claimed, can be regarded as mere detail. Many scientists would regard my friend's view as provincial. The relatively simple subatomic matter may have been created in the first moments of the Universe, but the more complex organized matter now surrounding us must have formed well after its start. Every dating technique developed by post-Renaissance science suggests that complexity steadily arises from simplicity, order from chaos.

Granted, the initial coagulation of matter from otherwise chaotic radiation shortly after the Universe flashed into existence was an event of incomparable significance. This emergence of matter as the dominant constituent is the first great transformation in the history of the Universe. But a second great transformation occurs when technologically competent, intelligent life emerges from that matter. Our civilization on Earth is now on the threshold of this second transformation.

EARLY UNIVERSE

To place the construction of all matter into perspective, consider figure 1, which summarizes the run of density and temperature throughout all time for a Big-Bang Universe. It represents the consensus of contemporary

1

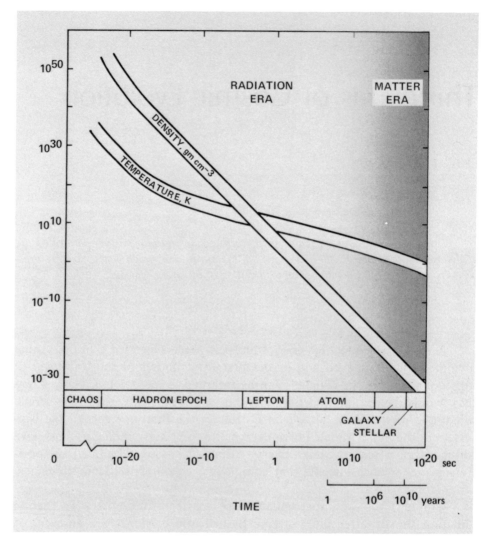

Figure 1. *Time variation of density and temperature for the Big-Bang model of the Universe.*

scientific thought in the broadest sense. Six major epochs are delineated, each corresponding to a major period in the history of the Universe. Specified across the bottom of this figure are the general names of the epochs, along with their time domains. Note that this plot is highly nonlinear, stretching from an incredibly small fraction of a second to the present time, 18 or so billion years after the origin of the Universe. The curves depict the *average* density and the *average* temperature of everything in the Universe at any point in time.

This figure suggests that in the beginning there was chaos. One cannot inquire about what happened at the exact moment of the Bang (precisely zero time). Some theorists argue, however, that it is possible to characterize the physical conditions at some extraordinarily small time after the Bang. For example, the currently known laws of physics specify a Universe younger than 10^{-23} sec to be characterized by an average density greater than 10^{50} gm/cm^3 and an average temperature greater than 10^{30} K. Of course, it is virtually impossible to appreciate such youth, for 10^{-23} sec is the amount of time it takes light to cross a proton. Equally difficult to comprehend are the large densities and temperatures characterizing this earliest epoch. The composition of the Universe at this time was indescribable, and its dominant action unimaginable.

The second major epoch is a bit closer to our limits of comprehension, although it is still characterized by severely nonterrestrial conditions — the *hadron epoch*. The name is derived from the fact that the heavy elementary particles such as protons, neutrons, and mesons, which were the most abundant type of matter at the time, are collectively known as hadrons. Calculations suggest that such particles existed as free unbound entities, considering the high temperature prevalent in the Universe well within its first second of existence. The hadrons unquestionably collided and interacted with one another and with other types of elementary particles, for the density was also extreme. The dominant action at this time is presumed to be the self-annihilation of hadrons into high-energy photons, thus creating a brilliant fireball of radiation. Lacking a solid understanding of elementary particle physics, scientists presently know very little more about this mystifying period.

As the Universe continued its rapid expansion, its contents cooled. A variety of models suggest that about a millisecond after the Bang, the conditions suitable for hadron annihilation had nearly subsided, thus allowing the initially less abundant, lighter elementary particles such as electrons, neutrinos, and muons to predominate. The average density and temperature of this *lepton epoch* had decreased to about 10^{10} gm/cm^3 and 10^{10} K. These physical conditions are still excessive by terrestrial standards, but they had diminished considerably compared to the chaotically dense and hot conditions extant a fraction of a second earlier. By the time the first second had elapsed, the leptons were self-annihilating into photons, much as had the hadrons earlier. The radiative fireball of the cosmic bomb was still being fed with new photons.

The radiation density exceeded the matter density by a large amount in these first few minutes; photons of radiation far outnumbered particles of matter. As soon as the elementary particles of matter began to coagulate, fierce radiation destroyed them. For this reason, the first three epochs are often collectively referred to as the *radiation era*. Whatever matter existed was merely an inconspicuous precipitate suspended in a sea of dense, brilliant radiation.

LATER EPOCHS

The fourth epoch — the *atom epoch* — extends in time from about 100 sec to about a million years after the Bang. Midway through this epoch, the average density had decreased to about 10^{-10} gm/cm^3, while the average temperature had fallen to about 10^6 K — values not terribly different from those in the atmospheres of stars today. A principal feature of the atom epoch was the gradual diminution of the original fireball, for the annihilation of hadrons and leptons had all but ceased.

Toward the beginning of the atom epoch, radiation still reigned supreme over matter, for the Universe remained flooded with photons. As the Universe expanded, however, the photon density decreased as the fourth power of the radius of the Universe, while the matter density decreased only as the third power. The early dominance of radiation thus gradually diminished. Sometime between a few minutes and a million years after the Bang, the charged elementary particles of matter were able to coagulate electromagnetically without being broken apart by radiation as quickly as they combined. This was a most important transformation in the history of the Universe. The dominance of radiation had subsided, for matter had gradually become neutralized, a physical state over which radiation has little leverage. Matter had, in a sense, overthrown the cosmic fireball. Henceforth it would dominate radiation as the principal constituent of the Universe. To denote this major turn of events, the last three epochs in figure 1 are collectively known as the *matter era*.

Once the matter era began, atoms appeared. The influence of radiation had grown so weak that it could no longer prohibit the joining of the leptons and hadrons that had survived annihilation. Hydrogen was the first element to form, since it required only that single electrons be electromagnetically joined to single protons. Copious amounts of hydrogen were synthesized in the early Universe, and it is thus the common ancestor of all things.

Hydrogen was not the only kind of atom formed early in the matter era. Indeed, at the start of the atom epoch, the average temperature of the Universe still exceeded the 10^7 K necessary to fuse two hydrogen atoms into helium via the proton-proton cycle. The Universe was cooling, but it took time for the average temperature to dip below this critical value. Consequently, some helium atoms must have been produced within the primordial fireball in the same way that they now form in the interior of stars.

Elements heavier than helium, on the other hand, could not have been produced in the early Universe. The synthesis of such elements requires temperatures even greater than 10^7 K. It also requires lots of helium atoms, for the heavier elements are constructed from lighter ones. The basic difficulty here is that, even though helium atom production was in high gear during the start of the atom epoch, the average temperature was falling quickly. Theoretical calculations suggest that, by the time there were sufficient helium atoms to interact with one another to produce the heavier elements, the temperature had fallen below the threshold value ($\sim 10^8$ K) required for the mutual penetration of doubly charged helium nuclei. In contrast to the rapid cooling of the early Universe, the dense interiors of stars in the present Universe are perfectly suited for the generation of hotter temperatures and thus heavier elements. The guts of stars are indeed where the heavies were created — and where they are still being created.

By the end of the atom epoch, matter was in firm control. Sometime during the fifth or *galaxy epoch*, gravity began to pull some of this matter together into enormous clumps. Galaxies were beginning to form. Indeed, they all must have originated long ago, for observations imply that no galaxies have formed within the past 10 billion years or so. Each galaxy contains substantial numbers of old stars, in addition to an often abundant complement of young stars. The quasars and remote galaxies must have formed in the earliest parts of this epoch.

The time scale of the last two epochs shown in figure 1 has been compressed enormously. An important and rapid series of events occurred immediately after the Bang, especially in the first few minutes that constitute the radiation era. However, once the Universe cooled sufficiently to allow atoms to form, subsequent events occurred more slowly.

By the middle of the galaxy epoch, the average density of the Universe had decreased by another factor of 10 billion, to 10^{-20} gm/cm^3. The average temperature of the entire Universe had also diminished to a relatively cool 3000 K. The Universe was becoming thinner, colder, and darker.

Finally, there is the present *stellar epoch*. Scientists can say with some assurance that it has been at least 10 billion years since the Bang. In fact, the Universe is probably older than that, perhaps as old as 18 billion years, although its precise age depends on the yet-to-be-determined change of the Hubble constant with time. The present average density is approximately 10^{-30} gm/cm^3, the critical value above which the Universe will eventually contract and below which the Universe will expand forever. The average temperature of everything in the Universe is presently 3 K. This then is the cooled relic of the incredibly hot fireball that existed eons ago, the fossilized grandeur of a bygone era.

The dominant action of the stellar epoch is the formation of stars, intermediate in size between atoms and galaxies. Research during the past several years has provided direct observational evidence that stars are actually forming within galaxies. Galaxies themselves are not forming in the present epoch, but stars within them apparently are — 18 billion years after the Bang.

An interesting by-product of star formation is the associated coagulation of matter into planets, life, and intelligence.

This history of the Universe is the prevailing view among most cosmologists. All theoreticians do not agree on specific events before about 1 sec. Depending on the intricacies of the model chosen, the density and temperature during the radiation era can change by several orders of magnitude. In virtually all models, however, the Universe is regarded to have been initially very hot and dense, after which it cooled and thinned.

COSMIC EVOLUTION

The history of the Universe can be viewed in another way, one that follows a more linear time scale. Figure 2 shows the arrow of time, along which are marked several important developments in the history of the Universe. Known popularly as the scenario of cosmic evolution, it links the development of galaxies, stars, heavy elements, life, intelligence, technology,

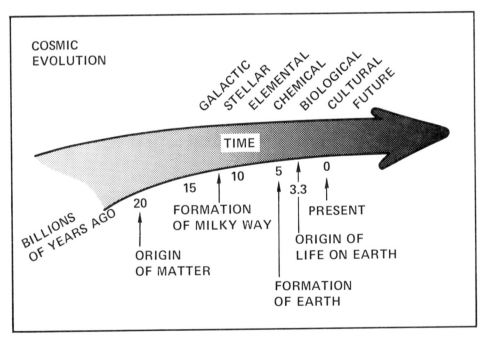

Figure 2. *Diagrammatic representation of the scenario of cosmic evolution.*

and the future. This diagram highlights the grand synthesis of a long series of gradual alterations of matter, operating over almost incomprehensible space and time, that have given rise to our Galaxy, our Sun, our planet, and ourselves.

Cosmic evolution is the study of the seemingly endless changes in the composition and assembly of various aggregates of matter and life throughout the Universe. It attempts to demonstrate that a clear thread links the evolution of simple atoms into galaxies and stars, of stars into heavy elements, of those elements into the molecular building blocks of life, of those molecules into life, of life into intelligence, and of intelligent life into culture and technology.

The scenario is supported handsomely by legions of experimental tests in physics, chemistry, biology, astronomy, geology, anthropology, neurology, sociology, and a spectrum of other disciplines. This support is general, however. We do not yet know all the details. Nor do we know much about either the starting or end points — the origin and destiny of the Universe. We do know that the Universe is not static: it is changing with time — it is evolving.

The scenario of cosmic evolution is a human invention. Despite seven major construction phases, it was not handed to us on a granite slab. Accordingly, it is subject to change as research progresses. As it stands now it is a broad guide to an understanding of the time evolution of matter, based on every available dating technique — not just methods utilizing radioactivity and fossilized life forms, but also self-consistent methods enabling us to date astronomical objects throughout the observable Universe.

Cosmic evolution stipulates that complexity arises from simplicity. It seems straightforward enough: light, quarks, atoms, stars, planets, life, intelligence — an entire hierarchy of material coagulations from radiation, to matter, to life. Yet this increase in complexity over time bothers some researchers because it seems to violate the second law of thermodynamics, which dictates that entropy (or disorder) should be increasing everywhere. Why should organization arise naturally from simplicity? In other words, why does entropy seem to decrease at certain selected locations within a universe where it is otherwise increasing? Frontier research suggests that the answer concerns the extent to which a system departs from thermodynamic equilibrium. A living organism, for example, is an open or unstable system, not in equilibrium with its environment. It resembles a heat engine that also concentrates energy. Consequently, life can construct and order itself by exchanging energy with the outside. Recent advances at the boundary between physics and chemistry suggest that classical thermodynamics, which predicts strict adherence to the second law, is restricted to systems in or near thermal equilibrium. These are closed systems, and their contents do indeed become disordered with time. Far from equilibrium, however, no system is stable, and this instability can lead to the occasional emergence of ordered macroscopic structures. It would seem then that the existence and organization of galaxies, stars, planets, and life are the result of energy having been captured by material systems far from equilibrium. Generally, destruction of structures occurs when they are near equilibrium, while construction of structures may occur when they are beyond some stability threshold.

SOME MISSING LINKS

Cosmic evolution is a broad working hypothesis that attempts to integrate all that is known into an overall framework of understanding. However, several of the details within that framework remain to be unraveled. These are important details, for without a specific understanding of each of the major evolurionary events, we can never hope to comprehend this all-encompassing view of our Universe.

Curiously, it seems that valuable insight into many of these unsolved, largely cosmic problems can be gained by adopting a highly interdisciplinary approach and studying phenomena almost completely out of context. Consider a few examples.

Galaxy Formation

The origin of the galaxies may constitute the biggest missing link in the entire scenario of cosmic evolution. Conditions at the present epoch of the Universe seem entirely inappropriate for the formation of galaxies. No observer has ever unambiguously reported evidence for galaxies forming at the present epoch, and no theorist can realistically suggest how they might do so given the present temperature and density throughout the Universe. Clearly, the hotter gas, more intense radiation, and greater turbulence of the early fireball were more conducive to galaxy formation; but specifically how they formed remains a mystery.

Contemporary researchers approaching this problem usually begin with the complex subject of hydrodynamics and examine the fate of density inhomogeneities in a turbulent medium. Since Earth's weather is a good example of turbulent gas flow, it is not inconceivable that studies of terrestrial phenomena may help us understand this extraterrestrial problem. Figure 3(a) shows the kilometer-sized swirling eddies that appear and disappear at random within Earth's atmosphere. These enhanced fluctuations in gas density become pronounced whenever air currents are particularly turbulent. Once in a while such an eddy can accumulate large quantities of moist air and grow into a full-fledged hurricane hundreds of kilometers across (fig. 3(b)). Although on a much smaller scale, this terrestrial phenomenon roughly mimics the overall morphology, the pancake shape, the differential rotation, and the concentration of energy within spiral galaxies (fig. 3(c)), which are thought to have formed by accumulating gargantuan quantities of hydrogen gas. Of course, meteorological conditions on Earth have no direct bearing on the origin of galaxies, but their several resemblances suggest that something can be learned about galaxy formation through the study of hurricane formation. In particular, since most meteorologists agree that some sort of turbulent "priming" is required to initiate a hurricane, the early stages of such storms could conceivably be used by astronomers to derive clues about the elusive density fluctuations that gave rise to protogalaxies in the early Universe.

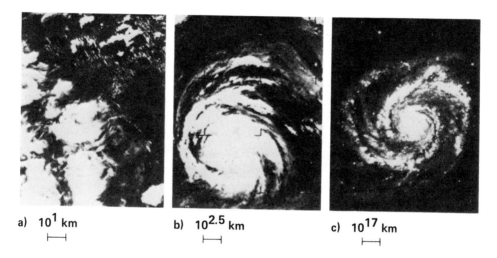

a) 10^1 km   b) $10^{2.5}$ km   c) 10^{17} km

Figure 3. *(a) A Skylab photograph of the top of Earth's cloud layers, showing the presence of several atmospheric eddies near the Canary Islands. (b) A full-scale hurricane is a collection of moisture hundreds of times larger than an atmospheric eddy. In 1967, Hurricane Beulah was photographed by one of the ESSA satellites hovering over the Gulf of Mexico. (c) A spiral galaxy is a collection of mostly gas, usually more than a trillion times larger than a hurricane (M51; Harvard Observatory).*

Origin of Life

Clues to the physical and chemical conditions on primordial Earth might similarly be gleaned from the study of cell-like organic ensembles synthesized under laboratory conditions. The production of some amino acids and bases from a primordial mixture of ammonia, methane, water, and energy has been known for some time. More sophisticated experiments in recent years have shown that repeated heating and cooling of simple organic molecules can yield spherical droplets that contain large concentrations of complex polymers. These are not proteins as we know them, but simpler proteinlike linkages of amino acids.

Figure 4 is a photomicrograph of a few of these so-called proteinoid coagulations. Although there is some dispute regarding the relevance to primordial Earth conditions of the laboratory experiments used to produce these proteinoids, the coagulations do seem to resemble morphologically some of the most ancient microfossils as well as modern blue-green algae cells. These curious chemical proteinoids appear to possess many of the attributes of bona fide living organisms: they are cell-like spheres a few microns across, each possessing a thick shell-like membrane; most even appear to exhibit a primitive metabolism, with some dissipating away while others swell and bud (fig. 4). This is not to suggest that the proteinoids should be in any way associated with life itself. Rather, because their physical and chemical properties so closely mimic those of procaryotic cells, the suggestion here is that researchers ought to be able to glean some insight into the ways and means of chemical evolution through further study of these proteinoid globules, despite the controversy over the appropriateness of the initial laboratory conditions. The laboratory proteinoids may be as far removed from real life as hurricanes are from galaxies. But their overall morphology and their microscopic kinetics suggest some resemblance to whatever were the progenitors of living organisms on Earth. Laboratory studies of such protocells may someday have reverse usefulness by demonstrating what Earth was like some 4 billion years ago.

Chemical Evolution

As a third example of how one research specialty might be taken somewhat out of context to study a seemingly unrelated problem of cosmic evolution, consider chemical evolution. The study of interstellar molecules may well yield some insight into the origin of life on Earth. This is not to suggest that interstellar molecules have any direct bearing on the onset of terrestrial life, but studies of galactic clouds could allow us to recover information about the early stages of chemical evolution lost forever on our planet.

Dark and dense interstellar clouds (fig. 5) may in fact rank on par with the Jovian atmosphere as the best place to study chemical evolution. These clouds are rich in a variety of small molecules such as carbon monoxide (CO) and formaldehyde (H_2CO), compounds typically five orders of magnitude less abundant than molecular hydrogen (H_2). These and other molecules are

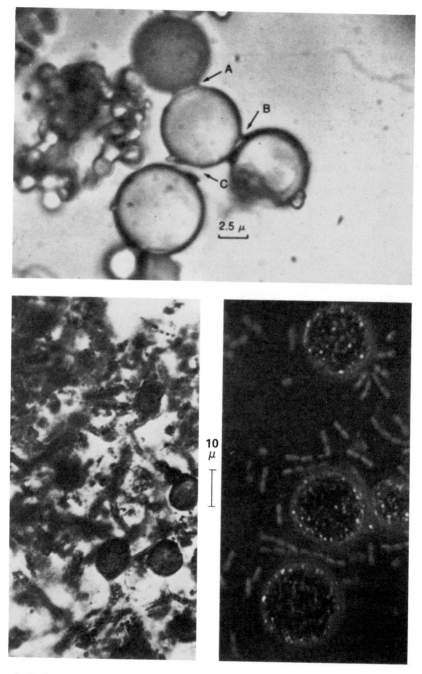

Figure 4. *A photomicrograph of proteinoid spheres, each containing a large concentration of amino acids (upper frame; from the research of S. W. Fox and his associates). Very old fossils, dated to be about 3 billion years old (left bottom frame; from the research of E. S. Barghoorn and his associates). Simple blue-green algae cells found almost anywhere on Earth (lower right frame).*

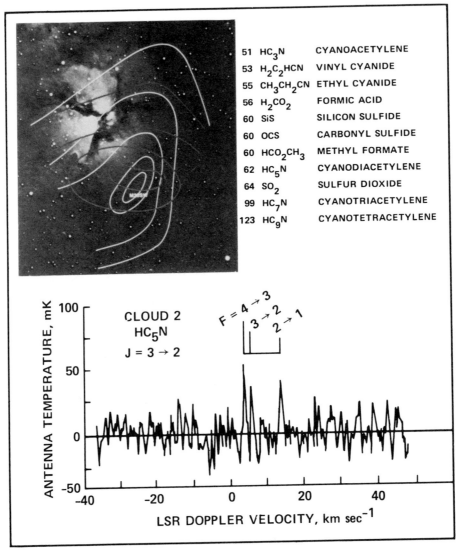

Figure 5. *Typical dark, dense, and dusty interstellar clouds can be seen outside the glowing Trifid Nebula. The contours show the distribution of several molecules within a particularly rich cloud known as M20SW (left top frame; photograph from Harvard Observatory; contours from author's research). Nearly a dozen interstellar molecules having masses greater than 50 atomic mass units had been observed by mid-1979 in the interstellar clouds of our Galaxy (right top frame). The unique hyperfine spectral features of the HC_5N molecule were observed at 8 GHz toward an interstellar cloud with the 1000-channel spectrometer of the Haystack Observatory (bottom frame; unpublished data by author and his associates).*

invariably found in regions containing large concentrations of dust, suggesting that dust plays either the role of catalyst in the formation of the molecules or the role of protector once the molecules form by some other mechanism. In addition, spectral lines characteristic of much heavier molecules have been detected in localized patches of these giant molecular clouds, which routinely span tens, sometimes hundreds, of light years. For example, about a dozen interstellar molecules are known that have a molecular weight exceeding 50 atomic mass units. These include many of the familiar products of the laboratory simulations of primordial Earth conditions: cyanoacetylene (HC_3N), formic acid (H_2CO_2), and several others listed in figure 5, presently topped off by cyanotetraacetylene (HC_9N). More than 50 interstellar molecules have been identified, and nearly 200 more as yet unidentified features have been observed, mostly in the millimeter-wave spectrum of interstellar clouds.

The greatest significance of these heavy molecules is that they exist in space. Apart from this, they are also significant because of their unexpectedly large abundance, often within 8 to 10 orders of magnitude (by number) of H_2. The observed spectra leave little doubt about their identification or their relative abundances. For instance, figure 5 shows a recently acquired high-resolution spectrum of the hyperfine transitions of the HC_5N molecule. Because the observed line strengths agree with the quantum-mechanical predictions for spontaneous emission, the measured intensities cannot be appreciably amplified by masing or other non-LTE processes. The very fact that such signals are detectable suggests that these heavy molecules are far more abundant than anyone would have guessed even after the rash of discoveries of interstellar molecules began about a decade ago.

While the consensus still maintains that the larger molecules are probably constructed from smaller atoms and molecules already extant in interstellar space, there is at present no satisfactory formation mechanism for the large organic coagulations now found there. Consequently, some researchers are beginning to consider seriously the possibility that some interstellar molecules could result from destruction rather than construction; that is, many of the interstellar molecules now observed may be fragments torn from much larger molecules yet to be detected. A statement once offered as a lark, namely, that the enigmatic interstellar dust grains have the same dimensions as virus particles, may in the end turn out to be prophetic. If so, the interstellar cloud could become the key to our understanding not only of the early stages of chemical evolution, but of the advanced stages as well.

Other Missing Links

It is not inconceivable that we could learn more about several of the other missing links of cosmic evolution by studying phenomena normally considered outside the realm of traditional investigation. New insight into the origin of our solar system is now being provided by radio and infrared observations of protostellar regions scattered throughout the Milky Way Galaxy. Major advances concerning the origin of human intelligence have been made by studying the behavior and learning abilities of the great apes. And the physical conditions close to the birth of the Universe itself may someday be appreciated by studying the death of such supermassive objects as black holes.

LIFE ERA

It is hard to condense 18 billion years of history into a few paragraphs. Figure 6 shows the broadest view of the largest picture. Radiation dominated matter in the earliest epochs of the Universe. The enormous number of

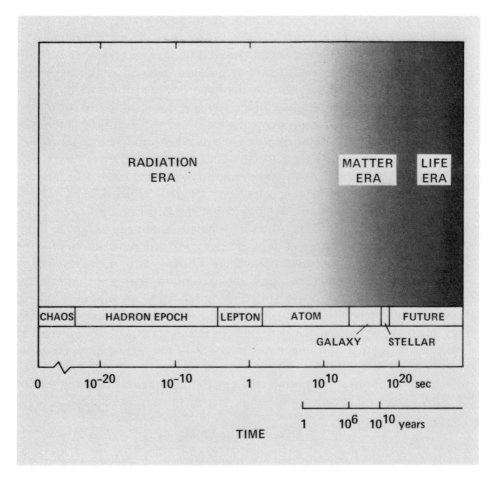

Figure 6. *Three principal eras of cosmic evolution.*

photons, and particularly the scattering of photons by the electrons, produced a fireball inside of which no atoms or molecules could have formed.

Slowly, as the Universe expanded, it cooled. Matter gradually began to coagulate into atoms and eventually into clusters of atoms. From the start of the matter era, matter dominated radiation, and it has dominated radiation ever since, successively forming galaxies, stars, planets, and life.

Life forms are most interesting pieces of matter, especially technologically intelligent ones. One can argue that technologically intelligent life is fundamentally different from lower forms of life and other pieces of matter scattered throughout the Universe. It is fundamentally different because it can tinker not only with matter but also with evolution. For example, whereas previously the gene (i.e., DNA) and the environment (be it stellar, planetary, geological, or sociological) had governed evolution, now we on Earth are suddenly gaining control of both the gene and the environment. We are now tampering with matter, diminishing the resources of our planet, often polluting it. And we are now on the verge of tampering with life, potentially altering the genetic makeup of human beings.

The emergence of technologically intelligent life heralds a whole new era — a *life era* — as suggested in figure 6. Technology enables life to begin to control matter, much as matter grew to dominance over radiation tens of billions of years ago. Matter is now losing its total dominance, at least at those isolated locations where technologically intelligent life resides.

The transformation from a matter era to a life era will not be instantaneous. Just as it took time for matter to dominate radiation in the early Universe, it will surely take a great amount of time for life to dominate matter. And, in fact, such domination may never be total, either because civilizations may never control resources on a truly galactic scale or because the longevity of technological civilizations may be inherently small. But one thing seems certain: we on Earth, as well as other intelligent life forms throughout the Universe, are now participating in a fantastically important transformation — the second most important transformation in the history of the Universe.

We now stand on an enormously significant threshold. We have come full cycle. We have become smart enough to reflect back upon the material contents that gave life to us. Life now contemplates life. It contemplates matter. It ponders its own origin and destiny. It explores the planetary system we call home. It searches for extraterrestrial life. It quests for new knowledge.

Provided civilizations remain curious, provided they are wise enough to survive, then it is not inconceivable that life could evolve sufficiently to overwhelm matter, just as matter overwhelmed radiation in the early Universe. Indeed, the destiny of matter in the Universe may well be controlled in part by the life that arose from it. Together with our galactic neighbors, should there be any, we may be in a position someday to gain control of the resources of much of the Universe, rearchitecturing it to suit our purposes and, in a very real sense, ensuring for our civilization a measure of immortality.

ADDITIONAL READING

Chaisson, E.: Cosmic Dawn: The Origins of Matter and Life. Atlantic, Little-Brown, Boston, 1981.

Prirogine, I.: From Being to Becoming: Time and Complexity in the Physical Sciences. W. H. Freeman, San Francisco, 1980.

Weinberg, S.: The First Three Minutes: A Modern View of the Origin of the Universe. Basic Books, New York, 1977.

I — Origin of Life

In his introductory remarks to the Conference, Tom Young told us of his faith, his conviction, that someday we will "find life out there." It is not unlikely, however, that we may not. And if we do not, we still have to acknowledge that life originated *someplace,* at least in one place. We currently assume that this occurred as a result of a continuing process of chemical evolution that is always going on in the Cosmos. Scientists have been exploring various aspects of how this may have happened for at least 50 years. I recall some very early experiments of the Nobel laureate, Otto Meyerhof, who was a biochemist, and who — just to play a long shot — used to put big bottles full of inorganic and organic chemicals on the shelf in his laboratory, sterilize them, and let them sit there while he went about his business for the next 20 or 30 years. He was hoping, of course, that some day the miracle would occur, that suddenly some life would appear in one of his bottles.

More sophisticated laboratory experiments of this type have been in full swing during the last 50 years, and these will be reviewed later by Sherwood Chang. The early steps in this process of chemical evolution are becoming fairly well understood, but the critical step in the formation of life, namely, the origin, the *actual* origin of a replicating system, is still very, very far away. We still do not have a good handle on that question. Partly, this is because we do not have adequate understanding of the actual local environment over the time frame within which this took place, and it is my belief that the kinds of space activities in which the United States and the Soviet Union are engaged, in exploring the Solar System, sooner or later will aid in this understanding by giving us a better picture of what the terrestrial environment was like at the time that these beginning processes were taking place.

Harold P. Klein
Director of Life Sciences
Ames Research Center

Organic Chemical Evolution

SHERWOOD CHANG

What were the chemical origins of earthly life, and what was the chronology of major events? A full understanding requires far more knowledge than we now have about the early history of the Solar System and Earth.

The study of organic chemical evolution represents a cosmic quest for an understanding of our chemical origins, starting from the Big Bang and proceeding through interstellar clouds, the solar nebula, the formation of the Sun and planets, to the origin of life on Earth. In this context, the exploration of environments across space and time is directed at understanding not only their present state, but also what they can tell us about their past, their origins, and their evolution. During this quest, we will learn more about our own origins on Earth and discover more about the constraints imposed by stellar, solar system, and planetary evolution on the origin and distribution of life in the cosmos. The latter knowledge then helps narrow future searches for life elsewhere in the Universe.

In an evolutionary sense, all life is a product of countless changes in the form and content of primitive matter wrought by processes of chemical and biological evolution. The course of biological evolution can be traced back to common ancestors in the Precambrian period. The prebiotic history of Earth and the Solar System remains much more obscure, however. The components of the Solar System are products of chemical evolution from interstellar matter, but the circumstances of this evolution and even the resulting present state of some bodies are very poorly understood. Fortunately, there are windows, though obscured, that permit us to look into the past and to discern some of the features of past events. These windows are provided by astronomical observations and by geological and geochemical studies of rocks from ancient Earth, the Moon, and meteorites from outer space. Data from these sources then provide bases for formulating working models that attempt to reconstruct environments and to describe the physical-chemical processes that shaped them in the past. These working models, in turn, will be modified by new observations.

BEGINNINGS OF CHEMICAL EVOLUTION IN THE SOLAR SYSTEM

Just as biological evolution implies that all organisms on Earth have a common ancestry, so chemical evolution implies that all matter in the Solar System has a common origin. Consider the following scenario: An interstellar cloud of dust and molecules collapses, perhaps triggered by the shock wave associated with a nearby supernova, thus beginning the chemical evolution of the Solar System. According to one current model, gravitational collapse of the interstellar cloud led to an enormous disk of gas and dust, the primitive solar nebula, shaped like a flying saucer with the proto-Sun at the center. To a first approximation, this disk has been pictured as having a chemical composition that was spatially uniform and similar to that of the present Sun, at least for the major elements. Detailed studies of meteorites, however, have revealed anomalies in the isotopic composition of some elements, indicating that the solar nebula may not have been as homogeneous as theorists have suggested. Gravitational collapse would have been accompanied by heating and the establishment of a pressure gradient, with pressure highest in the region of the proto-Sun and decreasing with radial distance. Similarly, temperatures would have been highest in the central region, possibly exceeding 1600 K. Temperatures would have decreased, with radial distance, falling steeply within several astronomical units, then much more gradually to the edges of the solar nebula, beyond the limits of the present orbits of the giant planets, where temperatures would have been less than 100 K.

According to the equilibrium condensation model, as the inner region of the solar nebula cooled, minerals formed from the hot gas, yielding solids at their various condensation temperatures. Some of these primary condensates may have undergone secondary transformations if they continued to react with the nebula gas as the system cooled. In the outer regions of the solar nebula, in and beyond the current orbits of the giant planets, where temperatures would have remained low, condensation and accretion of organic and inorganic interstellar material could have taken place at temperatures near 100 K, allowing volatile materials to be preserved. Because of turbulence and the pressure and temperature gradients in the solar nebula, some mixing of materials with high- and low-temperature histories would have occurred during the accretion of small bodies. Eventually the accretion of fine-grained condensed material led to larger and larger objects and ultimately to the formation of planets. Thus, from the solar nebula came the Sun, the planets and their satellites, comets, meteorites, and asteroids. The Sun continues to contribute matter to various objects in the Solar System through the injection of solar-wind particles. The dashed line in figure 1 suggests that comets may preserve intact material that originated in the interstellar medium. Some cosmochemists believe that comets are products of low-temperature condensation and accretion processes that occurred at the outer edge of the Solar System where the lack of heating permitted survival of interstellar matter.

The stage of condensation from the nebula gas was probably terminated by the so-called T-tauri stage in the proto-Sun's evolution, during which a very powerful solar wind swept the uncondensed gas out of the Solar System and into the interstellar medium. Generally, the material condensed in the

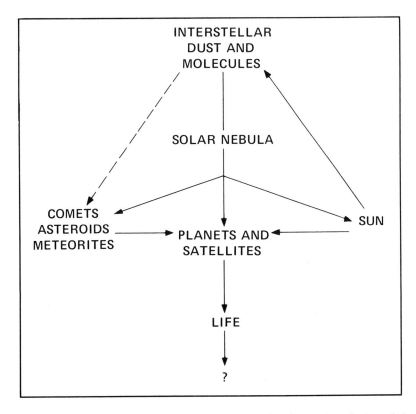

Figure 1. *Interrelationships between various bodies in the chemical evolution of the Solar System. Solid arrows indicate contributions of matter from one source to another. The dashed arrow signifies uncertainty regarding direct condensation of comets from interstellar matter. The arrow from "Life" implies its eventual dispersal from Earth.*

inner Solar System had a high-temperature origin and was depleted in volatiles (i.e., materials containing H, C, N, O, and S in volatile form and the
noble gases), but the materials that condensed and accreted in the outer
Solar System were rich in volatiles.

This very simple model has been criticized, and inevitably, as theory,
experiments, and observations progress, it will undergo changes, perhaps so
many that a new model will emerge. In the meantime, it provides a useful
framework for discussing various aspects of organic chemical evolution.
However, some cosmologists believe that the proto-solar system was initially
a partially ionized gas, in which case the notion of equilibrium condensation
would not be valid.

Sometime within 1000 million years of Earth's birth, life arose on its
surface and biological evolution began. Eventually, the death of the Sun may
be accompanied by an ejection of matter back into the interstellar medium
that spawned it. According to this scenario, the origin and evolution of life
on Earth were and will continue to be inextricably bound to the evolution of
both the Sun and Earth. It is somewhat ironic that life arose on Earth, a
planet that, relative to the Sun, is severely depleted in the volatile elements
that make up organic chemistry: hydrogen, carbon, nitrogen (see table 1).
On the other hand, the chemistry of the cosmos seems to be dominated by
these elements. From this knowledge springs the conviction that organic
chemistry constitutes an integral and fundamental part of cosmochemistry,
and from this comes the anticipation that, despite the seeming improbability

TABLE 1.– RELATIVE ABUNDANCES OF
SELECTED ELEMENTS (IN ATOM
PERCENT)

Element	Sun	Earth	Biosphere
Hydrogen	94	0.08	64
Helium	6	~0	0
Carbon	.04	.01	9.1
Nitrogen	.008	.00002	.1
Oxygen	.07	49	27
Neon	.004	~0	0
Sodium	.0002	.7	.005
Magnesium	.004	14	.02
Aluminum	.0003	4	.0004
Silicon	.004	14	.03
Argon	.0001	~0	0
Calcium	.002	.8	.008
Iron	.003	17	.002

of the origin of life on Earth, life may in fact be widely distributed through-out the Universe. With this introduction we shall proceed to a discussion of interstellar clouds, comets, outer planets, asteroids as represented by mete-orites, and the primitive Earth, and consider the organic chemistry of these various environments.

INTERSTELLAR CLOUDS

Interstellar clouds of dust and gas make up about 50% of galactic matter, and if the material in the clouds were spread uniformly over all space, the concentration of matter would amount to something like three hydrogen atoms per cubic centimeter. We consider two basic types of clouds: in very diffuse clouds, which contain little dust, concentrations of gas molecules are very low and hydrogen atoms are the dominant species; in dark, dense clouds, which are abundant in dust, molecular hydrogen is the dominant species and the gas concentrations range from about 10^3 to about 10^7 molecules/cm^3. A roster of molecules observed in the interstellar medium is given in table 2. The more complex molecules (triatomic or larger) occur in the dense clouds. The bulk composition of the interstellar gas is presumed to reflect cosmic elemental abundances. The dust in inter-stellar clouds is not well characterized; there is evidence to suggest the pres-ence of ice, silicates, graphite, macromolecular organic compounds, and mix-tures of these ingredients. The dust and molecules may have come from several sources: some of it may be a remnant of nebula condensation and solar-system formation, that is, material ejected into the interstellar medium by the T-tauri stage of stars, and some of it may have been ejected from the dense atmospheres of giant stars.

Estimates of the lifetimes of dense clouds before gravitational collapse exceed estimates of the lifetimes of molecules in the gas phase before freez-ing out on dust grain surfaces. Therefore, the fact that we observe interstellar molecules in the gas phase indicates a continuous production mechanism within the clouds themselves. In formulating a production mechanism, one must consider the environment in which it occurs. The temperatures are very low, 3 to 100 K, which means that chemical reactions in the clouds (except some reactions of hydrogen atoms) must occur with essentially zero activa-tion energy. In addition, the extremely low concentrations of molecules mean that all collisions between them (and therefore chemical reactions) are binary, that is, they involve only two species. These and other constraints have led to a model for the synthesis of interstellar molecules in dense clouds in which reactions are initiated by collisions of ubiquitous, high-energy, cosmic-ray particles with H_2 and He. Reactive ionic species are generated

TABLE 2.— INTERSTELLAR MOLECULES (1979)

Inorganic			Organic	
H_2	Hydrogen	Diatomic	CH	Methylidine
OH	Hydroxyl radical		CH^+	Methylidine ion
NO	Nitric oxide		CN	Cyanide radical
NS	Nitrogen sulfide		CO	Carbon monoxide
SiO	Silicon monoxide		CS	Carbon monosulfide
SiS	Silicon sulfide			
SO	Sulfur monoxide			
H_2O	Water	Triatomic	HC_2	Ethynyl radical
HNO	Hyponitrous acid		HCN	Hydrogen cyanide
N_2H^+			HNC	Hydrogen isocyanide
H_2S	Hydrogen sulfide		HCO^+	Formyl ion
SO_2	Sulfur dioxide		HCO	Formyl radical
			OCS	Carbonyl sulfide
NH_3	Ammonia	4-Atomic	HC_2H	Acetylene
			C_2CN	Cyanoethynyl radical
			H_2CO	Formaldehyde
			HNCO	Isocyanic acid
			H_2CS	Thioformaldehyde
		5-Atomic	CH_4	Methane
			HC_4	Butadiynyl radical
			H_2CNH	Methanimine
			H_2NCN	Cyanamide
			HC_2CN	Cyanoacetylene
			HCO_2H	Formic acid
			H_2CCO	Ketene
		6-Atomic	CH_3CN	Methyl cyanide
			CH_3OH	Methyl alcohol
			$HCONH_2$	Formamide
		7-Atomic	CH_3C_2H	Methylacetylene
			CH_3NH_2	Methylamine
			H_2CCHCN	Vinyl cyanide
			$H(C_2)_2CN$	Cyanodiacetylene
			CH_3CHO	Acetaldehyde
		8-Atomic	HCO_2CH_3	Methyl formate
		9-Atomic	CH_3CH_2CN	Ethyl cyanide
			$H(C_2)_3CN$	Cyanotriacetylene
			CH_3CH_2OH	Ethyl Alcohol
			$(CH_3)_2O$	Dimethyl ether
		11-Atomic	$H(C_2)_4CN$	Cyanotetracetylene

which can enter into binary reaction sequences that require little or no activation energy. Some general types of reactions are shown in table 3. In addition, chemical reactions may occur on dust grain surfaces that act as "collectors," and these would involve the recombination of free radicals, a type of reaction requiring little or no activation energy. In addition, atoms and free radicals may react with the grain itself.

Examination of the list of compounds in table 2 leads to two important observations. First, the compounds are chemically diverse and structurally complex. Second, many of them are known to be important intermediates in the production of organic matter in abiotic synthesis experiments (e.g., hydrogen cyanide, formaldehyde, and cyanoacetylene).

Clearly, the interstellar environment, as exotic and as seemingly inimical to chemical reactions as it may appear at first consideration, nonetheless exhibits a rich chemistry that manifests itself in the production of organic compounds that, for the most part, are familiar from terrestrial experience.

TABLE 3.– ION–MOLECULE REACTIONS IN INTERSTELLAR CLOUDS

Primary ionization:	$H_2 + CR \rightarrow H_2^+ + e + CR'$	
	$\rightarrow 2H^+ + e + CR'$	
	$He + CR \rightarrow He^+ + e + CR'$	
Secondary ionization:	$H_2^+ + H_2 \rightarrow H_3^+ + H$	
	$H_3^+ + A \rightarrow AH^+ + H_2$	$(A = O)$
	$He^+ + AB \rightarrow A^+ + B + He$	$(AB = CO)$
	$\rightarrow AB^+ + He$	
Hydrogenation:	$A^+ + H_2 \rightarrow AH^+ + H$	$(A = CH)$
Radiative association:	$A^+ + H_2 \rightarrow AH_2^+ + h\nu$	$(A = C)$
Insertion:	$AH^+ + CO \rightarrow ACO^+ + H$	$(A = CH_2)$
Hydrogen transfer:	$AH^+ + CO \rightarrow HCO^+ + A$	$(A = CH_2)$
	$AH + C^+ \rightarrow AC^+ + H$	$(A = NH_2)$
Dissociative recombination:	$AH^+ + e \rightarrow A + H$	$(A = HCN)$
Radical recombination:	$A + B \rightarrow C + D$	$(CH_3 + O \rightarrow H_2CO + H)$
CR = cosmic ray	e = electron	$h\nu$ = photon

Although many scientists believe there is little basis for speculating that interstellar organic molecules found their way intact and unchanged to the prebiotic Earth's surface, there is growing interest in the possibility that interstellar molecules may be preserved intact in comets and/or in altered form in carbonaceous meteorites. These possibilities stem from the idea that all solar-system matter had a common origin in an interstellar cloud of dust and molecules. To the extent that comets and carbonaceous meteorites contributed mass to early Earth and arrived at the surface *intact,* interstellar organic compounds could have survived to take part in subsequent chemical evolution.

COMETS

Comets occupy an especially interesting place in models of solar-system origin and evolution. They may have been a partial source of planetary atmospheres, and they are believed to have been the building blocks for the rocky cores of the outer planets. Present understanding places the origin of comets in the outer regions of the primitive solar nebula in and beyond the space now traversed by the giant planets. Perturbations of their original orbits by the formation of the giant planets are believed to have sent some into the inner Solar System to collide with the Sun and inner planets and others into orbits extending great distances from the Sun (up to 50,000 astronomical units (AU), where 1 AU = 150 million kilometers).

The components of a comet observable within several astronomical units of the Sun include the nucleus, the coma, and the tail (fig. 2). According to the "dirty ice" model, comet nuclei consist of simple and complex organic molecules and meteorite-like dust and rock embedded in a matrix of frozen water and possibly solid CO_2. As the comet approaches the Sun, heating by the Sun occurs and the ices sublime, ejecting volatile "parent" compounds (possibly H_2O, CO_2, CH_4, C_2H_2, NH_3, HCN, CH_3CN, etc.) and entraining nonvolatile dust and rock from the nucleus. In the coma, interactions of the parent compounds with solar radiation can lead to physical and chemical processes that result in the partial-to-complete breakdown of the so-called parent molecules. Ion-molecule reactions analogous to those occurring in interstellar clouds probably play an important role. The neutral daughter products are observed in the coma, while the positively charged ones are observed in the tail. According to an alternative view, all the observed species already exist in the nucleus and are simply released directly into the coma by evaporation. These two possibilities are summarized in figure 3. A third possibility is that both parent molecules and simpler species are released into the coma where they undergo reactions to yield the

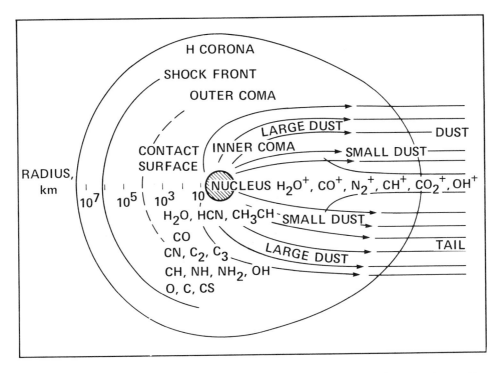

Figure 2. *Major features of a comet. The distance scale is logarithmic. Ions are observed in the tail, neutral species in the coma.*

observed ions, atoms, and molecules. In addition to the species indicated in figure 2, metallic elements (Fe, Si, Mg, Ca, Ni, Na, Cr) have been detected in spectroscopic studies of meteor showers associated with comets. The relative abundances of these elements suggest similarities between the chemical compositions of cometary dust and carbonaceous meteorites.

The comet nucleus is thought to be small, typically 1–10 km in diameter, but no direct observations have yet been made. It appears as a small point of light embedded within the bright, large, and extensive coma. The mass of the nucleus could range from 10^{15} to 10^{18} gm. The light from the visible coma and tail is emitted by atoms and molecules that have interacted with solar radiation. The size of the coma is remarkable, perhaps greater than 10^5 km in radius. The tail, composed of dust grains and ionic molecules, is even larger, possibly exceeding 10^7 km in some cases. When comets become visible in the inner Solar System, they may be spatially the largest objects in the sky.

As mentioned above, comets are believed to be material condensed and accreted at the outer edge of the primitive solar nebula. Thus a relationship may exist between comets and interstellar molecules; if one compares the molecules observed in comets (fig. 3) with interstellar molecules (table 2),

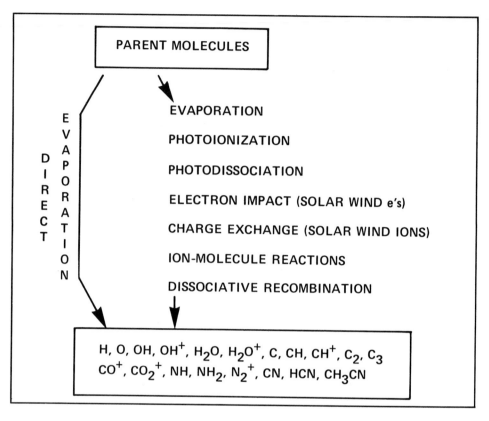

Figure 3. *Production of observed cometary molecules by direct evaporation from the nucleus or by evaporation of parent molecules followed by their interactions with solar radiation.*

there do seem to be similarities. Both populations contain cyanide derivatives with the CN group, and comet species can be produced by fragmentation of interstellar molecules. It has also been suggested that carbonaceous meteorites, which are rich in various forms of the volatile elements and organic matter, are remnants of volatile-depleted and moribund comets. If comets *do not* contain relatively unaltered interstellar matter, and if they formed at the outer edge of the solar nebula where temperatures were low enough to condense gases such as carbon dioxide and water, then the presence of parent organic molecules in comets is difficult to understand. There is no widely accepted model for chemical reactions in the solar nebula that could yield the chemistry of comets. Indeed, without direct observations of the nucleus, our knowledge of comet chemistry is exceedingly sparse and model-dependent. Since comets are poorly understood and may represent a chemical evolutionary link between the primitive solar nebula and the interstellar medium, their direct study by space probes constitutes a high-priority objective for many space scientists.

JUPITER

From the outer regions of the Solar System where comets originated, we move sunward to consider organic chemistry on Jupiter. Spectroscopic observations made by astronomers, theoretical considerations, and, more recently, direct study by space missions provide the basis for current models of Jupiter. The planet has approximately a solar-composition atmosphere consisting primarily of hydrogen and helium with minor to trace amounts of methane, ammonia, water, hydrogen sulfide, ethane, acetylene, phosphine, carbon monoxide, arsine, and the noble gases. The planet's compositional similarity to the primordial nebula makes it a critical object for cosmological study. Planetary processes that prevailed soon after its origin are probably still occurring now.

The model of the environment of the upper (and observable) Jovian atmosphere is depicted schematically in figure 4. Sunlight and extraplanetary particles are represented as entering at the top of the atmosphere. The changes in temperature (in °K) and pressure (in atmospheres) with depth (not drawn to scale) are shown on the far left-hand scale. The locations of a haze layer and the various observed and postulated cloud layers are also shown, as are the directions of motion of the latter (thick arrows) according to recent meteorological models. The depths (in terms of temperature and pressure) to which sunlight of various wavelengths penetrates into the atmosphere are indicated by the vertical arrows. The presence of gas constituents is also shown by arrows, starting at about the maximum altitude in the atmosphere where they may occur. Thus H_2, He, and CH_4 occur throughout the atmosphere, while H_2O does not occur above the 250 K level. The stratification of some atmospheric components results from the formation of their solid (or liquid) condensates at different temperatures (and altitudes). The jagged lines through the clouds indicate lightning flashes.

The distinct coloration of Jupiter's cloud cover and the variability of its patterns with time have been observed for over a century. These were the first indications of the occurrence of disequilibrium processes in the atmosphere. In 1952, Urey first suggested that complex organic molecules might be responsible for the cloud colors. Since then the hypothesis that Jupiter is at an advanced stage of organic chemical evolution has been widely promulgated. Support for this view has been marshaled from observations that complex organic compounds and colored organic polymers are produced when gaseous components like those in the Jovian atmosphere are subjected in laboratory experiments to electric discharges, thunder shock waves, high-energy proton irradiation, or ultraviolet irradiation. On the basis of theoretical models of Jovian atmospheric photochemistry, however, the contrasting view that the colors are attributable to inorganic substances photochemically

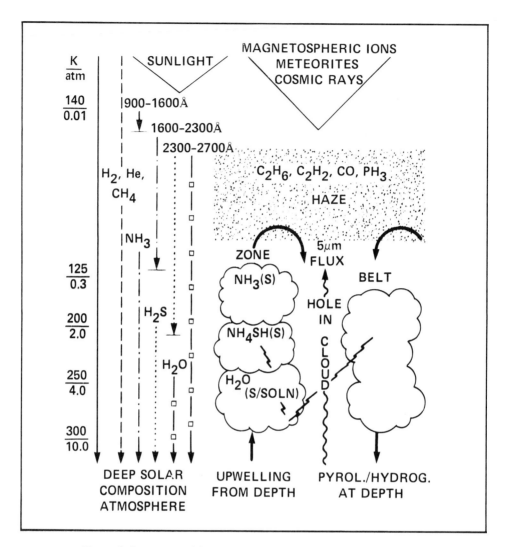

Figure 4. *Summary of features in a model of Jupiter's atmosphere.*

synthesized in the atmosphere has also been championed. Candidate inorganic species include red phosphorus (P_4), ammonium and hydrogen polysulfides ((NH_4)$_2$S$_X$ and H_2S$_X$), and elemental sulfur. No data yet obtained from planetary observations can verify either viewpoint to the exclusion of the other.

The hydrocarbons observed on Jupiter — acetylene (C_2H_2) and ethane (C_2H_6) — are believed to be produced by the interaction of sunlight with methane high in the atmosphere. There is also good reason to believe that acetylene is synthesized mainly during thunderstorms. The origin of the carbon monoxide is not clear. Two processes have been suggested which may act separately or jointly: an upwelling of material from deep in the atmosphere, where at high temperatures carbon monoxide is thermodynamically stable, or reactions of atmospheric methane with oxygen atoms injected into the upper atmosphere from extraplanetary sources. Images from the Voyager mission have shown lightning flashes on Jupiter near the ammonia clouds. Although the occurrence of lightning does not prove that the colored material is organic matter, it supports the view that organic matter is produced on Jupiter as it is in laboratory experiments designed to simulate Jovian phenomena. It thus lends plausibility to the mechanism and stimulates interest in providing future space probes of the planet with the ability to detect a variety of organic compounds.

The occurrence of organic compounds on Jupiter has been linked to the possibility of life on that planet. Although available evidence cannot eliminate this possibility, it does make it highly unlikely. The cloud features visible on the planet have a finite lifetime; that is, the various colored belts are observed to disappear with lifetimes ranging from weeks to years. The Great Red Spot has been observed for hundreds of years, but meteorologists believe that the material visible now is not the same material that was visible a hundred or even five years ago. Vertical atmospheric cycling transports material in the upper atmosphere down to some depth in the lower atmosphere where the temperatures and pressures are high. As a consequence, any organic compounds formed higher in the atmosphere by electric discharges, thunder shock waves, ultraviolet irradiation, and other processes are destroyed by hydrogenation and reconverted to the primary ingredients of the atmosphere — hydrogen, ammonia, methane, water, and hydrogen sulfide. Thus atmospheric circulation imposes severe constraints on the time available for simple atmospheric gases to be converted to complex organic macromolecules akin to proteins and nucleic acids, the critical biochemicals of life on Earth. For life to arise on Jupiter, the rate of chemical evolution must be extraordinarily fast, that is, fast compared to the rate of circulation of matter to hot regions deep in the atmosphere.

Many uncertainties thus exist about the nature and identity of the colored substances on Jupiter, but on the basis of available knowledge it appears that the dynamics of atmospheric circulation have limited the progress of Jovian organic chemical evolution.

METEORITES

Sunward from Jupiter lies the asteroid belt, samples of which are believed to find their way to Earth in the form of meteorites. Application of spectroscopic remote-sensing techniques to asteroids has revealed the presence of mineral assemblages on their surfaces analogous to those found to predominate in known types of meteorites. For our discussion, the most interesting meteorites are the ones called carbonaceous. (Objects with spectroscopic properties similar to those of carbonaceous meteorites appear to be quite common in the asteroid belt.) These objects consist of complex assemblages of relatively fine-grained mineral and organic matter that reflect a broad range of elemental compositions, textures, and petrologies, indicative of wide variations in the environment of origin for the various components.

According to a prevailing model for their origins, some of the mineral ingredients were formed primarily by equilibrium condensation from the cooling gaseous solar nebula. Other minerals resulted from secondary reactions of high-temperature condensates with lower temperature nebula gas while the condensates were still suspended as particles in the nebula. Presumably, the diverse ingredients were eventually assembled into rocky material on parent bodies, probably resembling asteroids, where compaction events and the environmental conditions further influenced their chemistry, mineralogy, and petrology. Later, disruption of the parent bodies (perhaps by collision with other bodies) yielded fragments representative of the various parts which, in time, fell under the influence of Earth's gravitational field. Carbonaceous meteorites therefore represent material formed very early in solar-system history and contain clues to the primitive environments and processes that produced them.

Figure 5 summarizes major and minor phases found in carbonaceous meteorites; gives their probable temperature of formation by equilibrium condensation from the gaseous nebula, by secondary alteration, or by non-equilibrium processes, either in a solar-composition gas or on a parent body; and shows their relative abundances and distributions in three types of carbonaceous meteorites. For present purposes, the major differences between the C1, C2, and C3 meteorites are their increasing content of volatile elements and their decreasing content of minerals of high-temperature origin. Accordingly, the amount of organic matter increases in the same order from about 0.5 to 5% by weight. High-temperature mineral phases occur most abundantly in C3 meteorites, along with metals and the mafic silicates, olivine and pyroxene, which comprise the bulk of their mass. These minerals exist only in low to trace amounts in C2 meteorites; all, except traces of mafic silicates, appear to be absent in the C1 meteorites.

A complex, chemically heterogeneous carbonaceous phase, characterized by insolubility in solvents and acids, occurs as the major carbon component in all three types of meteorites but is least abundant in the C3 meteorites. It is designated the solvent- and acid-insoluble carbonaceous (SAIC) phase. This material is especially noteworthy because it serves as host phase for several noble-gas components of the "planetary" type, the isotopic com-

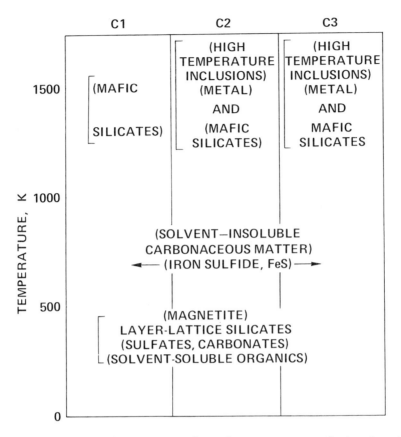

Figure 5. *Distributions and approximate formation temperatures of minerals and other phases in carbonaceous meteorites. Parentheses indicate low to trace amounts.*

position of at least one of which cannot be readily explained by solar or solar-system processes. Some of the SAIC material is probably relatively unaltered interstellar matter. The natural mechanism(s) responsible for producing the SAIC phases with their noble-gas contents is unknown, however. When carbonaceous matter is condensed in the presence of noble gases by passage of electric discharges through CH_4 and N_2 or by laser vaporization of carbon targets, the noble gases become trapped in a "planetary" elemental pattern; but the relevance of these specific mechanisms to the cosmochemical context in which the meteoritic SAIC phases were produced is unclear. In figure 5, the indicated formation temperature of SAIC material was arbitrarily chosen and may be viewed as simply the midpoint in a possible ± 500 K range of formation temperatures.

Organic synthesis promoted by Fischer-Tropsch-type (FTT) reactions, electric discharges, ultraviolet photochemistry, or other mechanisms must have occurred at temperatures sufficiently low to permit preservation of the variety of volatile and thermally labile organic compounds found in low abundances in C1 and C2 meteorites. Although it is uncertain about how these compounds were synthesized, some investigators favor production on a parent body rather than on mineral grains suspended in the solar nebula. An interstellar origin is also possible if the building blocks of meteorite parent bodies included material formed in regions of the solar nebula where cometary matter originated.

The predominant minerals in C1 and C2 meteorites (50–80%) are the layer-lattice silicates or phyllosilicates. These minerals resemble terrestrial clays in crystallographic structure, but exhibit elemental compositions remarkably similar to the pattern of cosmic abundances. This similarity suggested a solar nebula origin for this material, but recent observations and analyses indicate that a more likely mode of production involves hydrothermal alteration at about 350 K of previously formed silicates on a parent body. Also found in C1 and C2 (and rarely in C3) meteorites are minor amounts of magnetite, sulfates, and carbonates. Recent findings indicate that these, too, have a parent-body rather than nebula origin.

Table 4 shows the distribution of carbon in the Murchison meteorite, the most pristine and carefully examined carbonaceous meteorite. Note that the volatile organic compounds — the hydrocarbons, carboxylic acids, ketones, aldehydes, alcohols, and amines — constitute a small fraction of the total carbon and less than 0.05% of the total mass of the meteorite. Nonetheless, the variety of compounds indicates that the organic chemistry of meteorites is quite complex.

TABLE 4.– DISTRIBUTION OF CARBON IN
MURCHISON METEORITE

Species	Abundances
Acid-insoluble carbonaceous phase, %	1.3–1.8
CO_3, %	0.2–0.5
Hydrocarbons and lipids, %	0.07–0.11
Carboxylic acids, ppm	~350
Amino acids, ppm	10–30
Ketones and aldehydes, ppm	~17
Urea and amides, ppm	<2–15
Alcohols, ppm	~6
Amines, ppm	~2–3
N-heterocycles, ppm	<2–40
Sum, %	1.81–2.45
Total carbon, %	2.0–2.58

ppm = parts per million.

Clearly, organic chemical evolution prior to or on the meteorite parent body yielded substances which on primitive Earth may have constituted the building blocks of the first organisms. An intriguing question is how much the meteorite parent-body environment (or other source regions of meteorite organic matter) resembled that of prebiotic Earth. Continuing studies of the organic chemistry of meteorites integrated with inorganic, mineralogic-petrologic, and other types of investigation should yield new insights into the relationships between the organic chemical evolution represented in carbonaceous meteorites and that of prebiotic Earth.

VENUS AND MARS

The widespread occurrence of organic compounds in the cosmos and within our Solar System confirms the expectation based on cosmic elemental abundances that organic chemical evolution is a natural consequence of the evolution of matter in the Universe. But organic chemical evolution is inextricably intertwined with the evolution of environments, be they interstellar clouds, meteorite parent bodies, or planets, and its progress toward life may be terminated at different stages depending on the physical and chemical constraints imposed by the environment.

On Venus, Earth's nearest neighbor, the temperatures at the surface and in the lower atmosphere, planetwide, are so high that organic compounds such as amino acids, sugars, and nucleic acids cannot survive. Consequently, the probability that life or organic chemistry exists or could survive on Venus is virtually nil. However, there may have been organic chemical evolution on Venus early in its history, if its global environment resembled that of early Earth. Clearly, planetary evolution and thus chemical evolution followed different tracks on Venus and Earth.

On Mars, local environments exist today which are not totally inimical to life or which would have permitted relics of earlier organic chemical evolution or life to be preserved in ancient rocks and sediments. Since we only sampled a very limited number of Martian environments on the Viking missions, the existence of present and past organic chemical evolution, and even life, remains to be adequately confirmed or denied.

PREBIOTIC EARTH

In many respects our knowledge of early Earth is much like our knowledge of the early Solar System. It is model-dependent and relies on the reconstruction of an environment by extrapolation from a record preserved in, but deciphered only in fragmentary fashion from lunar rocks, meteorites, remotely discernible features of Venus and Mars, and very ancient rocks and sediments of Earth. As more of the record is unveiled, new evidence leads to new interpretations and revisions of models.

According to the Oparin-Haldane-Miller-Urey paradigm, a highly reducing atmosphere consisting of methane, ammonia, and water prevailed on primitive Earth. Passage of energy in various forms through this hypothetical atmosphere produced the reservoir of organic molecules from which life evolved. The existence of this reducing atmosphere required the presence of metallic iron in the upper mantle and crust, which appears to conflict with some geochemical observations. Recently, a case has been made for a primitive atmosphere composed predominantly of H_2O, CO_2, and N_2.

The background and basic features of this model are depicted schematically in figure 6. It starts with refractory minerals condensing from the cooling nebula and accreting to form the protoplanet. Rapid accretion was accompanied by melting and segregation into a molten metallic core and a fluid silicate mantle. The initial inventory of volatiles was driven to the surface. As the nebula gas continued to cool, metallic iron was converted to the ferrous state. Presumably, when the Sun passed through its T-tauri stage, the powerful solar wind blew the remaining nebula gas out of the inner Solar System, carrying Earth's primitive and possibly highly reducing atmosphere with it. (Because some doubt exists about the efficacy of the T-tauri wind, it is significant that another mechanism has been proposed that could achieve the same result: in a recent physical model of the primitive solar nebula, it has been suggested that tidal stripping of the atmospheric envelope of a giant, gaseous, inner protoplanet by the Sun could have occurred early, leaving behind a core of condensed matter.) Debris remaining from the nebula condensation was steadily accumulated by primitive Earth. This debris, presumably of carbonaceous meteoritic composition, contributed material to form the thin crustal veneer of Earth. Heating of the debris as it passed through the atmosphere, during its impact with the surface, or while it was embedded in the hot surface, released the volatiles to form the secondary atmosphere. As a result of Earth's continued cooling, a thin, solid crust probably existed about 4.1 to 4.0 billion years ago. The crust must have formed by about 3.9 billion years ago, because shortly thereafter aqueous environments and sedimentary processes had begun, as evidenced by the 3.8-billion-year-old metasedimentary rocks of Greenland.

According to the model, H_2O and CO_2 were the dominant constituents of the secondary atmosphere, N_2 occurred in minor amounts, and H_2 and CO were present only in traces, if at all. Traces of CH_4 and other hydrocarbons are presumed to have been oxidized readily to CO_2 by iron oxides in the crust. The composition of this steam atmosphere was determined by the

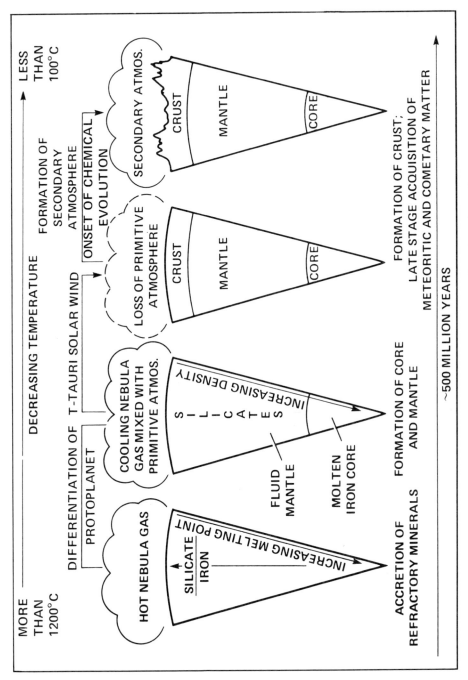

Figure 6. *Stages in a scenario for the Earth's early evolution.*

redox potential of the silicate crust and upper mantle and would have strongly resembled contemporary volcanic exhalations. Once the temperature of Earth's crust dropped below 373 K, water condensed to begin formation of the oceans, and the weathering of basic igneous rocks by CO_2 afforded carbonates. The prebiotic atmosphere that resulted would have closely resembled the present atmosphere minus oxygen.

We remain ignorant of the true composition of primitive Earth's atmosphere, but we can nevertheless explore the potential for abiotic organic synthesis in a variety of possible compositions ranging from highly reducing (H_2, NH_3, CH_4, H_2O) to nonreducing (CO_2, N_2, H_2O). Some of the results obtained with different gas compositions and energetic processes are summarized in table 5. Although other types of experiments have been conducted, the table is restricted to those involving electrical discharges, ultraviolet light, and the so-called Fischer-Tropsch-type synthesis. The last experiments involve passage of mixtures of H_2, CO, and NH_3 over mineral catalysts at 400 to 600 K to produce products.

The brief survey in table 5 shows that in abiotic syntheses the types of organic compounds formed, and probably their relative abundances, depend on both the composition of the reactant gases and the energy source. In a reducing atmosphere it appears to be relatively easy to produce some of the types of compounds that comprise either the molecular building blocks of life or potential precursors for them. For example, many biochemicals have been synthesized starting with the products observed in electric discharge experiments; thus, from simple aldehydes and nitriles it has been possible to produce amino acids, carbohydrates, purines, and pyrimidines. In a CO_2-N_2-H_2O atmosphere, however, only acids and aldehydes have been reported; the production of nitrogen-containing organic compounds appears to be inhibited. The selective absorption of ultraviolet light by various components of the atmosphere imposes limitations on the variety of organic compounds that can be produced by photochemical means. For example, experimental and computational studies have shown that ultraviolet irradiation of a CO_2-N_2-H_2O mixture yields formic acid and formaldehyde but no nitrogenous organic matter. Even when CO_2 is replaced by CH_4, no nitrogen is incorporated in organic compounds by photochemical means. When N_2 is replaced by NH_3, however, HCN, amino acids, and, presumably, other compounds are produced photochemically. The Fischer-Tropsch-type reactions produce a variety of organic compounds but require the presence of H_2 and NH_3.

In other experiments, the coupling of carbohydrates with purines has produced nucleosides of the purines, but it has not been easy to combine pyrimidine bases with carbohydrates to form the pyrimidine nucleosides. Phosphorylation of nucleosides to yield nucleotides has been achieved, as has formation of polynucleotides from mononucleotides. Amino acids have also

TABLE 5.– RESULTS OF EXPERIMENTS ON ABIOTIC
ORGANIC SYNTHESIS

Gases	Electrical discharges	Ultraviolet photochemistry	Fischer-Tropsch synthesis
CO_2, N_2, H_2O	HNO_3, etc (?)	Formic acid, formaldehyde	
CO_2, CO, N_2, H_2	HCN, amino acids, etc. (?)		
CO, N_2, H_2O		Hydrocarbons, alcohols, ketones	
CO, NH_3, H_2	HCN, amino acids, etc.		Hydrocarbons, aldehydes, ketones, nitriles, amino acids, purines, etc.
CH_4, N_2 (NH_3), H_2O	HCN, hydrocarbons, aldehydes, ketones, amino acids, carboxylic acids, etc.	Aldehydes, alcohols, ketones, hydrocarbons, (HCN),* (amino acids)*	

*Formed when N_2 is replaced by NH_3.

been converted to polypeptides of various lengths. Thus the view that proto-biopolymers could have been produced with relative ease on primitive Earth has been widely promulgated. Underlying these synthesis experiments, how-ever, is the assumption that biomonomers (e.g., amino acids, carbohydrates) or their simple precursors (e.g., hydrogen cyanide, formaldehyde) were readily produced in the atmosphere of primitive Earth. Furthermore, many of the syntheses of the complex biomolecules involve either single reactants or very simple mixtures of reactants and are carried out under laboratory conditions whose geological relevance to and plausibility in the primitive

Earth environment are not readily discernible. Despite these limitations, it is clear that, under certain conditions, simple atmospheric gases can be converted in stepwise fashion into complex molecules possibly related to primordial precursors of the proteins and nucleic acids of living systems.

Although production of the organic compounds necessary for chemical evolution would have proceeded readily in a reducing atmosphere (CH_4, NH_3, H_2O), the possibilities in a CO_2–N_2–H_2O atmosphere with traces of H_2, CO, and/or CH_4 appear to be limited, at least in the types of experiments carried out so far. However, reports have appeared recently of photochemical reactions involving gas-solid and gas-liquid-solid systems in which carbon and nitrogen "fixation" occurs with CO_2 and N_2 as the source gases. Investigations of organic synthesis in such heterogeneous systems are highly desirable if we intend to understand the possible pathways for production of organic matter in a nonreducing secondary atmosphere.

It is important to keep in mind that, as embodied in the Oparin-Haldane-Miller-Urey paradigm, the fundamental conception of abiotic organic synthesis as a necessary and natural prelude to the origin of life on Earth retains its validity. Changes in the chemical composition of model prebiotic atmospheres must be viewed as challenges to be met with fresh approaches to achieving abiotic organic syntheses compatible with changing environments. Furthermore, the nonreducing (or only slightly reducing) model, like the highly reducing one, remains poorly constrained in a number of ways and undoubtedly will undergo refinement and alteration with time. A realistic model probably lies somewhere between the two, with CO_2 and CH_4 both present and with nitrogen present predominantly in the form of N_2.

PROBLEMS AND PROSPECTS FOR FUTURE STUDY OF PREBIOTIC EARTH

An adequate understanding of the course of organic chemical evolution on primitive Earth and the chronology of major events involved requires far more knowledge about the early history of the Solar System and Earth than is available now. A listing of major issues is given in table 6.

The course of condensation and accretion of matter in the solar nebula would have been critical in establishing the chemical composition of material that formed the planets and the processes that initiated their geophysical and geochemical evolution. New knowledge of the geophysical and geochemical evolution of early Earth or other planetary bodies (including meteorite parent bodies) would lend insight into the timing of core differentiation and

TABLE 6.– SOME MAJOR ISSUES IN
ORGANIC CHEMICAL EVOLUTION
ON EARTH

Condensation and accretion in the solar nebula
Early geochemical/geophysical evolution
Origin and early evolution of the atmosphere
    composition
    thermal structure
    radiative transmission
    transport processes
Production rates for organics
    sources: half lives
    sinks: recycling mechanisms
    condensation mechanisms
Organic-inorganic interactions
    catalysis
    phosphate utilization
    sequestration
Environmental fluctuations
    day/night
    seasons
    tides
Phase separations
    membranes
    microstructures
Molecular selectivity
    biological subset of monomers
    genetic code
    chirality

crustal evolution and the thermal history of crust. From this knowledge could come information about the origin, composition, and early evolution of the atmosphere. The thermal structure would determine the rate at which H_2 (and therefore reducing power) was lost from the top of the atmosphere. It would also determine, in part, the atmosphere's composition and the kinds of chemical transformations that could occur in various parts of the atmosphere. For similar reasons, it is important to know the radiative transmission in the atmosphere, which determines the energy budget associated with incident sunlight. Transport processes in the atmosphere are also important. For example, if material is produced by ultraviolet light high in the atmosphere, how long does it take for it to get down to the surface of Earth? Is it a short enough time to preserve it against destruction or against conversion to a refractory material incapable of further involvement in chemical evolution? If material is produced on land, how can it be transported to bodies of water where it can encounter other materials? If it is produced in oceans, how is it concentrated?

Possible production rates for organic compounds, even in a reducing atmosphere, are not well known; synthesis could occur in the atmosphere, in the oceans, or at the interfaces between the atmosphere, bodies of water, and land. As mentioned above, there are also chemical reactions that act as sinks for organic matter. For example, amino acids and sugars would react to form water-insoluble products, which are thus removed from the reservoir of organic material that would have been available to continue organic evolution. How could material in sinks of this type have been recycled to make it available?

What were the environmentally plausible pathways available to convert simple monomeric amino acids and nucleotides to the biopolymers necessary for life? One pathway that has recently been receiving attention involves the interaction between inorganic matter (particularly clays) and organic matter. There is little doubt that organic chemical evolution occurred in a predominantly inorganic realm. An important question is how organic chemistry interacted with inorganic chemistry and mineralogy. Did the inorganic world provide catalysts for organic reactions? If so, what were they? How was phosphate utilized to produce nucleotides and polynucleotides? Did inorganic material sequester organic matter, thereby removing it from the realm of chemical evolution?

Environments on Earth are subject to all kinds of fluctuations: day and night, seasons, and tides. In some experiments, environmental fluctuations of temperature and moisture content have been successfully used to produce peptides from amino acids, for example. How important were fluctuations overall for organic chemical evolution? Were they necessary?

At some stage in organic chemical evolution it would have been necessary to achieve a phase separation between an evolving organic system and the external environment. Thus the origin of membranes must be considered. Could the organic microstructures produced in a variety of abiotic synthesis experiments serve as models for early membranes?

Finally, molecular selectivity during chemical evolution must be considered. In meteorites and in the products of model prebiotic organic synthesis experiments, a rich variety of compounds exists. Even within a single class — the amino acids — there are a large number of different molecular structures. How were the very limited number of amino acids now utilized by organisms selected from this larger abiotic set? How did the genetic code arise? And what was the origin of chirality, that is, the handedness prevalent in the amino acids and the sugars of all living organisms?

Partial answers may exist for some of these questions, but much remains to be learned. As a field of active scientific inquiry, the study of organic chemical evolution and the origin of life is still in its infancy, and contributions from a variety of disciplines, including astronomy, astrophysics, biology, geochemistry, inorganic chemistry, and organic chemistry, are essential. If research efforts continue on many of the issues that have been mentioned, we can anticipate significant progress for the future.

ADDITIONAL READING

Andrew, B. H., ed.: Interstellar Molecules. Reidel, Dordrecht, 1980.

Bar-Nun, A.: Acetylene Formation of Jupiter: Photolysis or Thunderstorms? Icarus, vol. 38, May 1979, pp. 180–191.

Bunch, T.; and Chang, S.: Carbonaceous Chondrites II: Carbonaceous Chondrite Phyllosilicates and Light Element Geochemistry as Indicators of Parent Body Processes and Surface Conditions. Geochim. Cosmochim. Acta, vol. 44, 1980, pp. 1543–1577.

Chang, S.: Comets: Cosmic Connections with Carbonaceous Meteorites, Interstellar Molecules and the Origin of Life. In: Space Missions to Comets, NASA SP-2089, 1979, pp. 59–111.

Dalgarno, A.; and Black, J. H.: Molecule Formation in the Interstellar Gas. Reports on Prog. Phys., vol. 39, June 1976, pp. 573–612.

Delsemme, A. H., ed.: Comets-Asteroids-Meteorites. Univ. of Toledo Press, 1977.

Flinn, E. A., ed.: Scientific Results of the Viking Project. J. Geophys. Res., vol. 82, Sept. 30, 1977, pp. 3959–4680.

Gehrels, T., ed.: Jupiter. Univ. of Arizona Press, Tucson, 1976.

Goldsmith, D.; and Owen, T. C.: The Search for Life in the Universe. Benjamin-Cummings Publishing Co., Menlo Park, Calif., 1980.

Gordon, M. A.; and Snyder, L. E., eds.: Molecules in the Galactic Environment. John Wiley & Sons, New York, 1973.

Miller, S. L.; and Orgel, L. E.: The Origins of Life on the Earth. Prentice-Hall, Englewood Cliffs, New Jersey, 1974.

Mission to Jupiter and Its Satellites, American Association for the Advancement of Science, 1979 (a special reprint collection); see also Science, vol. 204, no. 4396, June 1, 1979, pp. 913–921 and 945–1008.

Nagy, B.: Carbonaceous Meteorites. Elsevier Publishing Co., New York, 1975.

Windley, B. F., ed.: The Early History of the Earth. John Wiley & Sons, New York, 1976.

Sulfur: Fountainhead of Life in the Universe?

BENTON C. CLARK

Sulfur is ubiquitous in the Universe and essential to all life forms that we know. It supports the chemoautotrophic way of life and the photosynthetic. It may inhabit niches we cannot imagine, and the life zone about a star may therefore be wider than now estimated.

In July 1976, the first of two unmanned spacecraft, the Viking Landers, set down on the surface of Mars. Thus began the most ambitious search for extraterrestrial life to date, on the most Earth-like body in our Solar System. Disappointingly, the chemical tests for metabolic activity in Martian soil proved negative or ambiguous (Klein, 1978). There were no signs of lifelike forms in either lander pictures or the thousands of pictures taken by the dual orbiters. No warm, wet spots (oases) or organic compounds in the soil were found. Much was and is being learned of the geology and climatology of Mars, however, and one surprising finding has been the high sulfate content (~10–15%) of Martian soil (Clark et al., 1976), which has led one investigator to suggest the possibility of an indigenous biology, perhaps now extinct, whose energy metabolism is nonphotosynthetic in nature, but rather based on cycling of sulfur compounds through various redox states by enzymic catalysis of inorganic sulfur with atmospheric photolysis products (Clark, 1979).

Just over six months after the Viking adventure began, two other spacecraft began exploring another hostile (to man), yet unique environment. One craft, named the Angus, took 70,000 reconnaissance photographs of this never-before-explored zone. A second craft, the Alvin, was a manned submersible. The mission was the geological exploration of ocean-bottom thermal springs on the 2.5-km-deep center of the Galápagos rift zone. Hydro-thermal vents were indeed found, and although these are of considerable geochemical and geophysical interest, the most important discovery was the existence of previously unknown species of animals whose communal life is dependent on the primary productivity of sulfur-oxidizing bacteria (Corliss et al., 1979; Corliss and Ballard, 1977).

The parallels between these two explorations are striking, though coincidental. Three common threads are special environments, unique life forms, and the element sulfur. It is the purpose of this brief report to reemphasize the importance of sulfur and sulfur-containing compounds in biology as we know it, in planetary evolution, and in the Universe at large. Although it seems most likely that liquid water and organic compounds are essential ingredients for the vast majority of (if not all) biotic systems in the Universe, it will be my theme that sulfur compounds may be of equivalent rank and may well permit the proliferation of life in certain environments not otherwise considered hospitable.

COSMOCHEMISTRY OF SULFUR

The most common isotope of sulfur, ^{32}S, has a nuclear proton-neutron complement equivalent to eight alpha particles or two ^{16}O nuclei. Nucleosynthesis during explosive oxygen burning in the supernova phase of stellar evolution produces sulfur at high yield (Truran, 1973). Indeed, if the relative abundances of the elements in the Sun (Cameron, 1973) are a reasonable guide to element profiles in other main-sequence stars, then iron is the only element above sulfur in the periodic table that is more abundant than sulfur in the Universe.

Even the so-called chemically peculiar (CP) stars contain the "normal" complement of sulfur (Preston, 1974). At the time of the tabulation by Herbst (1978), over 1/6 of all molecular species identified in interstellar space were sulfur-bearing (COS, CH_2S, H_2S, CS, etc.). The omnipresence of sulfur in the Universe is summarized in table 1.

In our own Solar System, sulfur compounds are not only surprisingly common, but often play key roles in planetary processes. Besides the high level of sulfur in the Sun and widespread and diverse occurrences on Earth, the Venusian atmosphere is laden with sulfuric acid aerosol droplets — in spite of original expectations (Lewis, 1972) that sulfur would not accrete as condensed compounds at or within the orbits of Venus and Mercury. The Martian regolith is surprisingly rich in sulfur, as is almost certainly its core. Jupiter and the rings of Saturn contain sulfur, as does at least one comet

TABLE 1.– SULFUR IN THE UNIVERSE

Solar System	
Sun	$S/Si \approx 0.5$
Venus	Atmosphere (H_2SO_4, SO_2, COS, $S°$, CS_2?)
Earth	Core, mantle, evaporites, ocean, biosphere, ores
Mars	Regolith, core (SO_4^{2-}, FeS)
Io	Surface ($S°$, H_2S, SO_2, S_2O)
Jupiter	
Saturn	
Comets	
(Asteroids), meteorites	
Interstellar dust clouds	COS, CH_2S, H_2S, CS, SO, SO_2, SiS, NS
Main-sequence stars	
CP stars, T-tauri stars	
Supernovae and supernova remnants	
"Hot gas" in galaxy clusters	
Cosmic rays	

(A. Delsemme, private communication, 1979), probably most asteroids, and virtually all meteorites. The latter can contain up to 6% by weight of sulfur (Moore, 1971). The surface of Io is extremely sulfur-rich and may include several allotropes of native sulfur, and possibly frozen SO_2 (Hapke, 1979).

ROLE IN PLANETARY EVOLUTION

The core of Earth is believed to be predominantly iron, with minor alloyed nickel and up to 15% sulfur (Murthy and Hall, 1972). Sulfur is suspected of playing a key role during formation of the core since the Fe-S eutectic melts at a temperature several hundred degrees lower than pure Fe or Fe-Ni. Even though as much as 99% of Earth's total sulfur inventory may be sequestered in the core, the residual crustal average abundance of about 500 ppm is of extraordinary importance. Volcanic and fumarolic emissions are conspicuously rich in sulfurous gases, notably H_2S and SO_2, with much lower levels of $S°$, COS, and SO_3 (table 2). Many ore deposits consist of massive sulfides of several different elements; many evaporite beds are gypsiferous ($CaSO_4$). In sea water, sulfate ion is second only to chlorine as the most common dissolved anion. Total evaporation of the oceans would leave a layer of sulfates over 3 m thick. Earth's stratosphere is laced with sulfate aerosol particles. Both inorganic and organic forms of sulfur have

TABLE 2.– COMPOSITION OF REACTIVE
VOLCANIC GASES (PERCENT)*

	Showa-Shinzan Volcano		Kilauea Volcano
	194°C	750°C	1200°C
H_2	13.6	25.0	0.4
CO_2	76.4	65.0	48.6
CH_4	.2	.1	---
H_2S	4.3	.1	.04
SO_2	.5	1.7	49.6
CO	---	---	1.4
(HCl + HF)	5.1	8.3	.03

*H_2O- and N_2-free values. Data summarized in Carmichael et al. (1974).

apparently been crucial to the evolution of life on Earth (see below). The O_2-CO_2 balance of our atmosphere is intimately connected with the buffering action by the reduced and oxidized sulfur reservoirs (Garrels et al., 1976).

Planetary evolution need not invariably involve sulfur, even though the examples in table 1 appear to implicate its involvement as a rule. An interesting exception is the Moon, whose surface is low in sulfur as well as most other volatile elements. The Moon is not only sterile today, but is almost completely devoid of the resources required to sustain life as we know it and can conceive it.

SULFUR COMPOUNDS

The versatility of sulfur lies in its ability to take up numerous valence states, including –2, 0, +2, +4, and +6, and to form chain compounds and compounds in which sulfur atoms appear in more than one valence. Table 3 is a brief list of many of the known types of inorganic sulfur compounds. Most occur to one degree or another in nature, although some, such as dithionite (an extremely strong reducing agent) and peroxydisulfate (a powerful oxidant), are much too reactive to persist. The properties of the sulfur atom which render it of great importance in organic chemistry and biochemistry were first summarized by Wald (1962).

TABLE 3.– SULFUR COMPOUNDS

H_2S, CS_2, COS	SO_2
FeS (pyrrhotite, troilite)	Na_2SO_3 (sulfite)
$Fe_3S_2O_2$ (sulfomagnetite)	$Na_2S_2O_5$ (pyrosulfite)
–SH (sulfhydryl)	$Na_2S_2O_6$ (dithionate)
FeS_2 (pyrite)	
	SO_3
$S°$, S_8	Na_2SO_4 (sulfate)
	$Na_2S_2O_7$ (pyrosulfate)
SO	$Na_2S_2O_8$ (peroxydisulfate)
$Na_2S_2O_3$ (thiosulfate, "hypo")	
$Na_2S_4O_6$ (tetrathionate)	
$Na_2S_2O_4$ (dithionite)	
$Na_2S_3O_6$ (trithionate)	

ROLE IN BIOLOGY

In the chemistry of life, sulfur plays major roles in energy transduction, enzyme action, and as a necessary constituent in certain biochemicals. The latter include important vitamins (biotin, thiamine), cofactors (CoA, CoM, glutathione), and hormones. Table 4 summarizes the biological utilization of sulfur compounds. Prebiotic chemical evolution may also have involved sulfur (Raulin and Toupance, 1977).

TABLE 4.– BIOLOGICAL UTILIZATION OF SULFUR

1. Energy source (sulfate reduction, sulfide oxidation)	5. Energy storage (APS, PAPS)
2. Photosynthesis (non-O_2-evolving)	6. Enzyme Prosthetic group, (Fe-S proteins)
3. Amino acids (met, cys)	7. Unique biochemicals (CoA, CoM, glutathione, biotin, thiamine, thiocyanate, penicillin, vasopressin, insulin)
4. Protein conformation (disulfide bridges)	

BIOENERGETICS

Sulfur is continuously cycled through its valence states by the biogeo-chemical cycle in nature. Figure 1 shows some of the organisms that partici-pate in the biological component of these reactions. All valence states of sulfur are susceptible to biological attack.

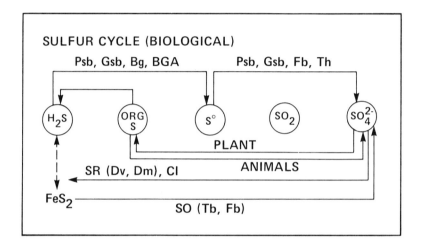

Figure 1. *Sulfur cycle in nature, as mediated by biological activity. Psb, purple sulfur bacteria; Bg, Beggiatoa; Fb, Ferrobacillus; Th, Thiobacillus; SR, sulfate reducers; SO, sulfate oxidizers; Dv, Desulfovibrio; Dm, Desulfotomaculum; Cl, Clostridium.*

Autotrophic energy metabolism, that is, the growth of organisms whose sources of chemical energy do not include organic compounds, involves sul-fur compounds in a surprising variety of cases. As shown in table 5, chemo-synthetic organisms exist for both oxidation and reduction of sulfur com-pounds: sulfur bacteria, sulfate reducers, and sulfur oxidizers. The sulfate reducers are not strict chemoautotrophs, however, in that not all of their internal organics can be synthesized from inorganic carbon (CO_2). A require-ment for small levels of either acetate or lactate as a partial carbon source has led to the classification of these organisms as mixotrophs. These organ-isms are broadly distributed despite their lack of satisfactory defenses against atmospheric O_2 (they are obligatory anaerobes).

Photosynthesis in the sulfur bacteria proceeds not by splitting H_2O, as in green plants and algae, but by splitting H_2S to obtain H atoms. This scheme of photosynthesis is simpler biochemically (involving a single stage, rather than the two-stage process for reduction of water), and it was only the

TABLE 5.– ENERGY METABOLISM INVOLVING SULFUR
COMPOUNDS*

Chemoautotrophy (sulfur oxidizers)
$$H_2S + 2O_2 \rightarrow H_2SO_4$$ Thiobacillus

Mixotrophy (sulfate reducers)
$$Na_2SO_4 + 4H_2 \rightarrow Na_2S + 4H_2O$$ Desulfovibrio

Photoautotrophy (purple and green sulfur bacteria)
$$H_2S + 2CO_2 + 2H_2O \rightarrow 2(CH_2O) + H_2SO_4$$ Chromatium
$$2H_2S + CO_2 \rightarrow (CH_2O) + 2S + H_2O$$ Chlorobium

*In sulfur oxidation, H_2S may be replaced by native sulfur, thio-
sulfate, or tetrathionate. For *T. denitrificans*, O_2 may be replaced by
nitrate. In sulfate reduction, SO_4^{2-} is replaceable by sulfite as well as
the H_2S alternates listed above. Molecular hydrogen can be substituted by organic hydrogen donors.

abundance of H_2O compared to H_2S that provided the evolutionary pressure
for the development of O_2-evolving photosynthesis.

Indeed, it is widely recognized that photoautotrophy arose with the
sulfur bacteria, possibly antedating the blue-green algae by as much as a
billion years. The reconstructed evolutionary sequence for microorganisms,
as proposed by Schwartz and Dayhoff (1978) and Hall (1979), indicates the
primitive nature of sulfur bacteria and the somewhat uncertain origin of the
sulfate reducers. Enzyme-sequence studies of Desulfovibrio species lead to
conflicting interpretations. Rubredoxin sequences place this organism close
to one of the primitive heterotrophs, Clostridium, but other molecules
(ferrodoxin and flavodoxin) indicate a later development. As shown below,
there is a plausible basis for the production of sulfites and sulfates even in
the primordial anoxic atmosphere. The presence of these substrates and
residual reducing power (H_2 and organics) provides the desired environment
for the sulfate reducers at very early times.

ENZYME CATALYSIS

The reactivity of the –SH group of cysteine has long been recognized as
important to the catalytic function of many enzymes. Additionally, the
–S–S– linkage to form cystine is a strong stabilizing force in establishing the
tertiary structure of proteins, which so often exerts a remarkable influence
on functional potency.

Sulfur plays yet another role in catalysis — perhaps its most fundamental contribution to living systems. That role is a "collaboration" with the iron atom to form Fe-S moieties that serve as prosthetic groups in the class of enzymes known as the "iron-sulfur proteins" (Lovenberg, 1973).

Iron-sulfur proteins are found in all living systems, including the highly unique methanogenic bacteria. These remarkable proteins are particularly suited to the catalysis of oxidation-reduction reactions, spanning the range +400 to –600 mV (scaled to the hydrogen electrode at –420 mV). Biological reactions catalyzed by Fe-S enzymes include uptake and evolution of molecular hydrogen (all known hydrogenases are Fe-S proteins), nitrogen fixation (nitrogenase), pyruvate metabolism, electron transport in bacterial photosynthesis (ferrodoxin and HiPIP), CO_2 fixation, $NADP^+$ reduction (chloroplasts), NADH oxidation (mitochondria), hydroxylation reactions, sulfite reduction, nitrite and nitrate reduction, xanthine oxidation, and probably many others (Hall et al., 1977). Indeed, they are the most numerous and most diversified components of the entire mitochondrial electron-transport system.

A particularly interesting subgroup of Fe-S proteins is the low-molecular-weight proteins, the ferrodoxins (Fd). They, too, apparently occur in all known organisms. The smallest and simplest Fd is found in Clostridium butyricum. It contains only 55 amino acid residues, 91% of which involve only 9 different amino acids (Hall et al., 1975). Ferrodoxins are thus interpreted as among the most primitive proteins available today for analysis. They represent the earliest electron-transport system and may even have formed the basis for the later evolution of heme-group binding in c-type cytochromes. Ferrodoxins are acid-labile and degrade under O_2. The Fe-S prosthetic group is particularly sensitive, and one purpose of the apoprotein is to protect this group from destruction by environmental reactants.

GALÁPAGOS DISCOVERY

The finding of small, isolated, and more or less complete ecosystems at the mouths of active hydrothermal submarine vents is important since such systems do not depend on photosynthesis for their primary productivity. The low $\delta^{13}C$ value of about –33 per mil for tissue from filter feeders collected at these vents is consistent with a chemosynthetic food source (Rau and Hedges, 1979). The presence of H_2S and CO_2 in the vented water, the abundance of sulfur-oxidizing bacteria, and the scarcity of heterotrophs all point to an energy source other than sunlight (Corliss and Ballard, 1977). However, these extraordinary communities mainly involve oxidizers and hence require dissolved O_2 from seawater. This O_2 is *not* generated by the vent, but mainly by photosynthesis in the surface layers.

It is interesting to speculate, however, that submarine volcanic emissions *could* provide all the necessary ingredients for a self-sustained ecosystem. As indicated in table 2, volcanic emissions include a number of reactive gases, including both reduced compounds (H_2, H_2S, CH_4, CO) and oxidized species (CO_2, SO_2, SO_3). Proportions vary considerably, even at the same volcanic area, but almost invariably include most of these species. Note that O_2 is so small as to be negligible. Quenching of high-temperature equilibrium compositions by rapid cooling provides a supply of gases that can further react with one another at seawater temperatures. For example, H_2 can react with either CO or CO_2 (to produce CH_4 and H_2O). These reactions are the sole energy source for growth by the methanogens.

Another energy source is the linked reactions shown in table 6. Here the sulfurous acid produced by volcanic SO_2 and H_2O ultimately yields sulfates by weathering reactions with igneous minerals such as olivine. These sulfates, which can include Mg, Ca, Na, or Fe salts, are then reduced by sulfate-reducing organisms with the aid of volcanic H_2. Once core formation had approached completion on the primitive Earth, and volcanic gases had settled into a redox balance that permitted formation of SO_2 and CO_2, the substrates for sulfate reduction and methane production became available.

TABLE 6.– A SULFUR–BASED ENERGY SOURCE FOR
SUPPORT OF A BIOLOGICAL SYSTEM*

1. Volcanic:	$SO_2 + H_2O \rightarrow H_2SO_3$
2. Weathering:	$2H_2SO_3 + Fe_2SiO_4 \rightarrow 2FeSO_3 + 2H_2O + SiO_2$
3. Disproportionation:	$4FeSO_3 \rightarrow 3FeSO_4 + FeS$
4. Biological:	$FeSO_4 + 4H_2 \rightarrow FeS + 4H_2O$
Net:	$2SO_2 + 6H_2 + Fe_2SiO_4 \rightarrow 2FeS + SiO_2 + 6H_2O$

*Step 4, the reduction of sulfate, is an energy-yielding reaction (as are the preceding ones) commonly employed by *Desulfovibrio* and other sulfate reducers as a metabolic energy source. Step 3, the disproportionation of sulfite to sulfate and sulfide, is not necessarily a required part of the reaction sequence since the sulfate reducers can also utilize sulfite as a substrate.

OTHER WORLDS

Hart (1979) presented calculations of climatological evolution for Earth-like planets orbiting other stars. His surprising conclusion is that the continuously habitable zone (CHZ), defined as the orbital distance at which such planets enjoy moderate temperatures (i.e., neither runaway glaciation nor runaway greenhouse heating occur), is narrow about G stars (±3% of the optimum distance for our own Sun) but probably nonexistent about K and M stars. It is wider about F stars, but such stars evolve rapidly into red giants. The likelihood of the terrestrial environment is thus lower than here-tofore expected. Certainly none of the 13 other large bodies (Moon-sized or larger) in the Solar System possess atmospheres or climates similar to Earth's. As the space exploration program has clearly shown, the only safe generalization is that planetary bodies cannot be generalized. They are *all* different in very fundamental respects, including bulk and atmospheric com-position, thermal history, and climate.

The assumption that only Earth-like environments qualify as CHZs is not a secure one. We have been biased by the idea that photosynthesis is of such fundamental significance that advanced biotic systems can persist only in environments coupled to illumination. However, the existence of niches only very indirectly coupled with the solar photon flux, such as the Galápagos vent communities, other benthic and marine mud ecosystems, and salt-marsh environments, emphasizes that the most fundamental requirement is energy flow to provide recycling of, or a fresh supply of, chemical poten-tial energy.

The basic requirements of life may simply be (1) a flux of energy, (2) a stable temperature regime compatible with the biochemistry of the organ-isms, (3) a liquid milieu, and, of course, (4) an initial supply of building-block elements, such as C, H, N, O, P, S, and transition metals. These ele-ments need not, in principle, be replenished since they can be recycled. Under these conditions, certain non-Earth-like environments may not only be conducive to life, but may be available in far greater abundance in the Universe.

Consider H_2O-rich bodies. In our Solar System, this includes not only Earth, but quite possibly Mars and Triton, and certainly Ganymede, Callisto, and Europa. Liquid water does not exist at the surface of any of these bodies, except Earth, but we should not discount the existence of "buried" liquid water reservoirs. All planet-sized bodies were once hot. Possible sources of this heat include gravitational potential energy, impact heating during accretion, short-lived radioisotopes (^{26}Al), and long-lived radioiso-topes (K, U, Th). For most bodies, the loss of this heat by radiation to space is not yet complete, even after 4 billion years. In some cases, especially bodies very rich in H_2O, much of this heat may have been lost by transfer to the surface by liquid and solid-state convection (Reynolds and Cassen, 1979), though three-body tidal interactions of the type apparently respon-sible for continuous heating of Io are also possible (Peale et al., 1979). Regardless of the manner in which they are formed, there is good reason to consider buried liquid water reservoirs as possible life-supporting environ-ments. The probable availability of dissolved salts, including sulfur com-pounds, and the existence of energy flow, in the form of a planetary heat flux, satisfy the life-support requirements listed above. Coupling to stellar or planetary luminosity may be completely unnecessary. For example, the scheme in table 6, whereby high-temperature reactions deep within the planet release upward-moving gases, which then may undergo additional, biologically catalyzed reactions, is possible. Recycling of the FeS produced by convective consumption into deeper layers could rejuvenate SO_2 and hence continuously support the process until the heat source is finally exhausted. This is not necessarily any more limiting than photosynthesis, since our own Sun will become a red giant in about another 6 billion years, in the process destroying not only terrestrial life, but all traces of its prior existence.

CONCLUSIONS

Sulfur is ubiquitous and probably plays several important roles in any exobiological organization. Although it can participate in photosynthesis, it also permits the chemoautotrophic way of life. Habitable zones include not only the surface ocean environment, but also the much more probable sub-surface oceanic regions. Earth-like environments as abodes for life may be the exception rather than the rule. Occupation of the more abundant buried zones is possible, and these should ultimately become an object of exploration. Whether such environments can support life long enough and at a sufficient level of activity to permit the evolution of highly encephalized forms (intelligent life) is conjectural. But the exotic and diversified nature of the mere handful of planetary bodies in our own Solar System substantiates the point of view that life in other parts of the Universe may inhabit environments we are not even capable of conceptualizing today.

Future accomplishments in many research fields will bear on the speculations presented here. These fields include chemical evolution, biological evolution, sulfur chemistry, enzyme mechanisms, chemoautotrophic ecological niches, planet and satellite formation, and the thermal history of planet-sized bodies. Of special interest are the large planetary satellites (e.g., the Galilean satellites of Jupiter) — their thermal history and the longevity of subsurface liquid water (Reynolds and Cassen, 1979). Another key question is the likelihood of totally chemoautotrophic systems surviving at scales and for durations adequate to permit higher evolution. Further elucidation of the evolutionary rise of microorganisms on Earth will shed light on this problem, as will additional study of special environments, such as submarine volcanic vents and their colonization by microorganisms.

REFERENCES

Cameron, A. G. W.: Abundances of the Elements in the Solar System. Space
Sci. Rev., vol. 15, Sept. 1973, pp. 121–146.

Carmichael, I. S. E.; Turner, F. J.; and Verhoogen, J.: Igneous Petrology.
New York: McGraw-Hill, 1974.

Clark, B. C.; Baird, A. K.; Rose, H. J.; Toulmin, P.; Keil, K.; Castro, A. J.;
Kelliher, W. C.; Rowe, C. D.; and Evans, P. H.: Inorganic Analyses of
Martian Surface Samples at the Viking Landing Sites. Science, vol. 194,
1976, pp. 1283–1288.

Clark, B. C.: Solar-Driven Chemical Energy Source for a Martian Biota.
Origins of Life, vol. 9, July 1979, pp. 241–249.

Corliss, J. B.; and Ballard, R. D.: Oases of Life in the Cold Abyss. National
Geographic, vol. 152, Oct. 1977, pp. 441–453.

Corliss, J. B.; Dymond, J.; Gordon, L. I.; Edmond, J. M.; von Herzen, R. P.;
Ballard, R. D.; Green, K.; Williams, D.; Bainbridge, A.; Crane, K.; and
van Andel, T. H.: Submarine Thermal Springs on the Galápagos Rift.
Science, vol. 203, 1979, pp. 1073–1083.

Garrels, R. M.; Lerman, A.; and Mackenzie, F. T.: Controls of Atmospheric
O_2 and CO_2: Past, Present, and Future. Amer. Scientist, vol. 64,
May-June 1976, pp. 306–315.

Hall, D. O.: Solar Energy Use Through Biology — Past, Present, and Future.
Solar Energy, vol. 22, 1979, pp. 307–328.

Hall, D. O.; Lumsden, J.; and Tel-Or, E.: Iron-Sulfur Proteins and Super-
oxide Dismutases in the Evolution of Photosynthetic Bacteria and
Algae. In: Chemical Evolution of the Early Precambrian, C. Ponnamperuma, ed., New York: Academic Press, 1977, pp. 191–210.

Hall, D. O.; Rao, K. K.; and Cammack, R.: The Iron-Sulfur Proteins: Structure, Function, and Evolution of a Unique Group of Proteins. Science
Progress (Oxford), vol. 62, 1975, pp. 285–317.

Hapke, B.: Io's Surface and Environs: A Magmatic-Volatile Model. Geophys.
Res. Lett., vol. 6, Oct. 1979, pp. 799–802.

Hart, M. H.: Habitable Zones Around Main Sequence Stars. Icarus, vol. 37, Jan. 1979, pp. 351–357.

Herbst, E.: The Current State of Interstellar Chemistry of Dense Clouds. In: Protostars and Planets, T. Gehrels, ed., Tucson: Univ. of Ariz. Press, 1978, pp. 88–99.

Klein, H. P.: The Viking Biology Experiments on Mars. Icarus, vol. 34, June 1978, pp. 666–674.

Lewis, J. S.: Metal/Silicate Fractionation in the Solar System. Earth and Planetary Science Letters, vol. 15, July 1972, pp. 286–290.

Lovenberg, W., ed.: Iron-Sulfur Proteins. 3 volumes, New York: Academic Press, 1973.

Moore, C. B.: Sulfur. In: Handbook of Elemental Abundances in Meteorites, B. Mason, ed., New York: Gordon & Breach, 1971.

Murthy, V. Ramo; and Hall, H. T.: The Origin and Chemical Composition of the Earth's Core. Phys. Earth Planet. Interiors, vol. 6, Dec. 1972, pp. 123–130.

Peale, S. J.; Cassen, P.; and Reynolds, R. T.: Melting of Io by Tidal Dissipation. Science, vol. 203, Mar. 1979, pp. 892–894.

Preston, G. W.: The Chemically Peculiar Stars of the Upper Main Sequence. Ann. Rev. Astron. Astrophys., vol. 12, 1974, pp. 257–277.

Rau, G. H.; and Hedges, J. I.: Carbon-13 Depletion in a Hydrothermal Vent Mussel: Suggestion of a Chemosynthetic Food Source. Science, vol. 203, Feb. 1979, pp. 648-649.

Raulin, F.; and Toupance, G.: The Role of Sulfur in Chemical Evolution. J. Molec. Evol., vol. 9, Aug. 1977, pp. 329–338.

Reynolds, R. T.; and Cassen, P. M.: On the Internal Structure of the Major Satellites of the Outer Planets. Geophys. Res. Lett., vol. 6, Feb. 1979, pp. 121–124.

Schwartz, R. M.; and Dayhoff, M. O.: Origins of Prokaryotes, Eukaryotes, Mitochondria, and Chloroplasts. Science, vol. 199, Jan. 1978, pp. 395–403.

Truran, J. W.: Theories of Nucleosynthesis. Space Sci. Rev., vol. 15, Sept. 1973, pp. 23–49.

Wald, George: Life in the Second and Third Periods or, Why Phosphorus and Sulphur for High Energy Bonds? In: Horizons in Biochemistry, M. Kasha and B. Pullman, eds., Academic Press, N.Y., 1962, pp. 127–142.

Role of Interfacial Water and Water in Thin Films in the Origin of Life

DUWAYNE M. ANDERSON

Molecular precursors of life probably evolved in a wet environment. Interfacial water and thin water films appear to form environments that are more favorable for precursor generation over a wider temperature range than water in the liquid phase.

It is widely accepted that some, perhaps many, of the abiotic chemical reactions and processes leading to the origin on Earth of replicating microorganisms occurred very early in the history of Earth in close proximity to the surfaces of clay minerals and other inorganic substrates (Bernal, 1967; Fox, 1965; Rutten, 1971; Anderson and Banin, 1975; Shimoyama et al., 1978). The mineral substrate is believed to have had the following functions:

1. Concentration of chemical reactants by adsorption
2. Preferential orientation of precursor molecules
3. Initiation and reaction rate enhancement by catalysis
4. Stabilization of intermediates
5. Suppression of randomness in recombination, addition, polymerization, and condensation reactions.

The last function is of particular significance for, "if the association of amino acids into a polymer were a completely random phenomenon, there would not be enough mass in the Earth — assuming it were exclusively amino acids — to make one molecule of every possible combination of units in a low molecular weight protein" (Dixon and Webb, 1964, as cited by Katchalsky, 1974). In considering these functions it is usually recognized, but not sufficiently appreciated, that the reactions of interest must have occurred in an aqueous environment and that aqueous interfacial phenomena were involved.

It has been observed that since the temperature coefficients of certain destructive decomposition reactions that would tend to inhibit the abiotic formation of RNA and DNA are quite large, low temperatures would have been advantageous and, in general, the lower the better (Miller and Orgel, 1974). It is of interest therefore to examine the properties of clay-water interfaces at low temperatures from the point of view of their possible involvement in chemical reactions important in abiotic syntheses of chemical precursors to cellular life.

This topic was reviewed by Anderson (1967) and subsequently by Anderson and Morganstern (1973). Anderson and Banin (1975) also covered essential aspects in their review of the role of soil and water in the origin of life. The nature of the various interfacial environments present in clay-water systems will be reexamined here with particular emphasis on their properties and behavior at low temperatures.

At temperatures below 0°C, the normal freezing point of water, five distinct interfacial environments are possible: ice-ice (grain boundaries), ice-air, silicate-water-air (SWA), silicate-water-silicate (SWS), and silicate-water-ice (SWI).

For the ice-ice interface, an order-disorder transition caused by ice lattice mismatch from one crystal to another causes the appearance of a transition zone with liquid-like properties between grains as temperatures approach 0°C. When impurities or solutes are present, they are concentrated at the interface, lowering the freezing point of the residual water and thickening the interfacial zone.

A liquid-like transition layer at the ice-air interface has long been suspected. Estimates of its thickness are as much as 100 Å (10^{-6} cm) near 0°C to about a monomolecular layer at –30°C and below. Again the presence of contaminants and solutes acts to increase the thickness of the interfacial liquid at any given temperature by depressing the freezing point of water in the interfacial liquid. Nuclear magnetic resonance (NMR) spectra of snow indicate the existence of a liquid-like component, but the width of the NMR signal compared to that of pure water indicates that in the interfacial phase it becomes more viscous and less free to move as a consequence of its interaction with the solid phase.

Silicate-water-silicate and silicate-water-air interfaces have been relatively well characterized at temperatures above freezing; they are less well characterized at lower temperatures. Electrical conductance measurements, NMR data, and protonation reactions indicate that the interfacial water in closest proximity to a clay surface and its exchangeable counter ions is dissociated to a much greater degree than normal (Anderson and Morganstern, 1973). This makes the interfacial liquid characteristically highly acidic and explains why protonation reactions are facilitated in this environment. This characteristic is of fundamental significance in considering the influence and possible roles of the SWS interface in primitive abiotic organic reactions.

The silicate-water interface may also be bounded by air at temperatures below freezing, but usually it is bounded by ice. Thus, depending on the quantity of water per unit of substrate surface, one may have distinct silicate-water and water-air interfaces and the SWS, SWA, and SWI interfaces, each with characteristic properties that vary with temperature, pressure, and solute concentration. The range of properties and behavior of these interfaces can thus be very large.

Until recently not much was known about the SWI interface, but certain basic information has now been obtained. When water-containing layer lattice clays (smectites) are frozen, the SWS interface decreases immediately from its initial value to about 9 Å. At about –10°C, the lattice spacing decreases further to about 6 Å, with no further noticeable decline at lower temperatures. For the SWI interface, a characteristic relationship between interface thickness and temperature can be derived from thermodynamic theory, and this has been confirmed by experimental measurements. The relationships for two homoionic forms of montmorillonite and for kaolinite, a nonswelling clay, are given in figure 1.

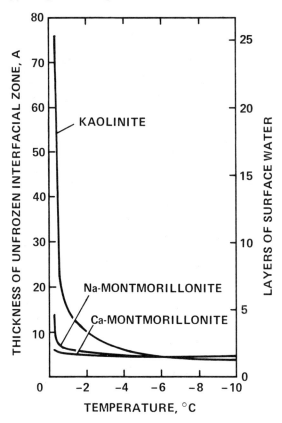

Figure 1. *Unfrozen water content plotted against temperature for three representative clays. Nature, vol. 216, Nov. 11, 1967.*

The nature of the silicate-water interface at temperatures below 0°C is illustrated in figure 2. Adsorbed water molecules (open circles) are shown bound to the clay surface together with some chemisorbed anions. Other specifically adsorbed molecular species and tightly bound cations might also be present. Hydrated cations are found separated from the mineral surface by their hydration sheath (dotted circles). They are shown symmetrically surrounded by water molecules, although asymmetrical arrangements are more likely and proton delocalization must be accounted for. Farther out is

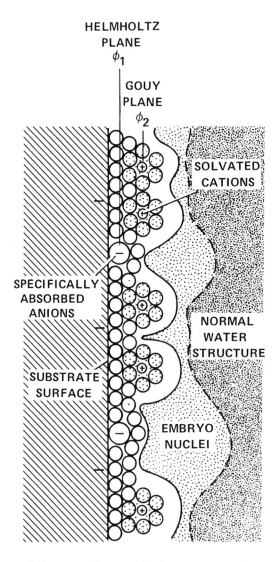

Figure 2. *Schematic of the zone of ice nuclei formation in soil water. Israel Journal of Chemistry, vol. 6, 1968, p. 350.*

the transition region separating the strongly adsorbed water from that which can be said to exhibit the properties and configuration of water in bulk. This zone is shown possessing bulges that fit in between the water domains of cationic perturbation. It is these bulges that are possible natal sites for embryo nuclei; consequently, they are shown as also extending outward into the bulk liquid. Once formed in regions like this, nuclei may be relatively stable. They are easily incorporated into the bulk ice on freezing, and on subsequent melting many may remain intact and active, provided the temperature is not raised too far above the melting point. This explains the common observation that subsequent freezings require little or no undercooling.

Although the influence of surface forces is somehow involved in their formation, the embryo nuclei do not grow by attachment to mineral surfaces, nor are they torn out of an extended network of ordered, chemisorbed water. Initially, adsorbed water is so disrupted by interaction with the surface and its adsorbed ions that embryo ice crystals are not able to form there. At a distance somewhat farther removed from the surface, however, the various adsorption forces may combine to exert a stabilizing influence that allows extensive ordering effects in the interfacial water. Depending on location relative to surface heterogeneities and distance outward from the surface, domains of ordered water, perhaps of more than one type, may exist. The location, extent, and lifetimes of these domains shift as the temperature is raised or lowered. Critical aspects of the environment created in proximity to a substrate surface thus involve reduced molecular motion and a tendency toward more open hydrogen-bonded regions in the fluid. As the temperature is lowered, the coherence and extent of hydrogen bonding must increase. Thus embryos form, and as the temperature is lowered, the favorable environment provided by the interplay of adsorption forces facilitates their enlargement into active nuclei. If one chooses to adopt the flickering cluster, mixture model of water, embryo nuclei may be said to form by the growth to critical size of one of the normal components of the liquid. Viewed in this way, heterogeneous nucleation becomes satisfyingly similar to homogeneous nucleation; it is only a matter of the temperature lowering required, in one case compared to the other, to permit the embryo to grow, by molecular accretion, to critical size.

Drost-Hansen (1967) has explored the nature of the water-ice interface in connection with the measurement of freezing potentials. As he visualizes it, six zones may be distinguished, ranging from the bulk liquid (consisting of clusters of clathrate cages and monomers) to normal ice. Separating these boundary phases is a disordered transition layer. Next is a zone consisting of a highly ordered layer of four coordinated water, termed "polar ice" inasmuch as many of the protons in this region are believed to be aligned perpendicular to the interface. The existence of this zone was postulated primarily to account for selective ion incorporation in ice during freezing.

Immediately adjacent are two zones within the solid phase in which dipole orientations are in various stages of relaxation and the selectively incorporated anions are being neutralized by migrating protons.

The main features of the SWI interface are illustrated in figure 3. The zone of embryo formation mentioned earlier, synonymous with the zone of "enhanced order" in figure 3, is believed to be more structured than free

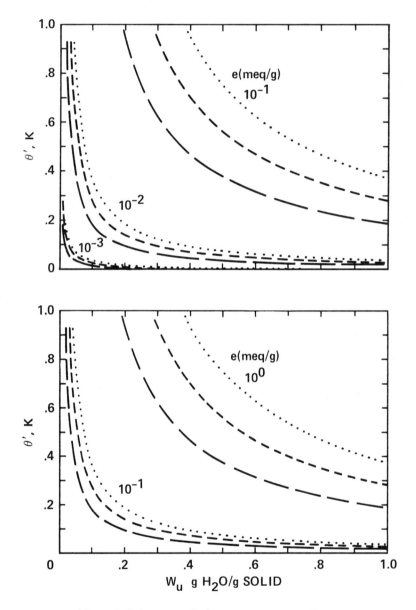

Figure 3. *Schematic of silicate-water-ice interface.*

water in the sense that the speed of formation and dissolution of the clusters of hydrogen-bonded water cages is thought to be diminished and the size of clusters and number of cages are thought to be larger than for free water. On the other hand, the number of clusters or cages per unit volume is believed to be reduced in the zones of disorder; monomers predominate there. Moreover, monomers closest to sites of surface charge, and the adsorbed cations, and water coordinated to ions located in the interfacial region undergo proton delocalization that, according to conductance and NMR data, leads to a proton activity in this zone many orders of magnitude higher than that of free water. It seems a paradox that, although the viscosity of water in the silicate-water and the silicate-water-ice interfaces is higher than that of normal water, proton mobility is greatly increased; nevertheless the evidence available at present indicates that this is the case.

Two zones shown in figure 3 are termed "disordered." For the zone adjacent to the silicate surface, this is based on the weight of evidence against hydrogen-bonded water networks and in favor of increased proton motion. For the zone adjacent to the ice surface, the argument is based on the principle of microscopic reversibility, which requires that the addition of water molecules to the ice lattice during interface advance must be caused by the same mechanism as the removal of water molecules during melting. If groups of water molecules were added on as "ice-like" units, it would be difficult to explain the phenomena of supercooling and the large entropy of fusion. If the units were not ice-like but clusters or cages of some different geometry, then there would be difficulty in simultaneously incorporating large numbers of these into a continuous hexagonal lattice. We conclude therefore that the dominant mechanism is the addition or subtraction of monomers. Furthermore, on the molecular level there must be continual exchange of water molecules within and between zones. The species most compatible with the phases present is the monomer; hence addition or subtraction of monomeric units is the most probable mechanism by which the interface is thickened and thinned.

NMR data show the SWI interface to be more structured than either bulk water or the silicate-water interface, and the above arguments require the presence of two zones in which monomers are dominant. It follows then that clusters and/or cages predominate in the intervening region. Measurements of the partial specific volume of water in an SWI interface indicate a water density less than that of free water, in harmony with the concept of increased hydrogen bonding, but it appears that hydrogen bonding proximate to clay surfaces is less well developed than that in free water. Since monomers are held to predominate in the two disordered zones, tending to increase the average interfacial water density, a net decrease in the average interfacial water density requires the existence of an intermediate zone where hydrogen-bonded clusters and/or cages predominate.

In exercising this tentative model of the SWI interface, it should be kept in mind that cations predominate in numbers over anions at the silicate surface and that, during advance of the ice-water boundary, anions probably predominate next to and within the polarized "pseudo-ice" transition layer. It is generally conceded that all cations and most anions tend to be water structure breakers, whereas F^- and OH^- seem to be net "structure makers" in water. It is thus possible that OH^- may play an important role in promoting local orientational order if delocalized protons are attracted to and tend to reside near the negatively charged silicate surfaces.

Protonation reactions have not yet been attempted at the SWI interface. However, the evidence cited above leads one to believe that attempts to carry out such reactions will be successful. Freezing soil water is equivalent to drying the soil; in both cases water leaves the mineral matrix. Since proton delocalization is greatest at the lowest hydration states, we conclude that protonation reactions are facilitated by a thin interfacial region. Protonation reactions in the SWI interface therefore may be possible at quite low temperatures in systems where the water (ice) content is high enough to inhibit reaction at normal temperatures.

When electrolytes are present, the general relationship illustrated in figure 1 is shifted to the right due to the freezing point depression effect (Banin and Anderson, 1974). Figure 4 illustrates typical effects.

The effect of a progressive lowering of the temperature of clay-water mixtures is illustrated in figure 5. Each curve represents a different mixture. As the temperature is lowered (moving along a given curve from right to left), nucleation occurs at a characteristic temperature (freezing-point depression) and a large exotherm expressed in electrical units of intensity appears. This exotherm is due to the evolution of the latent heat of freezing; it falls gradually to zero as freezing progresses. For drier clay-water mixtures, this exotherm appears at progressively lower temperatures and is progressively smaller. As the temperature drops further, one or more additional exotherms appear. These are associated in some as yet imperfectly understood manner with individual interfacial water domains. Since kaolinite and halloysite are nonexpandable clays and therefore exhibit only SWA and SWI interfaces, and since they exhibit only one low-temperature phase change at about $-40°C$, this exotherm is probably associated with a phase change in the SWI interface. The fact that montmorillonite possesses both SWA and SWS interfaces in the unfrozen condition and SWI and SWS interfaces in the frozen state, combined with the appearance of two or three low-temperature exotherms depending on the exchangeable ion present, is indicative of the different characteristics of these interfacial domains and the fact that they are altered to a significant degree by the presence of various ionic species.

The preceding discussion illustrates the great physical and chemical diversity of the interfaces at which various abiotic organic chemical reactions

LIQUID-LIKE INTERFACIAL ZONE   ICE

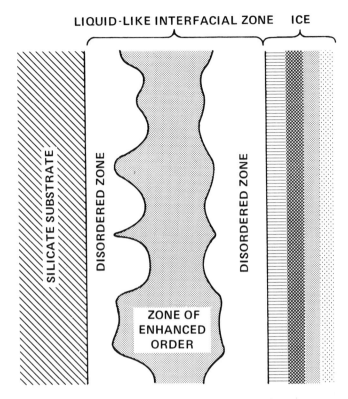

Figure 4. *Freezing-point shifts of pore solutions remaining in freezing porous bodies, due to increasing salt concentration of the initial pore solution. (Long dashes represent 1:1 electrolytes; short dashes 1:2 electrolytes; dots 1:3 electrolytes.)*

important in chemical evolution could have occurred. Superimposed on this diversity is the heterogeneity of types of clay substrate surfaces made possible by variations in mineralogy characteristic of the many different clay types. It should not be overlooked that curved surfaces as well as planar surfaces exist. Most clays change shape during hydration, dehydration, freezing, and thawing. Halloysite, for example, rolls into tubular configurations and unrolls during such cycles, and single smectite sheets have been observed to fold and unfold during wetting and drying. Curved interfaces such as these may have been crucial at certain stages of nucleotide synthesis. Little work has been done as yet to investigate such possibilities, but recent experimental and mathematical techniques have rendered these once formidable problems tractable.

As Miller and Orgel (1974) have observed, "All of the template-directed reactions that must have led to the emergence of biological organization take place only below the melting temperature of the appropriate organized polynucleotide structure. These temperatures range from 0°C, or lower, to perhaps 35°C in the case of polynucleotide-mononucleotide helices." They also

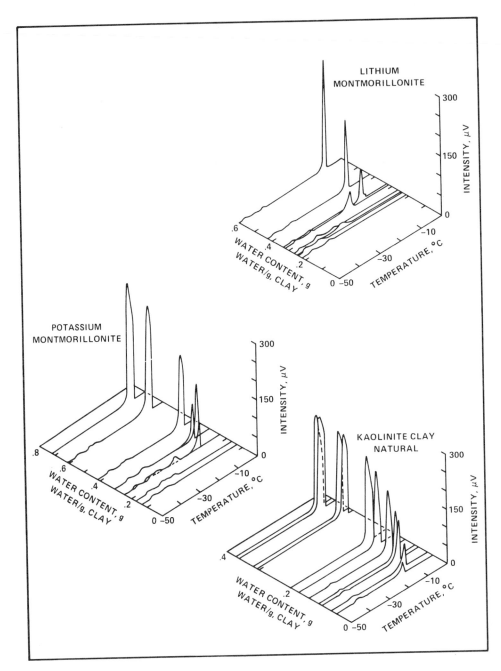

Figure 5. *Low-temperature differential thermal analysis data for representative clay-water mixtures (after Anderson and Tice, 1971). Soil Sci. Soc. America, vol. 35, Jan.-Feb. 1971.*

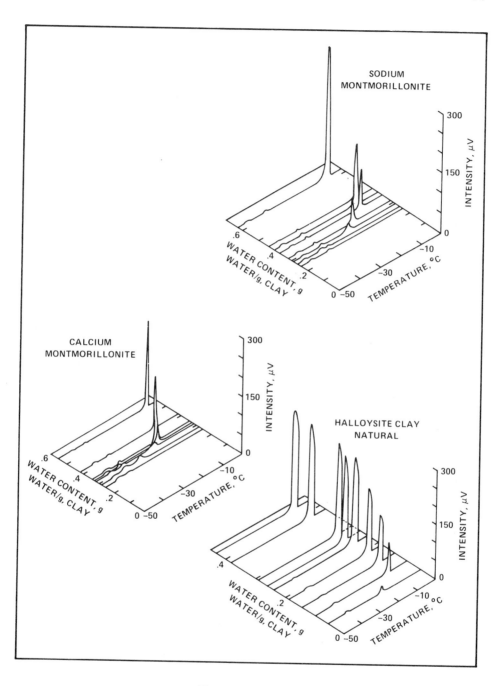

Figure 5. *Concluded.*

observed that the instability of the various organic compounds and polymers important at other stages "makes a compelling argument that life could not have arisen in the ocean unless the temperature was below 25°C. A temperature of 0°C would have helped greatly and –21°C would have been even better" (–21.8°C is the eutectic point for NaCl solutions).

At more advanced stages, low temperatures would inhibit membrane transport. Kushner (1976) has suggested that the physiological functioning of cells requires a semiliquid state for the lipids within the membrane and suggests that the lowest temperature permitting growth of primitive organisms might be determined by the melting point of the lipids they are capable of producing. Regulation by enzymatic systems is also impaired by temperatures near 0°C, although the work of DeVries (1970) on antarctic fishes has shown that, if ice nucleation can be prevented, cellular process can proceed normally several degrees below 0°C.

Horowitz (1976) has noted the necessity of considering the thermodynamic activity of water whenever one considers the water requirements of living cells. He gives an activity of 0.62 as the probable lower limit to growth as we presently know it. Figure 1 is interesting in this connection since this general relationship has been derived from thermodynamic theory (Low et al., 1968). For montmorillonite clay, the activity of the unfrozen interfacial water approaches 0.62 at a temperature of about –25°C. At this point, the interface is only one or two monomolecular layers thick; therefore it must also be highly acidic in nature. Between –5° and –1°C, the interfacial water activity ranges from 0.95 to 0.99 and the thickness of the interfacial zone varies from two to six monomolecular layers of water. Thus it appears that this interface possesses properties compatible with some living organisms down to temperatures far below 0°C. As indicated in figure 5, the limit is probably reached somewhere between –25° and –30°C, where the first low-temperature exotherms due to a change of phase in the interfacial water appears.

The preceding discussion has been confined to a demonstration that the properties of the unfrozen interfacial water films present in frozen clay-water systems are such that a wide variety of organic reactions of importance in the origin of life could have been facilitated: by the stabilizing presence of clay substrates; by the catalytic properties of the acidic clay surfaces; by the various dynamic geometrical characteristics of the surfaces presented by different clay lattices, which could, for example, have been important in facilitating ordering and coding in primitive polynucleotides; and by the presence of a liquid phase at relatively high water activities at temperatures quite far below the normal freezing point of water. This, in effect, makes the temperature range 0° to –35°C, even without the presence of brines, compatible with the other requirements for the abiotic synthesis of many of the organic molecules required for the natural appearance of life on primitive Earth and opens up several fruitful new lines of experimentation and theoretical investigation.

Water and ice are known to exist on Mars, Europa, Ganymede, Callisto, and the moons of Saturn. If the abiotic synthesis of organic molecules important in the evolution of life is occurring on any of these planetary bodies, it is likely that a significant portion of the reactions of interest are taking place in interfacial environments such as those described here. Future unmanned landers should include instrumental and sampling capabilities to search for evidence of these processes.

I wish to thank Dr. James MacAlear for many interesting and stimulating discussions on the possible influence of the dynamic character of clay surface geometry.

REFERENCES

Anderson, D. M.: The Interface Between Ice and Silicate Surfaces. J. Colloid Sci., vol. 25, Oct. 1967, pp. 174–191.

Anderson, D. M.; and Banin, A.: Soil and Water and Its Relationship to the Origin of Life. Origins of Life, vol. 6, 1975, pp. 23–26.

Anderson, D. M.; and Morganstern, N. R.: Permafrost: North American Contribution. Proceedings 2nd Int. Conf. on Permafrost, Yakutsk, Siberia, 1973. Washington, D.C.: National Academy of Science.

Anderson, D. M.; and Tice, A. R.: Low Temperature Phases of Interfacial Water in Clay-Water Systems. Soil Science Society of America Proceedings, vol. 35, no. 1, Jan.-Feb. 1971, pp. 47–54.

Banin, A.; and Anderson, D. M.: Water Resources Res., vol. 10, 1974, pp. 124–128.

Bernal, J. D.: The Origin of Life. Cleveland: World Publishing Co., 1967, p. 345.

DeVries, A. L.: In: Antarctic Ecology, vol. 1, W. M. Holdgate, Ed. New York: Academic Press, 1970.

Drost-Hansen, W.: The Water-Ice Interface as Seen From the Liquid Side. J. Colloid Interface Sci., vol. 25, 1967, pp. 131–160.

Fox, S. W., ed: The Origins of Prebiological Systems and of Their Molecular Matrices, Academic Press, N.Y., 1965, p. 167.

Horowitz, N. A.: Life in Extreme Environments: Biological Water Requirements. In: Chemical Evolution of the Giant Planets, C. Ponnamperuma, ed., Academic Press, N.Y., 1976, pp. 121–128.

Katchalsky, A.: Chemico Diffusional Coupling in Heterogeneous Peptide Synthesis. In: Irreversible Thermodynamics and the Origin of Life, G. F. Oster, I. L. Silver, and C. A. Tobias, eds., Gordon & Breach, New York, 1974, pp. 1–9.

Kushner, D. J.: Microbial Life at Low Temperatures. In: Chemical Evaluation of the Giant Planets, C. Ponnamperuma, ed., Academic Press, N.Y., 1976, pp. 85–93.

Low, P. F.; Anderson, D. M.; and Hoekstra, P.: Water Resources Res., vol. 4, 1968, pp. 379–394.

Miller, S. L.; and Orgel, L. E.: The Origins of Life on the Earth, Prentice-Hall, Englewood Cliffs, N.J., 1974.

Rutten, M. G.: The Origin of Life by Natural Causes. Elsevier, Amsterdam, 1971.

Shimoyama, A.; Blair, N.; and Ponnamperuma, C.: Synthesis of Amino Acids Under Primitive Earth Conditions in the Presence of Clay. In: Origin of Life, 1978, pp. 95–99.

II — Life Supporting Environments

It would be fine if astronomers would identify for us which stars could be accompanied by planets, and it would be staggering if through some technologic *tour de force* they were able to demonstrate the positive presence of planets in a few such cases. Yet the next step — to determine whether life as we know it could exist there — involves many parameters whose interplay is poorly understood even under optimum observational circumstances and which seem utterly indeterminate at interstellar distances. Everyone recognizes that this is so. Yet the fact that this Conference is being held is evidence of our belief or hope that somehow or someday the subject may open up, probably in a way no one here can anticipate.

That may be too slender a thread for most scientists to hang their professional energies on, but the subject of the viability of planetary surfaces in our Solar System is fully worth our attention. The exploration of the Solar System by spacecraft is opening up a vast field of investigation which did not exist as a serious subject 15 years ago. We cannot help but gain a better feeling for the possibilities — not the probabilities — of life near other stars if we are able first to understand the processes that have operated on the surfaces of our own planets. We begin with an examination of what may or may not be a very special case, our own Earth.

Dr. George H. Herbig
Professor of Astronomy
University of California
at Santa Cruz

Atmospheres and Evolution

LYNN MARGULIS and JAMES E. LOVELOCK

Bacteria have dominated Earth's biosphere since early Precambrian times, altering surface and atmosphere, maintaining and evolving the environment of all life. To understand the phenomena on which our survival depends, we must view Earth in a unified context.

Our purpose is to summarize some work of many people concerning regulation of the atmosphere and its relation to the evolution of microbial life. Remarkably, looking back, one recognizes a strange unconscious kind of cooperation among millions of species of organisms over at least 3 billion years. This cooperation, itself a product of evolution, has altered the entire surface of planet Earth.

We will have five main points. First, from a planetary perspective — that is, comparing Earth, Mars, and Venus — Earth's atmosphere is totally improbable. It has too much oxygen in the presence of too many gases that react with that oxygen. Furthermore, for a long time — hundreds of millions of years at least — Earth's atmosphere has not been obeying the rules of physics and chemistry alone. Perhaps we are beginning to understand why not.

Second, life on Earth is far older than most people had reckoned until recently. Life extends as far back as the oldest sedimentary rocks, perhaps as far back as the oldest rocks on Earth, which are metamorphic (Schidlowski et al., 1979).

The third point is that the living world is not divided primarily into plants and animals. Green plants and animals are in fact closely related organisms. The living world is rather divided into two other groups: *procaryotes* such as bacteria, whose cells lack nuclei, and *eucaryotes* such as animals, plants, and fungi, which are nucleated organisms. The differences between these groups are far greater than the differences between animals

79

and plants. The oldest life on Earth consisted of bacterial cells, which tend to be small. Bacteria have dominated Earth since the early Precambrian, and despite our anthropocentrism, they dominate Earth today. Bacteria far more than eucaryotes have altered the planet and its atmosphere. We must understand the interactions of bacteria with the atmosphere and the surface of Earth to perceive why Earth is so different from Mars and Venus.

Fourth, regarding the origins of the eucaryotic cells, they probably evolved from bacterial cells by mechanisms other than the simple accumulation of random mutations and that the establishment of intracellular symbioses played an important part in this process. Different species of bacterial cells established associations and coevolved, forming new kinds of units that became ancestors to animal, plant, and fungal cells. The appearance of organisms composed of nucleated cells led to many changes on Earth and was a prerequisite for the superficial "explosion" of invertebrate animals at the Phanerozoic boundary.

Finally, we will discuss the idea that the gases of the atmosphere are kept in balance mainly by bacterial cells. The argument will be that species diversity is absolutely required for current maintenance of conditions hospitable to life. Furthermore, without species diversity, the living system on the surface of Earth would not have originated and been maintained for more than 3 billion years.

The results of the Viking mission to Mars are having a profound effect on our thinking about Earth, which now appears to be the only planet in the Solar System that harbors life (Mazur et al., 1979). An important scientific goal is therefore to understand the factors that make our planet and its living system unique. It is our belief that, unless scientists of various disciplines employ their tools in a joint effort to look at Earth in a planetary context, we will never understand the phenomena upon which our survival depends.

Let us start by comparing the atmospheres and surfaces of Mars, Earth, and Venus. The surface of Mars, pockmarked with craters, probably records very ancient events in the history of the Solar System. The Viking lander found a rubble-strewn, extraordinarily dry terrain. The atmosphere is primarily composed of carbon dioxide, although it contains almost 2% nitrogen and less than 1% oxygen. Mars now is dry, oxidized, and, relative to the Earth, unchanging.

Before the hot, acidic conditions on Venus burned up the camera in less than a minute, Venera 9 took photographs of another dry, rubbly surface. The Soviet missions to Venus have studied the surface of that planet. They reveal a place in some respects very much like Mars. The atmosphere of Venus, too, is composed of carbon dioxide and small quantities of nitrogen, traces of oxygen, and very little water.

In what way is Earth's surface improbable? Superficially, it appears to be a normal planet between Mars and Venus. Without life, Earth's atmosphere would probably be far more like that of its neighbors, containing carbon dioxide, nitrogen, trace amounts of oxygen, and water vapor. But as it is, Earth's atmosphere has far too little carbon dioxide and far too much oxygen.

Earth's atmosphere contains the hydrogen-rich gases that are found in the atmospheres of the outer planets, but it also contains oxygen, which tends to react with hydrogen and hydrogen-rich gases. There are other chemical anomalies in Earth's atmosphere: it contains sulfur compounds such as dimethyl sulfide as well as highly improbable compounds such as terpenes, disparlure, eugenol, myoporum, nepetalactone, and butyl mercaptan. Unlike any of the inner planets, Earth's atmosphere also contains phosphate, but it is in particulate forms we call spores, seeds, birds, and bats! The presence of these complex organics can only be understood on the basis of their informational role among the biota (Margulis and Lovelock, 1975).

Table 1 documents the chemical improbability of the atmosphere by answering the question: Given that the atmosphere is 20% oxygen, what quantities of the other reactive gases would we expect to be simultaneously present?

Hydrogen and oxygen, of course, react to form water; nitrogen and oxygen react to form nitrate. Ammonia and oxygen also react, as do methane and oxygen. Earth's atmosphere seems to be composed of a combustible mixture. From a chemical point of view, it is far from equilibrium, yet it is maintained in a steady state with a composition unlike that of the atmospheres of Mars and Venus. It is rather more like fuel than like exhaust. Given 20% oxygen and standard conditions, table 1 shows what the approximate equilibrium concentration of the other gases should be. Nitrogen is too abundant by about a factor of 10, but methane is too abundant by a factor

TABLE 1.— THE ATMOSPHERE PROBLEM

	Present concentration	Expected equilibrium concentration[a]	Approximate discrepancy	Residence time, yr	Output, 10^6 tons/yr
Nitrogen	0.8	$<10^{-10}$	10^9	3×10^6	1000
Methane	1.5×10^{-6}	$<10^{-35}$	10^{29}	7	2000
Nitrous oxide	3×10^{-7}	$<10^{-20}$	10^{13}	10	300
Ammonia	10^{-8}	$<10^{-35}$	10^{27}	.01	1500
Methyl iodide	10^{-12}	$<10^{-35}$	10^{23}	.001	30
Hydrogen	5×10^{-7}	$<10^{-35}$	10^{28}	2	1000

[a]Given that oxygen comprises about 0.20 and is released into the atmosphere at the rate of about 110×10^9 tons per year.

of about 10^{35}! Scientists often make errors of a few percent, and are occasionally off by factors of 100, but when we make errors of 10^{35}, we are very poor scientists indeed. Models that attempt to explain the composition of Earth's atmosphere by physics and chemistry alone tend to sustain errors of 10^{35}, a strong argument that such models do not suffice. The atmosphere also contains far too much nitrous oxide, ammonia, hydrogen, and hydrocarbons, given the quantity of atmospheric oxygen. Furthermore all gases listed in table 1 move through the atmosphere. From a geological point of view they have very short residence times: they cycle in years, months, or even days in some cases. Of the nonnoble gases, the residence time of nitrogen is the longest, of the order of a million years. Those of methane, nitrous oxide, and the others are far shorter — only 10 years or less.

The annual output of these gases — the production rates — are enormous, measured in billions of tons per year. These large production rates and short residence times imply that gaseous emissions occur continuously. What produces and maintains such reactive gases at anomalously high concentrations? The tectonic and volcanic contribution is minor. In fact, the vast quantity of cycling gas is produced by microorganisms and plants, with animals contributing very little. Furthermore, neither plants nor animals qualitatively contribute to the production of many gases, such as methane and nitrous oxide; however, for each gas listed in table 1 there is always a microbial source (Margulis and Lovelock, 1974).

Our second point is that life is very much older than previously thought. Figure 1 shows a "time line" measured in billions of years. Little is known about Earth before 4 billion years ago. However, starting about 3.5 billion years ago, a fossil record of life is abundantly available. Most scientific effort has gone to develop a paleontological perspective that does not begin until the Phanerozoic, only about a half a billion years ago. Extensive fossil reconstructions of life during the past half billion years are available, as museum lovers are well aware. But what occurred prior to half a billion years? Why is the record for life so different during the first 3 billion years of Earth's history?

By half a billion years ago, there were well-developed communities of invertebrate animals and algae. The "dawn of fossil records" is marked by trilobite fossils that appear about 600 million years ago at the Cambrian boundary. By the lower Paleozoic era, many kinds of skeletalized animals had appeared, some squid-like, some starfish-like. Most look generally familiar even though they were members of species now extinct. An immense body of paleontological evidence exists for communities of reptiles, tropical forests, aquatic invertebrates, coral reefs, and so forth by Mesozoic times. Ecosystems during the Phanerozoic, the last half billion years, although differing from modern ones in detail, are recognizable and changed in familiar ways. They are quite different from those of the pre-Phanerozoic.

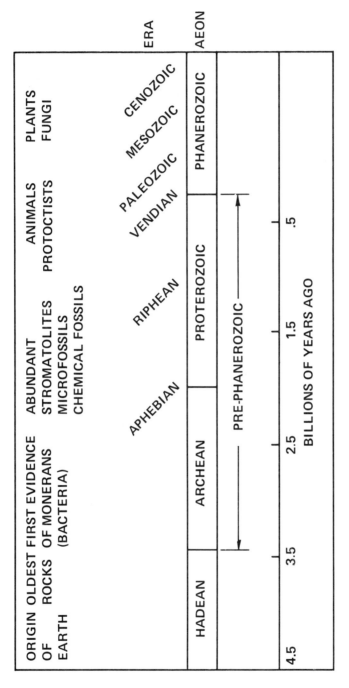

Figure 1. *Times of origin of major events during past eons.*

During the late Cenozoic, populations of woolly rhinoceroses and mammoths and other familiar but now extinct beasts roamed the Earth. Forest, lake, savannah, alpine, reef, and a multitude of other well-known ecosystems to support these animals had long since been established. In the last few million years, human-like australopithecines and members of our genus *Homo* have thrived. On the scale of billions of years, whatever else he is, man is but an upstart on Earth, some 0.004 billion years old. Furthermore, *Homo sapiens* came into a well prepared scene in which complex interacting microbial and plant ecosystems were highly developed.

Bacterial ecosystems preceded those dominated by animals, plants, and fungi (Reimer et al., 1979). Familiar animal, plant, and fungus species are, in fact, closely related to each other, and the eucaryotic cells of which they are composed appear relatively recently in the fossil record, during the Proterozoic. Some differences between them and the smaller, nonnucleated bacterial cells that greatly preceded them are indicated in figure 2.

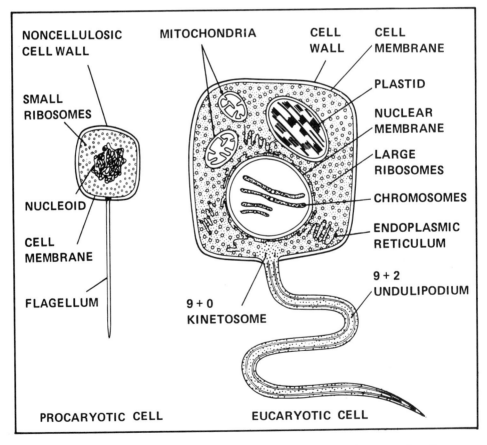

Figure 2. *Comparison of the structure of procaryotic and eucaryotic cells.*

All bacteria, including the blue-green algae, are morphologically relatively simple. Chemically, however, bacterial cells are extremely complex and diverse. From the point of view of gas metabolism, bacteria are far more versatile than animals, plants, and fungi, which take in oxygen and release carbon dioxide. Plant cells also take in carbon dioxide for photosynthesis, but animal, plant, and fungal aerobic metabolism is generally rather uniform. The point we wish to make here is that nearly every major metabolic feat performed by these eucaryotes is also represented in the bacterial world. Moreover, bacteria are able to carry out many other gas transformations such as nitrogen fixation and methanogenesis. It is generally conceded that bacterial cells initially changed the atmosphere; for example, the blue-greens first produced large quantities of atmospheric oxygen (Cloud, 1974). We will argue that the metabolic uniformity of eucaryotes can be understood in the context of a symbiotic theory of the origin of animal, plant, and fungal cells, which asserts that certain partnerships among bacterial cells eventually led to the origin of nucleated cells (Margulis, 1981).

Bacterial cells lack chromosomes, whereas plants and fungi tend to have chromosomes so like those of animal cells that often the three types of cells are difficult to distinguish. The ultrastructure of bacterial cells, including the blue-greens, as revealed by electron microscopy is so different from that of nucleated organisms that there is no ambiguity in the distinction.

What was the course of evolution during the enormous stretch of time from 3.5 until about 5 billion years ago? The Precambrian era was once considered to be nonfossiliferous, that is, to have left scant or no evidence of life at all. But in the last 20 years or so, it has been shown that life was abundant during the Archean and especially during the Proterozoic. Fossiliferous rock samples from all over the world and from various points in time before the Phanerozoic are now available. Part of the difficulty that hampered recognition of pre-Phanerozoic life is related to techniques and to scientific expectations. One aspect of this is the gradualistic view of cell evolution that prevailed before the symbiotic theory was considered.

A comparison between an ancient rock sample, a carbon-rich cryptocrystalline silicate rock called a chert, and a modern sediment is shown in figure 3. Until recently such commonplace reservoirs of life have been ignored by biologists and paleontologists. In fact, upon close examination, both the ancient and modern sediments are filled with evidence for life. A great part of the revolution in our thinking about rocks and soft sediments such as these in the context of the early history of life is attributable to the insight of Elso Barghoorn of Harvard and his late colleague Stanley Tyler. These investigators realized that the conspicuous fossil record of large organisms must have been preceded by something smaller. Such ideas led them to microscopic studies of thin sections of unaltered sedimentary rocks.

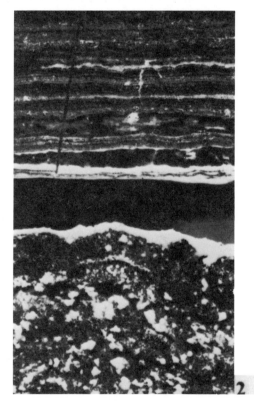

Figure 3. *Comparison between an ancient thinly laminated chert from the 3.4-billion-year-old Swaziland system in South Africa with banded sediments from a live community of microbes at Laauna Figueroa, Baja California, Mexico.*

If some of the black and smooth areas of cherts such as those shown in figure 3 are sectioned with a diamond knife and examined with optical microscopy, inclusions interpreted to be remnants of the oldest organisms on Earth can be observed. Whether they are indeed fossil bacteria is still debated; we think that the evidence for fossils in rocks older than 3 billion years is excellent (Knoll and Barghoorn, 1977). By 2 billion years ago, a multitude of microbes thrived; these are detected as fossils in thin sections and macerations of rocks (Schopf, 1975).

A further important clue to ancient microbial life is given by the stromatolites, which are layered sedimentary rocks produced by the trapping, binding, and precipitation of sediments by actively metabolizing communities of microorganisms. They may be thought of as the ruins of ancient "cities" of microorganisms, which began as microbial mats but then became lithified and preserved. Examples from all over the world are now available (Walter, 1976).

Some ancient cherts reveal upon microscopic examination large numbers of well-preserved, fossil microorganisms, including filamentous ones that resemble modern blue-greens so closely that they probably belonged to genera that persist today and that still produce atmospheric oxygen by photosynthesis. Direct microfossil evidence for the potential to produce atmospheric oxygen 2 billion years ago and probably far earlier is abundantly available.

Are there living ecosystems dominated by microorganisms that might tend to lithify and thus might be used for comparison with ancient stromatolites and fossiliferous cherts? Figure 4 shows Laguna Figueroa in Baja California del Norte, Mexico, some 250 km south of San Diego. Very little life seems to prevail in such hypersaline lagoons; in the absence of fresh water input, animals and plants are virtually absent. However, we are beginning to recognize that not only are such sediments thriving with microbial life, but they may provide realistic analogs to fossilized microbial communities. Laguna Figueroa is a closed, generally dry lagoon dominated in some regions by conspicuous evaporatic polygons (Horodyski, 1977). When these are sectioned, sediments of differing colors and textures are revealed. The colored laminations are due in part to stratified microbial communities, for example, mats dominated at the surface by the sheathed cyanobacterium, *Microcoleus vaginatus*, and underlain by purple photosynthetic bacteria (Margulis et al., 1979a). Smelly gases are emitted from the sediments; gas bubbles are seen everywhere. Whatever scums these mats and sediments contain, they certainly are sources of some of the "out-of-equilibrium" gases presented in table 1. The mud-bound microbes produce hydrogen sulfide, ammonia, methane, and many other gases — none of which have yet been measured in the field.

Figure 4. *Panoramic views of the microbial mat communities at Laguna Figueroa. (Courtesy of Robert Horodyski)*

In dry seasons this lagoonal complex extends for about 8 km in the north-south direction and 1 km across. Is life depleted in this hypersaline hot harsh region? No. Complex patchy colored surfaces prevail, composed of dense communities of microbes, the vast majority of which have not been identified and cultured. Some of the microbes in the mats at Laguna Figueroa resemble those preserved in cherts several billion years old (Knoll and Barghoorn, 1977); these communities are also similar to the inferred environment of deposition of the carbon-rich shales of South Africa, which are over 3 billion years old (Reimer et al., 1979).

In communities such as those at Laguna Figueroa, which may serve as analogs to Precambrian microbial ecosystems, the photosynthetic bacteria are usually well represented. Sheets of purple anaerobic photosynthetic bacteria — organisms that remove hydrogen sulfide but do not produce oxygen — may even form extensive surface colonies. These bacteria produce food for other forms of bacteria thriving below the surface. The conspicuous layers, at least near the surface, represent the stratifications of the naturally growing microbial ecosystem (fig. 3). In many places the surface layer is made of evaporites such as gypsum and halite; sand and sometimes aragonite are present. The second layer often tends to be dominated by blue-greens, which produce oxygen. The lower layers usually harbor anaerobic photosynthetic bacteria, with nonphotosynthetic bacteria of many kinds below them. Figure 5 shows some photosynthetic and other bacteria isolated from the microbial mats. Electron micrographs of material from surface scums of these mats reveal many photosynthetic bacteria, some probably new to science (Margulis et al., 1979a).

The sediments of the Laguna Figueroa community are compared in figure 3 with a 3.4-billion-year-old banded chert from the South African Swaziland system. Again, the community of live microorganisms is hypothesized to be analogous to some ancient one that was silicified. The modern microbial community probably lives under conditions of higher salinity than the Archean one. In today's world, if there were less salt and less extreme conditions generally, animals and plants would tend to dominate. This sort of invasion by eucaryotes was seen in the spring of 1979 after a long period of heavy rains and an influx of fresh water. Conspicuous microbial communities such as these are restricted to quiet waters in hot saline tropical and subtropical locations such as the Bahamas, Baja California, and western Australia, and to hot springs and geysers. In the Precambrian, when competition by animals and plants was lacking, they were probably far more widespread.

Evidence for ancient communities of microorganisms abound in the Precambrian fossil record from 3.5 until 0.5 billion years. Stromatolites made of calcium carbonate or, less frequently, of silica are especially common from 1.0 to 0.5 billion years ago. Some have striking modern counterparts. Examples of stromatolites from tropical western Australia have been described in detail (Walter, 1976). If these laminated calcium carbonate rocks are sectioned again, stratified communities of microbes will be found with them. The species comprising the community in large stromatolites differ from those found in Laguna Figueroa sediments. In carbonate-accreting environments, sheathed filamentous blue-greens grow, trapping sand, calcium carbonate particles, and other bits of sediment. By trapping particulates in their mucous polysaccharide sheaths, blue-greens often cover themselves with clasts until the light they require for photosynthesis no longer penetrates (Golubic, 1973). Many of these filamentous blue-greens exhibit a kind of gliding motility. Thus they are capable of active movement, emerging from beneath the surface layer of particles and growing up through it. This type of growth pattern forms layers rich in the remains of organisms:

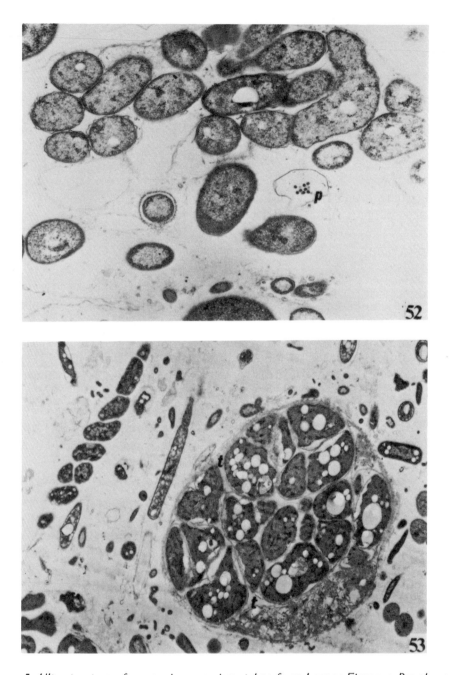

Figure 5. *Ultrastructure of some microorganisms taken from Laguna Figueroa. P = phage. (Courtesy of David Chase)*

organic debris alternating with layers rich in clastics. Such recent stromatolites found at Hamelin Pond in Shark Bay, western Australia, are similar to ancient ones formed some 2.3 billion years ago found near the Arctic Circle in northwest Canada (Walter, 1976).

The direct evidence for pre-Phanerozoic life also includes "chemical fossils," refractory organic complexes found in the keragen fraction, material, by definition, not extractable from rocks by standard aqueous or organic solvents. Thus there is substantial stromatolite microfossil and geochemical evidence that early Earth was dominated by bacterial communities.

From where, then, did communities of animal, plants, and fungi come? Eucaryotic cells contain organelles ("little organs" or cell parts) called mitochondria that are the sites of oxygen respiration (fig. 2); the mitochondria are absent from all bacterial cells. In addition, plant and algal cells contain plastids, membrane-bound compartments within which photosynthesis takes place. It is our opinion that both the photosynthetic plastids of plant cells and the mitochondria of plant, animal, and fungal cells were once independent bacteria. New biochemical data on the sequence of amino acid residues in certain proteins are in agreement with the concept that plastids and mitochondria were once free-living bacteria, thus suggesting a symbiotic origin of eucaryotic cell components (Schwartz and Dayhoff, 1978). A plant cell can be thought of as a composite formed by partnerships among bacteria that could convert sunlight into food, bacteria that became plastids, and aerobic bacteria that generated energy by burning oxygen and became mitochondria.

One striking difference between eucaryotic and procaryotic cells is the nature of the whiplike organelle of motility, the flagellum. In eucaryotes, the flagellum has a peculiar complex ninefold symmetry. It is made of nine sets of doublet tubes of protein and two central tubes, all called microtubules. These so-called 9+2 structures (fig. 2) are absent in procaryotes. The organelle of flagellated bacteria is far smaller, solid, and simpler, and is composed of an entirely different protein than those found in eucaryotes. The name "undulipodium" (Smagina in Corliss, 1979) has been suggested to refer to the eucaryotic but not the procaryotic flagellum. The uniformity of the 9+2 undulipodia is remarkable; for example, the ultrastructure of both the gill cilia of marine animals and the oviduct cilia of women display this ninefold symmetrical pattern of microtubules. Furthermore, not only do the sperm tails of mammals and amphibians, indeed of nearly all animals, have the 9+2 motif, even male ginkgo trees have the same detailed structure in the tails of their sperm! This sort of observation has led to a suggestion that both animals and plants evolved from cells bearing undulipodia, which may themselves also have been free-living once (Margulis et al., 1979b). The hypothesis we are now testing is that undulipodia were once highly motile and originally on their own as spirochaete bacteria that formed symbioses with the rest of the cell.

Regardless of the precise details, if the symbiotic theory is correct, all animal cells — including our own, of course — formed from partnerships of two or maybe three members of different procaryotic species. Plant cells with their photosynthetic organelles evolved from these same partnerships with yet another set of symbionts — photosynthetic ones that became plastids. In summary, then, the concept is that bacteria of several very different species came together in certain sequences to form new complex units: eucaryotic cells. Natural selection acted on the complexes — associated microbes left more offspring than their unassociated free-living counterparts. The theory states that the first step in the origin of animal, plant, and fungal cells was an association between two types of bacteria, one fermenting and one respiring. The respiring bacteria were efficient in using oxygen. A second step involved the acquisition of motility through the association with spirochaete symbionts that eventually evolved to be organelles of motility, the undulipodia. The third step, which occurred only in the ancestors of algae and plants, was the acquisition of the capacity for photosynthesis, that is, the formation of hereditary symbioses with procaryotes that became plastids. Thus animals, plants, and fungi — complex organisms made of complex cells — have not only evolved by natural selection acting on organisms in which favorable single mutations have accumulated (neo-Darwinian evolution) but are also products of symbiotic associations.

The power of associations can be impressively illustrated by analogy to live modern symbioses. New symbioses still occur that lead rather abruptly to new complexes with traits different from either of the individual partners. For example, bacteria are known that always invade other bacteria (Stolp, 1979); this may explain how bacteria that became mitochrondria first entered fermenting host cells. Live examples of motility symbioses are also known: some spirochaete bacteria attach to the surface of host cells, for example, to *Mixotricha paradoxa*, and move them around.

Mixotricha is a huge eucaryotic microorganism that lives in the hindgut of dry wood termites. The single cell is about half a millimeter in length. It ingests pieces of wood and digests the cellulose. It has four small undulipodia that it uses as rudders to change its direction. But in order to swim it uses its symbionts: about 500,000 highly motile surface spirochaete bacteria anchored to its surface. There are other types of symbiotic bacteria inside the *Mixotricha* cell (Cleveland and Grimstone, 1964). *M. paradoxa* is a symbiosis of at least three different kinds of microorganisms, and living inside termites, it is itself a symbiont. Without the wood-digesting abilities of microbial complexes such as *M. paradoxa*, wood-eating termites could not survive, since only a few bacteria, protists, and fungi contain cellulases and other enzymes required to digest wood.

Some motility symbioses can be seen on films of microbes from *Pterotermes occidentis* and *Kalotermes schwartzi*, dry wood termites from

southern Arizona and southern Florida (Margulis et al., 1978). These termites break wood into manageable pieces and ingest it. Their soldiers, however, have mouthparts so modified for defense of the colonies that they
cannot even ingest wood: they must therefore be fed by other termites
through the rear. Soldier termites of these species thus depend on hindgut
microorganisms supplied by members of other castes. E. O. Wilson of
Harvard University has suggested that termites became social insects primarily because they must transmit their wood-digesting hindgut microbes to
each other.

The microbial community is sensitive to oxygen. In dry wood termites,
the anaerobic microbial community digests wood while the hypertrophied
termite hindgut maintains the appropriate environmental conditions for
microbial growth and cellulose digestion.

The dry wood termites can live on a dry, hard wood diet — billiard
balls or cellulose fibers and water! If by chemical or heat treatment they are
deprived of their community of microbes, they die within weeks. Dry wood
termites, like all organisms, need sources of nitrogen and sulfur in their diets.
Wood has very little nitrogen and sulfur; cellulose fibers have none. At least
some termites therefore harbor other bacterial symbionts that fix the nitrogen from the air. Only a few other bacteria are capable of fixation of atmospheric nitrogen. Thus, a single termite is in fact a complex ecosystem har-
boring, in some cases, over 30 different species, many of them anaerobic.

The dry wood termite hindgut community usually contains a variety of
species of spirochaete bacteria. The role of these spirochaetes in the community is unknown; despite enormous effort, it has not been possible to
grow them (and indeed most of the other termite microbes) outside the
hindgut ecosystem. Production and removal of gases is probably involved in
providing the proper milieu for the microbes, but which gases and in what
quantity are not known.

Spirochaetes of termites tend to form symbiotic associations, which is
why we have studied them (To et al., 1978). Some large spirochaetes contain
microtubules, an extremely rare feature in procaryotic cells (Hollande and
Gharagozlou, 1967; Margulis et al., 1978). The presence of microtubules at
least superficially like those found in undulipodia suggests that spirochaetes
and undulipodia may have common ancestors.

Spirochaetes often move or beat together, generating forces that can
move other, larger organisms. Some show behaviors such as might be
expected of organisms ancestral to undulipodia, if indeed undulipodia were
once free-living (Margulis et al., 1979b). Individual spirochaetes beat in complete synchrony, apparently by virtue of their physical proximity.

Is the protein making up the spirochaete microtubules the same protein, tubulin, found in the microtubules of animals, plants, and fungi? Microtubules composed of tubulin have been discovered in nerve and sensory cells,

and in the mitotic spindles of all animals, plants, and fungi. Where did micro-tubules come from? Were they brought into cells symbiotically by once free-living spirochaetes that became undulipodia? Within a decade perhaps we may be able to prove or disprove the hypothesis of spirochaetal origin of undulipodia and other organelles composed of microtubules.

Microbial symbioses often occur in anaerobic microenvironments. Many wood-digesting protists in termite guts are covered with hundreds of surface bacteria. In lake muds, in the digestive tracts of animals, and in the anaerobic environments found below the evaporitic surface in the microbial mats at Baja California, for example, symbiotic relationships between microorgan-isms seem to be common. Without such complex microbial ecosystems, there would be no breakdown of wood and many other organic products. Without microbial degradation and gas release, there would be immense accumula-tions of dead and diseased carcasses and a failure to cycle through the gas phase the elements required for life to continue.

Some extant symbioses provide analogies for the acquisition of photo-synthesis. In many distinctly different cases, animals or other nonphotosyn-thetic but motile organisms have established stable symbioses with photosyn-thetic partners; the complex tends to survive, grow, and leave more offspring because it has the double advantage of motility and photosynthesis. In nutrient-poor waters such associations are particularly common. For exam-ple, coral animals and giant clams of the south Pacific are routinely sym-biotic with photosynthetic algae called dinoflagellates. Several species of worms such as *Convoluta paradoxa* routinely harbor algae in their tissues. Some snail-like molluscs have acquired photosynthetic partners, including many if not most of the radiolarians and foraminiferans. In all of these cases, nutrient transfer occurs from the photosynthetic to the nonphotosynthetic partner. For example, *Convoluta paradoxa* does not feed through its closed mouth; it sunbathes. The algae living between the cells of the worm's tissues produce food for the worm by photosynthesis. Furthermore, the continuity of the partnership between algae and worm is assured. When the flatworm dies, its tissue algae are released into the water. The swimming algae are attracted by some chemical signal to the surface of the worm eggs, and all normal members of the species *Convoluta paradoxa* are green due to the chlorophyll in the symbiotic algae. The partnership has been selected because it has advantages relative to each separate partner alone.

The origin of such algae-animal associations is analogous to the origin of plant cells. Plant cells can be thought of as animal cells that early in their history trapped photosynthetic microbes inside them. Such photosynthetic, oxygen-releasing microbes evolved into organelles and with time became the totally dependent plastids.

How does this discussion of cells relate to our earlier remarks about atmospheres? The major relationship between the larger groupings of organisms, the kingdoms plotted as a function of time, is shown in figure 6. The outlines of the history of life over time can be roughly drawn for the last 3.5 billion years. A remaining mystery is how life in the form of anaerobic bacterial cells arose from nonlife so quickly. The details of the origin of life, especially of the immediate precursors to procaryotic cells, are still obscure. However, from about 3.5 billion years ago until the present, we can trace the unbroken threads of life. The earliest organisms — anaerobic bacteria of many kinds — are still with us today. All produce and remove atmospheric gases, and many strongly interact with the clastic particles, minerals, and humic acids of the sediments. These and other bacteria, fungi, and protists make soil what it is: a far more complex material than the martian or lunar regolith. Bacteria originated early, have diversified a great deal on the metabolic level, and have modulated Earth's surface environment until the present.

After the diversification of anaerobic bacteria, another revolution occurred: oxygen became plentiful in the environment, at first because it was a waste product of the photosynthetic activities of the blue-greens. Since oxygen in large quantities is required by virtually all eucaryotes, it was probably plentiful in the environment by the time eucaryotic cell organization

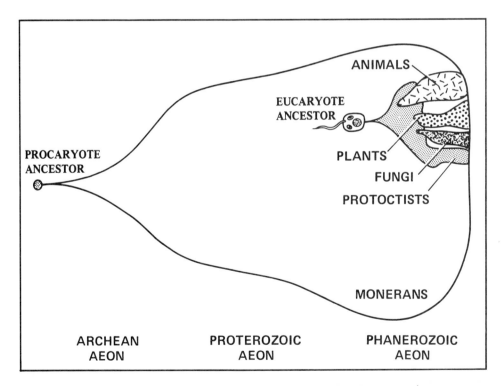

Figure 6. *Temporal relationships between the five kingdoms of organisms. (Schwartz and Margulis, 1981)*

emerged. Many new sorts of cells probably originated by symbiosis, and some populations of cells that bore undulipodia with microtubules eventually evolved. If cell symbiosis theory is correct, serial microbial symbioses preceded the origin of skeletalized invertebrates with hard parts. In any case, eucaryotic cells must have evolved before animals and plants composed of this type of cell.

In fact, the dramatic appearance of calcium carbonate, phosphate, and other mineralized hard parts was probably only a very conspicuous manifestation of many other evolutionary innovations at the cell and tissue levels that are not preserved in the fossil record (Lowenstein and Margulis, 1980).

The frontispiece of Sachse de Lowenheimb's book, *Oceanus Macromicrocosmicus,* published in 1664, is shown in figure 7. De Lowenheimb was a champion of Joseph Harvey's concept of circulation of the blood. Apparently, literate people at that time agreed that water flowed through the environment as through a closed circulatory system. The water raining into the rivers and flowing down the mountains eventually entered lakes and oceans. Water coming down as rain after evaporation was not lost: it cycled. Sachse de Lowenheimb used the concept as an analogy. Blood leaving the heart and passing through the arteries came back through the veins; it must therefore flow in a closed system.

Now we use the analogy again, but with an extension. All scientists agree that vertebrate blood vascular systems are closed. The only time blood is in equilibrium with its surroundings is when a person or animal is dead! Furthermore, the blood is a highly modulated fluid system controlled by and for a whole functioning living system — the animal. In blood there are many regulated deviations from equilibrium, including, for example, temperature and the amounts of bicarbonate, salt, and oxygen. These deviations are purposeful in that they maintain the organism. They are products of relentless and continuous natural selection. To regulate deviations from equilibrium at the ultimate expense of solar energy is an intrinsic property of living systems.

We have named our extension of the analogy between the blood vascular system of animals and the lower atmosphere of Earth, the *Gaia hypothesis* (Margulis and Lovelock, 1974). The Gaia concept is that Earth's atmosphere is a highly modulated circulatory system produced by the biosphere for the biosphere. What aspects of the atmosphere are modulated? Probably the composition and related features such as the acidity. Our atmosphere is too alkaline, relative to the other inner planets, and it harbors far too much oxygen. Earth's temperature may be actively modulated as well. That certain aspects of the atmosphere are actively regulated by life is relatively easy to show, although how this regulation works in detail is an extraordinarily difficult problem to solve. Recently, we have considered the role of anaerobic gas release processes such as methanogenesis in the stabilization of the oxygen

Figure 7. *Frontispiece to Sachse de Lowenheimb* Oceanus Macro-microcosmicus *(1664).*
(Courtesy Trustees, Wellcome Institute for the History of Medicine, London.)

content at its current 20% (Watson et al., 1978). Far more work needs to be
done. The analogy to the blood circulatory system is again instructive.
Although there is little doubt that blood ion ratios, protein composition,
temperature, pressure, and so forth are not in equilibrium with the environment of the organism but must be actively regulated to support the organism, the physiological mechanisms by which regulation is achieved are still
not entirely understood. It has taken more than 300 years to get this far

with vertebrate blood regulation. The atmosphere is even more complex and even less is known. However, until we recognize that Earth's atmosphere is out of equilibrium for a purpose, we doubt if a profound understanding of either climates or paleobiology is possible.

The composition and other aspects of the lower atmosphere are purposefully regulated in the same way that the atmosphere inside a beehive is purposefully regulated. Neither the disequilibria of the troposphere nor the atmosphere in a beehive is constructed according to a divine plan; both are products of many years of evolution. Natural selection has acted strongly to optimize complex systems by optimizing their component parts. This has led to Earth's dynamically stable yet chemically improbable Gaian atmosphere in which the temperature and oxygen content are optimal for the microbial animal and plant ecosystems on the planet's surface. Understanding how Earth's atmosphere and surface are maintained in disequilibrium is a crucial scientific goal whose achievement is aided inordinately by comparison with our flanking planets Mars and Venus. Earth's surface has apparently ·followed a peculiar course for hundreds of millions of years at least.

As we have discussed mud-dwelling bacteria, termite hindgut microbes, and the atmosphere of Earth, we have shown how symbioses and other interactions between organisms of very different species have been continuously important in environmental regulation. Without such interactions in which some organisms produce gas and carbon compounds that others utilize and remove, for example, Earth might also have a hot or cold dry planetary surface with a dead CO_2 atmosphere. It might be much more like Mars and Venus. In fact, without the gas-exchanging microorganisms we would never have evolved at all.

We are grateful to NASA (NGR-004-025), the Boston University Graduate School, and the Guggenheim Foundation for support of this work. Some of this paper was based on an oral presentation given to the Boston University community (by L. M.) as a part of The University Lectureship, Boston University, 1978.

REFERENCES

Corliss, J. O.: The Ciliated Protozoa. Characterization, Classification and Guide to the Literature, 2nd edition. London and New York: Pergamon Press, N.Y., 1979.

Cleveland, L. R.; and Grimstone, A. V.: The Fine Structure of the Flagellate *Myxotricha paradoxa* and Its Associated Micro-organisms. Proc. Roy. Soc. London B, vol. 159, March 17, 1964, pp. 668–686.

Cloud, Preston: Evolution of Ecosystems. Amer. Sci., vol. 62, Jan.-Feb. 1974, pp. 54–66.

Golubic, S.: The Relationship Between Blue-Green Algae and Carbonate Deposits. In: The Biology of Blue Green Algae, N. S. Carr and B. A. Whitton, eds., Oxford, 1973, pp. 434–472.

Hollande, A.; and Gharagozlou, I.: Morphologie Infrastructorale de *Pillotina calotermitides* praecox. C. R. Acad. Sci. Paris, vol. 265, Oct. 1967, pp. 1309–1312.

Horodyski, R. J.: *Lyngbya* mats at Laguna Mormona, Baja California, Mexico — Comparison with Proterozoic stromatolites. J. Sedimentary Petrology, vol. 47, 1977, pp. 1305–1320.

Knoll, A. H.; and Barghoorn, E. S.: Archean Microfossils Showing Cell Division from the Swaziland System of South Africa. Science, vol. 198, Oct. 1977, pp. 396–398.

Lowenstein, H. W.; and Margulis, L.: Calcium Regulation and the Appearance of Calcareous Skeletons in the Fossil Record. Proceedings, Third International Symposium on the Mechanism of Biomineralization in the Invertebrates and Plants. Kashikoyima, Mie, Japan, 1980, pp. 289–300.

Margulis, L.: Symbiosis in Cell Evolution. San Francisco: W. H. Freeman Co., 1981.

Margulis, L.; Barghoorn, E. S.; Banerjee, S.; Giovannoni, S.; Francis, S.; Ashendorf, D.; Chase, D.; and Stolz, J.: The Microbial Community in the Banded Sediments at Laguna Figueroa, Baja California: Does It Have Precambrian Analogues? Precambrian Research, vol. 11, 1979a, pp. 93–123.

Margulis, L.; Chase, D.; and To, L. P.: Possible Evolutionary Significance of Spirochaetes. Proc. R. Soc. Lond., vol. B204, April 11, 1979b, pp. 189–198.

Margulis, L.; and Lovelock, J. E.: Biological Modulation of the Earth's Atmosphere. Icarus, vol. 21, April 1974, pp. 471–489.

Margulis L.; and Lovelock, J. E.: The Atmosphere as Circulatory System of the Biosphere: the Gaia Hypothesis. Co-Evolution Quarterly, Summer 1975, pp. 30–40.

Margulis, L.; To, L.; and Chase, D.: Microtubules in Prokaryotes. Science, vol. 200, June 1978, pp. 1118–1124.

Mazur, P.; Barghoorn, E. S.; Halvorson, H. O.; Jukes, T. H.; Kaplan, I. R.; and Margulis, L.: Biological Implications of the Viking Mission to Mars. Space Science Reviews, vol. 22, June 1979, pp. 3–34.

Reimer, T. O.; Barghoorn, E. S.; and Margulis, L.: Primary Productivity in an Early Archean Microbial Ecosystem. Precambrian Res., vol. 9, 1979, pp. 93–104.

Schidlowski, M. P.; Appel, W. U.; Eichmann, R.; and Junge, C. E.: Carbon Isotope Geochemistry of the 3.7×10^9-year-old Isua Sediments, W. Greenland: Implications for the Archean Carbon and Oxygen Cycles. Geochim. Cosmochim. Acta, vol. 43, Feb. 1979, pp. 189–199.

Schopf, J. W.: Precambrian Paleobiology: Problems and Perspectives. Ann. Rev. Earth Planet. Sci., vol. 3, 1975, pp. 213–249.

Schwartz, R. M.; and Dayhoff, M. O.: Origins of Prokaryotes, Eukaryotes, Mitochondria and Chloroplasts. Science, vol. 199, Jan. 1978, pp. 395–403.

Schwartz, K. V.; and Margulis, L.: Phyla of the Five Kingdoms: a Guide to the Diversity of Life on Earth. San Francisco: W. H. Freeman Co., 1981.

Stolp, H.: The Interaction Between *Bdellovibrio* and Its Host Cell. Proc. R. Soc. Lond., vol. B204, April 11, 1979, pp. 211–217.

To, L. P.; Margulis, L.; and Cheung, A. T. W.: Pillotinas and Hollandinas: Distribution and Behavior of Large Spirochaetes Symbiotic in Termites. Microbios, vol. 22, 1978, pp. 103–134.

Walter, M. R., ed.: *Stromatolites* (Developments in Sedimentology, vol. 20). Amsterdam: Elsevier, 1976.

Watson, A.; Lovelock, J. E.; and Margulis, L.: Methanogenesis, Fires and the Regulation of Atmospheric Oxygen. BioSystems, vol. 10, Dec. 1978, pp. 293–298.

Origin and Evolution of Continents and Oceans

KARL K. TUREKIAN

Continental growth began soon after the formation of Earth and most of its hydrosphere. Perhaps it was only when decay processes became less effective that the continents could maintain integrity and a selectively depleted mantle could begin to develop.

The geometry and dynamics of the ocean basins and continents over the past 200 million years are comprehensible within the framework of the plate tectonic theory. The fundamental features of the theory are that Earth's outer layer, called the lithosphere, is divisible into a number of plates that are moving relative to each other and are being created and destroyed at the plate boundaries. The plates diverge by the creation of new ocean floor, and the resulting convergences with other plates cause destruction of the lithosphere along the edges. The divergence invariably leads to the formation of an ocean basin or an increase in the area of an existing basin. Most convergent plate boundaries have at least one oceanic side — this means that ocean basins are destroyed even where the other boundaries are continental. There are, however, a number of convergences of plates that are both continental, as in the case of the Indian plate and the Asiatic plate. In such cases, continental crust must be destroyed, but by mountain-building and erosion. Thus, in principle, if no other processes were acting, Earth's surface would after a time consist of ocean basins covered with a layer of sediment eroded from the elevated mountains at the continental convergences.

The geological record tells us that this process did not occur efficiently for most of the 3.8 billion years recorded in the rocks. Obviously other processes have been acting to protect the continents, at least since 3.8 billion years ago.

The two processes acting on behalf of continental perpetuity are: (1) the transformation of sediment piles deposited adjacent to the continent or brought to the continental margin by the convergence of ocean- and continent-bounded plates into metamorphic rocks welded to the continental block and (2) the addition of new continental material from the mantle. The geological record tells us that both of these processes have been active throughout the history of Earth.

To the extent that new continental material is formed from the mantle, there has been net continental accretion. The question then becomes: What was the pattern of increase of continental crust with time? A companion question is: In what way was the growth of the volume of oceans related to the growth history of the continental crust?

ISOTOPIC CONSTRAINTS

The best model for the formation of Earth, for a long time, was based on the accretion of material of chondritic, and more specifically carbonaceous chondritic, composition. The subsequent redistribution of the elements was supposed to give rise to the present configuration of hydrosphere-atmosphere, crust, mantle, and core.

Two important papers in the 1960s comparing certain properties of terrestrial materials with the supposed chondritic precursor showed that this assumption was too simple. The late Paul Gast, in a classic paper on "the limitations on the composition of the mantle" (Gast, 1960), clearly showed that the $^{87}Sr/^{86}Sr$ ratio of mantle-derived materials required the mantle source region to have a considerably lower Rb/Sr ratio than any chondrite analyzed. Ringwood (1966) showed that an Earth partitioning from a carbonaceous chondrite bulk composition could not yield the highly oxidized state of the upper mantle and the high concentrations of siderophile elements, such as nickel, found there if the mantle had been formed in equilibrium with an iron-nickel core produced by reduction and segregated by gravity.

These ideas clearly required new attitudes with regard to the composition and formation history of Earth's four major inorganic spheres. Since then results on rare gases and their isotopes, the rare earth elements and their isotopes, and the isotopes of oxygen have added additional constraints. In particular, the use of radioactive ^{147}Sm and its daughter ^{143}Nd in a manner similar to the use of the $^{87}Rb-^{86}Sr$ couple has added much to our knowledge of the development of Earth's crust. The decay constants for ^{87}Rb and ^{147}Sm are 1.42×10^{-11} yr^{-1} and 6.54×10^{-12} yr^{-1}, respectively.

Several groups have been involved in these studies: the La Jolla group (beginning with Lugmair et al., 1975), the Paris group (beginning with

Richard et al., 1976), the Caltech group (beginning with De Paolo and Wasserburg, 1976), and the Lamont-Doherty group (beginning with O'Nions et al., 1977). A review of the results to date appears in O'Nions et al. (1979).

The relationship between the $^{87}Sr/^{86}Sr$ ratio of mantle-derived materials and the $^{143}Nd/^{144}Nd$ ratio indicates oceanic basalts, specifically the midocean ridge basalts (MORB), are derived from a mantle source that has been depleted for a long time in Rb relative to Sr and in Nd relative to Sm. Since both Rb and Nd are known to be highly concentrated in continental magmatic rocks relative to Sr and Sm, respectively, the assumption is that the depletion is due to the formation of continental crust early in Earth's history.

If one assumes that the bulk Earth has the same relative abundances of the rare earth elements as chondrites (and most other meteorites) and that the initial isotopic composition at least of Sm and Nd was the same throughout the Solar System, then it is possible to extrapolate the correlation line of $^{87}Sr/^{86}Sr$ vs $^{143}Nd/^{144}Nd$ to intercept the $^{143}Nd/^{144}Nd$ expected in the bulk Earth today. This should be the same value found in chondrites. This intercept yields a value of the $^{87}Sr/^{86}Sr$ ratio for the bulk Earth which can be translated into a Rb/Sr ratio for the bulk Earth, assuming an initial $^{87}Sr/^{86}Sr$ ratio and the age of Earth based on meteorite studies. The $^{87}Rb/^{86}Sr$ ratio of the bulk Earth by this model is about 0.09, compared to chondritic values ranging from 0.73 to 0.78, reaffirming Gast's initial observation.

O'Nions et al. (1979) and McCulloch and Wasserburg (1978) show that continental crust, back through the Archean, has been forming from material of bulk Earth composition as far as the rare earth complexion is concerned and not from depleted mantle as for the ocean basalts.

We can then divide Earth's mantle into two zones: a "depleted" mantle produced from bulk Earth composition by the melting and extraction of continental crust and a mantle retaining its initial bulk Earth composition.

Based on these results, one can argue that continental crust always formed from bulk Earth material and, based on the MORB results, one can argue further that the dominant source of material supplying basaltic rock to the diverging plate boundaries is the "depleted" mantle.

If we extend our definition of "depleted" beyond relative Nd and Rb depletion to the class of elements to which they belong, sometimes called the "large ion lithophiles," then the depleted mantle should be lower than the bulk Earth mantle in U, Th, Pb, K, and other elements that can be grouped with these. Indeed, these elements are low in MORB and are therefore inferred to be low in the MORB mantle source, which is depleted in Rb and Nd.

The oceanic islands have values for $^{143}Nd/^{144}Nd$ and $^{87}Sr/^{86}Sr$ intermediate between the MORB value and the bulk Earth value and are also higher in their U and K concentrations. They can therefore be inferred to be a mixture of the two different types of mantle reservoirs.

The close association of Ca, Sr, the rare earth elements, Th, and U during the condensation of solid phases from a solar composition nebula (see Grossman and Larimer, 1974, for a summary) has led various people to estimate the bulk planetary concentration of these elements based on heat flow (O'Nions et al., 1979; Turekian and Clark, 1975). Table 1 shows the concentrations of Nd and Sm for the bulk Earth based on such a calculation together with the continental crust values taken from O'Nions et al. (1979). One can calculate the amount of bulk Earth that had to have been processed to yield the continental crust with its characteristic Sm/Nd ratio if one knows the Sm/Nd ratio of the depleted mantle. If we assume that the rare earth elements are delivered to the midocean ridge basalts from the depleted mantle source with no further fractionation of the rare earths, then the ratio observed in these rocks will provide us with a suitable estimate. If any fractionation occurs during the formation of the MORB magma, then the MORB source (depleted mantle) would have to have a smaller Sm/Nd ratio.

A plot of the Sm/Nd ratio of depleted mantle against the mass of undepleted mantle processed to form depleted mantle relative to the mass of the continental crust is shown in figure 1, based on the appropriate data in table 1. The MORB Sm/Nd ratio is taken to be 0.38 (Sun et al., 1979). This corresponds to a mass of processed mantle relative to continental crust of 30. Since the continental crust is about 2×10^{25} gm, about 60×10^{25} gm of depleted mantle must exist. The model predicts a depleted mantle of 60×10^{25} gm resulting from the formation of the continental crust. Since the mass of the total mantle is about 400×10^{25} gm, about 15% of the mantle is depleted. This calculation is based on one set of data. The amounts of the different reservoirs could change with a different choice of data, but the fact remains that a significant but not overwhelming amount of depleted mantle is consistent with observations.

TABLE 1.— SAMARIUM AND NEODYMIUM DISTRIBUTIONS IN EARTH BASED ON ISOTOPIC SYSTEMATICS AND HEAT FLOW (DATA IN PART FROM O'NIONS ET AL., 1979; SUN ET AL., 1979)

Zone	Mass, 10^{25} gm	Sm, ppm	Nd, ppm	Sm/Nd, weight ratio
Undepleted mantle	$\leqslant 343$	0.47	1.44	0.33
Continental crust	2	3.7	16	.23
MORB	.5	2.0	5.3	.38
Depleted mantle	$\geqslant 60$	$\geqslant .33$	$\geqslant .86$	$\leqslant .38$
Bulk Earth	597.6	.32	.97	.33

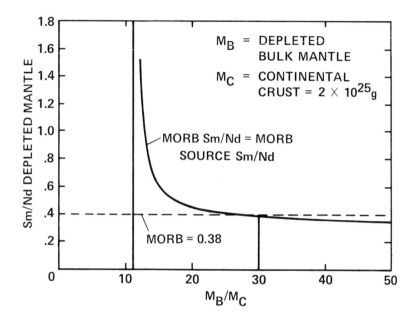

Figure 1. *Plot of Sm/Nd of depleted mantle vs the ratio of mass of the processed depleted mantle (M_B) to the mass of continental crust ($M_c = 2 \times 10^{25}$ gm).*

Because MORB comes from shallow parts of the mantle and continental crust from deeper parts, the assumption seems inevitable that Earth is zoned. This is also compatible with models of terrestrial heat production (Clark and Turekian, 1979).

There is a problem with this zone scheme. If the large ion lithophiles and generally the low-melting fraction are extracted from bulk mantle materials, then one has the feeling that the residual mantle should be denser than the original mantle and thus not be in the shallower parts of the mantle. But our observational data seem to require just such a conclusion. One way out of this dilemma is to add "lighteners" to depleted mantle material. The most obvious lightener is water. It could be added directly to the processed mantle material from an external source (the ambient ocean?). It could act as lightener either as a molecular additive or as an oxidant with the loss of released hydrogen by diffusion. The latter mechanism is preferable. Whatever the mechanism, it implies the existence of an extensive aqueous phase capable of entering the depleted mantle. Earth in that sense may not have been degassing water over time but rather using it up as an oxidant!

The driving force for all this activity must be the heat generated in Earth by radioactive decay of U, Th, and K (and possibly other shorter-lived radionuclides in its early history). With the decrease, by decay, of the heat-generating radionuclides, the process must slow and the rate of production of new continental crust must decrease.

As undepleted mantle is transformed into depleted mantle by the formation of continental crust, the depth of the boundary between the two increases. This could inhibit the formation of additional continental crust by decreasing the efficiency of mass transport from this depth to effect heat loss. The loss of Earth's heat at the present time may be more efficiently mediated by convection and volcanic activity in the upper depleted zone than by the eruption of new continental material from the deeper undepleted zone. The eruptions called "hot spots" or "plumes" as inferred from the location of nonridge oceanic islands may result from release of substances from the deep undepleted zone. In this they may resemble, in a small way, the mechanism for the production of continents at an earlier time.

IS THE GROWTH OF CONTINENTS ACCOMPANIED BY THE GROWTH OF THE HYDROSPHERE?

Are oceans increasing in volume, or did they appear essentially in their present volume at Earth's surface early in its history? This question has troubled geologists for a long time. The case for continuous degassing was detailed by Rubey (1951). The case for instantaneous degassing has been based mainly on the antiquity of rocks with sedimentary features on cratons (Armstrong, 1968) and on calculations based on ^{40}Ar in the atmosphere (Fanale, 1971).

The case for continuous degassing of some gases from the interior, notably the rare gases, has received support from a recent ^{3}He measurement in rocks, hot springs, and ocean water (for reviews see Craig et al., 1978; Tolstikhin, 1978).

The fact that oceanic basalts actually add more gas than they lose during hydrothermal alteration (Dymond and Hogan, 1973) eliminates degassing as a new source of gases for all but ^{3}He and Ne, for which the calculation had been made anyway.

There are therefore few constraints on devising a model for the way in which the ocean arrived at Earth's surface. My prejudice, shared with others (e.g., Anders and Owen, 1977), is that the hydrosphere arrived as part of a low-temperature veneer added to the accreting planet and that it was released mainly at the time of infall as the result of gravitational heating either as it passed through the growing atmosphere or during impact.

If this is the case, then the growing (undepleted) mantle could have started out more anhydrous than the bulk material arriving at Earth. Certainly any water that became a part of the mantle would later be involved in the segregation of continental crustal material from the mantle.

There is no reason to assume a priori that the volatile-bearing material infalling to Earth had to be coupled to the alkali-metal-bearing phases required for continental crust formation. The data for chondrites (from Minster and Allegre, 1979) show that as one proceeds from the least-equilibrated H-group chondrites (H-3) to the most equilibrated (H-8), the Rb/Sr ratio and the K concentration increase. This is in exactly the opposite sense of the distribution of the volatile components such as carbon and the rare gases (Marti, 1967). Clearly, K and Rb do not follow the "low-temperature" volatiles in chondrites, and there is no reason to expect them to have done so during the accretion of Earth.

I contend therefore that water and some other volatiles were released in large part to Earth's surface early in its history, perhaps at the time of accretion, and that there need be no correlation with continental crust formation and the volumetric growth of the oceans. Indeed, as implied in the previous section, the availability of a supply of water in Earth's outer spheres may have provided the opportunity to oxidize Earth's iron distributed in the mantle as the result of accretion. This would elevate the oxidation state of the rocks of the upper, depleted mantle while guaranteeing the retention there of the otherwise siderophile elements such as Ni. I confess that the exact mechanism of planetary oxidation is not completely clear to me, but certainly it involves the incorporation and dissociation of H_2O with subsequent loss of H_2 by diffusion from the upper mantle and ultimately from Earth altogether.

WHY DO CONTINENTS APPEAR TO DATE BACK ONLY 3.8 BILLION YEARS?

If the continent-making process is coupled to heat production in Earth, then one would expect to see continental material dating back to the beginning of Earth or at least rocks with a radiogenic isotopic imprint that remembered the early days. As we have seen, neither of these have been found, and so another explanation must be sought. I shall conclude by offering one that may be difficult to prove in light of the paucity of data for the early days of Earth's history.

At the present time, the mantle convective cycle involves release of basalts from the depleted upper zone of the mantle. In the past, when continental crustal material was forming, it was forming from bulk Earth composition (or undepleted) mantle. Just as at the present time at the convergence areas the oceanic basalts are injected back into the mantle, so in the early days when the continental crust was forming it may have been destroyed by reinjection into the mantle. It would then essentially return the depleting mantle to its primitive undepleted state. Only when the process of resorption became less efficient did the continents attain their integrity and the process of formation of depleted mantle begin in earnest.

REFERENCES

Anders, E.; and Owen, T.: Mars and Earth: Origin and Abundance of Volatiles. Science, vol. 198, Nov. 4, 1977, pp. 453–465.

Armstrong, R. L.: A Model for the Evolution of Strontium and Lead Isotopes in a Dynamic Earth. Rev. Geophys., vol. 6, May 1968, pp. 175–199.

Clark, S. P., Jr.; and Turekian, K. K.: Thermal Constraints on the Distribution of Long-Lived Radioactive Elements in the Earth. Phil. Trans. R. Soc. Lond., Ser. A, vol. 291, 1979, pp. 269–275.

Craig, H.; Lupton, J. E.; and Horibe, Y.: A Mantle Helium Component in Circum-Pacific Volcanic Gases: Hakone, the Marianas, and Mt. Lassen. In: Terrestrial Rare Gases, E. C. Alexander, Jr., and M. Ozima, eds., Center for Academic Publications, Tokyo, Japan, 1978, pp. 3–16.

De Paolo, D. J.; and Wasserburg, G. J.: Nd-Isotope Variations and Petrogenetic Models. Geophys. Res. Lett., vol. 3, May 1976, pp. 249–252.

Dymond, J.; and Hogan, L.: Noble Gas Abundance Patterns in Deep-Sea Basalts — Primordial Gases from the Mantle. Earth Planet. Sci. Lett., vol. 20, Sept. 1973, pp. 131–139.

Fanale, F. P.: A Case for Catastrophic Early Degassing of the Earth. Chem. Geol., vol. 8, Sept. 8, 1971, pp. 79–105.

Gast, P. W.: Limitations on the Composition of the Upper Mantle. J. Geophys. Res., vol. 65, Apr. 1960, pp. 1287–1297.

Grossman, L.; and Larimer, J. W.: Early Chemical History of the Solar System. Rev. Geophys. Space Phys., vol. 12, Feb. 1974, pp. 71–101.

Lugmair, G. W.; Scheinin, N. B.; and Marti, K.: Search for Extinct [146]Sm, 1: The Isotopic Abundance of [142]Nd in the Juvinas Meteorite. Earth Planet. Sci. Lett., vol. 27, Aug. 1975, pp. 79–84.

Marti, K.: Trapped Xenon and the Classification of Chrondites. Earth Planet. Sci. Lett., vol. 2, 1967, pp. 193–196.

McCulloch, M. T.; and Wasserburg, G. J.: Sm-Nd and Rb-Sr Chronology of Continental Crust Formation. Science, vol. 200, June 1978, pp. 1003–1011.

Minster, J-F.; and Allegre, C. J.: [87]Rb-[87]SR Chronology of H Chrondites: Constraint and Speculations on the Early Evolution of Their Parent Body. Earth Planet. Sci. Lett., vol. 42, 1979, pp. 333–347.

O'Nions, R. K.; Carter, S. R.; Evensen, N. M.; and Hamilton, P. J.: Geochemical and Cosmochemical Applications of Nd Isotope Analysis. Ann. Rev. Earth Planet. Sci., vol. 7, 1979, pp. 11–38.

O'Nions, R. K.; Hamilton, P. J.; and Evensen, N. M.: Variations in [143]Nd/ [144]Nd and [87]Sr/[86]Sr Ratios in Oceanic Basalts. Earth Planet. Sci. Lett., vol. 34, Feb. 1977, pp. 13–22.

Richard, P.; Shimizu, N.; and Allegre, C. J.: [143]Nd/[146]Nd, a Natural Treasure: An Application to Oceanic Basalts. Earth Planet. Sci. Lett., vol. 31, 1976, pp. 269–278.

Ringwood, A. E.: Chemical Evolution of the Terrestrial Planets. Geochim. Cosmochim. Acta, vol. 30, Jan. 1966, pp. 41–104.

Rubey, W. W.: Geologic History of Sea Water. Bull. Geol. Soc. Am., vol. 62, 1951, pp. 1111–1147.

Sun, S.-S.; Nesbitt, R. W.; and Sharaskin, A. Y.: Geochemical Characteristics of Mid-Ocean Ridge Basalts. Earth Planet. Sci. Lett., 1979.

Tolstikhin, I. N.: A Review: Some Recent Advances in Isotope Geochemistry of Light Rare Gases. In: Terrestrial Rare Gases, E. C. Alexander, Jr., and M. Ozima, eds., Center for Academic Publications, Tokyo, Japan, 1978, pp. 33–62.

Turekian, K. K.; and Clark, S. P., Jr.: The Non-homogeneous Accumulation Model for Terrestrial Planet Formation and the Consequences for the Atmosphere of Venus. J. Atmos. Sci., vol. 32, June 1975, pp. 1257–1261.

Climatic Stability

DONALD M. HUNTEN

Over geological time, the Martian climate has changed drastically, but Earth's only substantially. Since we cannot explain why in either case, we should keep an open mind about our climatic history and about the factors that can affect it.

This paper is at best a brief summary of a complex subject. For more detail, the recent paper by Pollack (1979) is warmly recommended. For Earth, another excellent source is Appendix A in the United States Committee for the Global Atmospheric Research Program (GARP) report (1975). A perspective over the whole life of Earth is given by Walker (1977).

OBSERVED CHANGES

Figure A.1 in the GARP report (1975) illustrates a notable change in the Argentière glacier of the French Alps over a period of only 115 years — it has retreated far up the valley from its 1850 position. It is clear that substantial changes in climate occur over the span of a long human lifetime. Larger changes are found — the "Little Ice Age" of 1400–1750 and the ice ages themselves, the latest from about 125,000 to 15,000 years ago. These fluctuations, along with many others, are illustrated in figure A.2 of the same report.

It was demonstrated by Mariner 9, and amply confirmed by the Viking Orbiters, that many areas of Mars have been affected by a running fluid. Although liquid water is unstable on Mars today, all the proposed alternatives are far more objectionable. Therefore, Mars must have once had a very different climate, somewhat warmer and with a considerably greater atmospheric pressure. Figure 2 in Masursky et al. (1977) shows two different

111

channel types, indicating different types of flow. The "fluviatile" channels seem to be ordinary rivers with tributaries, and the simplest way to generate them is by the collection of rainfall over a long period. The more spectacular channels are best explained by a single enormous flow, as by the breaking of a dam. The polar caps show another kind of evidence (Pollack, 1979): a "laminated terrain" that suggests some sort of cyclic variation. The age of these features is unknown, but the channels are thought (through crater counts) to be one to several billion years old.

FEEDBACK AND BISTABILITY

The existence of ice ages on Earth and the evidence just summarized for Mars have suggested to many people that climate may be hypersensitive to small changes in the forcing factors. In other words, the system may contain positive feedback mechanisms that are significant enough to allow two alternative states, stable to small changes but susceptible to being "flipped" by somewhat larger ones. An obvious mechanism could be an increase in cloudiness, ice cover, or both with reduced temperature. The increased albedo to solar radiation could reduce the net heat input and lead to a further cooling.

INTERNAL INFLUENCES

Albedo changes fall into the category of internal influences. Others are changes in the ocean circulation, continental drift, veils of volcanic dust, and the evolution of the atmosphere. Models of the climate can be constructed and run on a very large computer, but they must specify all these factors at an observed value. They are essentially useless for predicting a response to changed conditions.

EXTERNAL INFLUENCES

Variations in the output of the Sun are of obvious interest. At very short wavelengths, large variations are well documented and show a strong correlation with solar activity as measured by sunspot area. These wavelengths carry, however, less than 10^{-4} of the total solar energy, and it is hard to imagine any effect on surface climate even if they disappeared altogether. Rapid variation of the Sun's total output is inhibited by its huge mass — even the outer convective zone has a thermal time constant of 10^5 years.

Eddy (1976) has shown that sunspots were absent from 1645 to 1715 and has named this period the "Maunder minimum." Other indirect indicators of solar activity are in agreement and suggest several earlier such periods. There is a rather good correlation with periods of reduced temperature on Earth. As pointed out above, our present understanding of the energetics does not support the idea of a causal connection, but the evidence is nevertheless suggestive.

On a much longer time scale, theories of the Sun's evolution indicate an increase by a factor of 1.37 in output over its life. Curiously, both Mars and Earth appear to have been warmer, if anything, in the remote past. Proposed explanations usually have involved an increased greenhouse effect, perhaps due to NH_3 or CO_2 in the early atmosphere. This solution has its own problems since NH_3 is rapidly photodissociated.

The quasi-periodic changes in orbital eccentricity and axial inclination, generally called Milankovitch variations, comprise a quite different class of external influences. After the polar laminated terrain on Mars was discovered, Ward (1974) computed obliquity changes of $20°$ peak to peak. Although they are fairly large for this planet, it has been impossible to make a correlation because the time scale for Mars is unknown. The same difficulty held up terrestrial studies until Hays et al. (1976) used deep-sea cores to obtain a chronology over nearly 500,000 years. A Fourier analysis revealed periods of around 22,000, 42,000, and 100,000 years. A similar analysis of the driving functions shows 22,000 years (precession), 41,000 years (obliquity), and 97,000 years (eccentricity). Although the 100,000-year peak is the largest, there is no obvious reason why the eccentricity of the orbit should affect climate at all! More recently, Kominz and Pisias (1979) did a coherence analysis. They found none at the 100,000-year period, but significant coherence at the two shorter periods. Only about 15% of the variance, however, is accounted for in this way. Nevertheless, there seems to be a clear indication that small external changes do have a measurable effect.

ATMOSPHERIC COMPOSITION

The current rate of escape of hydrogen from Earth, continued over geologic time, would leave behind just the amount of free oxygen now in the atmosphere. Photosynthesis and burial of carbon are now much more important sources of oxygen. Excess oxygen presumably converts ferrous iron in the crust to ferric. Hydrogen loss at present is limited by the very low stratospheric humidity, which seems to be a sensitive consequence of the particular structure and circulation of the atmosphere. A primitive atmosphere might support a much larger loss rate and a correspondingly larger generation of oxygen (see, e.g., Hunten, 1973). It can also be argued that crustal iron is directly oxidized by H_2O, with H_2 coming out in volcanoes and supporting the escape (Walker, 1977; see also Kasting and Donahue, page 149 in this volume). The oxygen would then never appear in the atmosphere.

CONCLUSION

The principal conclusion is that we should keep an open mind about past climates. They are more likely to be different from than similar to modern ones.

REFERENCES

Eddy, J. A.: The Maunder Minimum. Science, vol. 192, no. 4245, June 18, 1976, pp. 1189–1202.

Hays, J. D.; Imbrie, J.; and Shackleton, N. J.: Variations in Earth's Orbit — Pacemaker of Ice Ages. Science, vol. 194, no. 4270, 1976, pp. 1121–1132.

Hunten, D. M.: The Escape of Light Gases from Planetary Atmospheres. J. Atmos. Sci., vol. 30, Nov. 1973, pp. 1481–1494.

Kominz, M. A.; and Pisias, N. G.: Pleistocene Climate: Deterministic or Stochastic? Science, vol. 204, April 13, 1979, pp. 171–173.

Masursky, H.; Boyce, J. M.; Dial, A. L.; Schaber, G. G.; and Strobell, M. E.: Classification and Time of Formation of Martian Channels Based on Viking Data. J. Geophys. Res., vol. 82, Sept. 30, 1977, pp. 4016–4038.

Pollack, J. B.: Climatic Change on the Terrestrial Planets. Icarus, vol. 37, Mar. 1979, pp. 479–553.

Understanding Climatic Change, A Program for Action. United States Committee for the Global Atmospheric Research Program, National Academy of Sciences, Washington, D.C., 1975.

Walker, J. C. G.: Evolution of the Atmosphere. Macmillan, New York, 1977.

Ward, W. R.: Climatic Variations on Mars. I — Astronomical Theory of Insolation. J. Geophys. Res., vol. 79, Aug. 20, 1974, pp. 3375–3386.

Stellar Influences on the Emergence of Intelligent Life

MARTIN COHEN

Where would intelligent life most likely arise? Around stars not too different from our Sun: Single, of mass 0.4 to 1.4 times the Sun's mass, and inhabiting a calm galactic neighborhood for three or more billions of years.

We live in a galaxy containing some hundred billion stars. Where would we expect to find intelligent life? It is appropriate to ask where planets may have formed. In one viable picture of the formation of our Solar System, the planets formed at roughly the same time as the Sun and represent the agglomerated dusty debris of a gigantic solar nebula. Such debris would have a temperature of a few hundred degrees and would be readily detectable in the infrared region of the spectrum. Present-day infrared investigations reveal potential planetary systems around many young stars, some quite different from the Sun in mass and temperature. How can we restrict the nature of the central star in a solar system that is to contain intelligent life?

Let us examine some crucial stellar parameters, namely, surface temperature, mass, evolutionary history, and rotational velocity. Two vital observational attributes of stars are their total energy output and their surface temperature. A plot of these two parameters against one another for stars around us in the Galaxy is not a scatter diagram, however, but reveals a broad swath through the figure. This we term the main sequence, and it represents the locus of mature stars. Stars evolve to the Main Sequence in time scales very short compared with their sojourn in middle age on the sequence. Likewise, they evolve rapidly away from the sequence as they die. Time spent on the main sequence is a time of stability for stars, a period of virtual constancy of radiation output and temperature. Gone are the massive flares and violent stellar winds that probed their planetary systems (if any) in early evolution. The length of this period of stability is a strong function of stellar mass. Our Sun will reside on the main sequence for 13 billion years, of which it has so far expended only 6 billion years. A star of 10 solar masses spends only 30 million years there. A star only 1/3 solar mass spends an essentially infinite time on the main sequence.

Stars more massive than the Sun are hotter, emit more strongly in the blue and ultraviolet, and would input more energy to orbiting planets at a given distance. Oceans on the planets would be very warm. The hard radiation would accelerate the rate of mutation of species in the water, hastening the emergence of life from the oceans onto the land. On land, increased hard radiation would shorten the life span of creatures and accelerate genetic alterations. This might shorten the time required to evolve intelligent life. However, a hotter star would have a stronger stellar wind and more energetic flares than the Sun has. The increased ionizing radiation (and subsequent ground penetration by ultraviolet radiation) could confine life to the oceans (or caves) where any harmful effects would be minimized. Consequently, a hot star could also hinder the emergence of intelligent species. Stars less massive than the Sun are cooler and emit much of their energy at long wavelengths (the red part of the spectrum and the near-infrared) which penetrate readily to planetary surfaces. For planets at a common distance, however, a cool star yields a much cooler planetary climate than does a hot star, with a presumably lowered rate of mutation.

There seems to be a consensus that the emergence of intelligence on a planet requires about 3 billion years of relative constancy of stellar output. Therefore, massive hot stars are unfavorable for life because of their very short main-sequence lifetimes, despite their possibly rapid induced rates of mutation. Cool stars afford unlimited time but with lowered mutation rates. We can narrow our focus to stars with mass less than about 1.4 $M_\odot$ ($M_\odot$ denotes the mass of the Sun) and temperature about 7000 K, to ensure a minimum time of 3 billion years on the main sequence.

A habitable planet demands a specific environmental temperature range. The volume of space around a star in which this constraint is satisfied is large for hot stars but greatly diminished for cool ones. Life-bearing planets would have to be very close to a 3000 K star. This proximity leads us to another consideration — planetary rotation. Planetary spin affects many factors that bear on the likelihood of life, including surface gravity across the planet, diurnal temperature variations, global climatic patterns, and the force of winds. Consider temperature alone. A high rotation rate would smooth out day-to-night changes (although too high a spin could also promote a zero-gravity region near the equator from which matter would be lost). A lower rotation rate would enhance the differences between day and night and could strain the capacity of plants to survive. In the lower limit of rotation, the planet's day and year become equal and the rotation becomes locked to that of the star, a situation inimical to life. All water and carbon dioxide could be lost to ices on the cold face, while an eternal desert would cover the hot face. There is a slowing down of rotation that comes about as a result of the tidal braking effect of the star on the body of its planet. Since tidal torque depends very steeply on an inverse power of the orbital distance, a planet very close to its star will be rapidly braked and its rotation locked. In fact, we cannot meet the incompatible constraints of minimum planetary temperature and limited tidal braking for too cool (i.e., too low mass) a star. This implies a limit of at least 0.7 $M_\odot$ (4000 K) for a star that sustains an intelligent species on a planet.

We can lower this limit on stellar mass somewhat for planets with sufficiently massive and/or close satellites. We can lock a planet to its satellite's rotation yet still have it rotate with respect to the central star. The limit on stellar mass is thereby weakened to greater than about 0.4 $M_\odot$ (3000 K).

Speaking of massive satellites suggests a comment that should be made about life on massive planets. We have so far considered the star as the primary source of energy for the planetary environment. This condition can be relaxed in special circumstances. Jupiter (and to a lesser extent Saturn) radiates into space more energy than it receives from the Sun at its orbital distance, given its albedo. The excess energy is easily accounted for by postulating a minuscule, slow internal contraction of Jupiter in which gravitational energy is radiated away as infrared flux. This makes it more fruitful to regard Jupiter as a failed star than as a giant planet, with a likely central temperature of order 30,000 K, which is too low for nuclear burning. It is not impossible, then, that at levels where the Jovian atmosphere is comfortably warm for life, the gaseous mix could have the required pressure for complex chemistry. Of course, this could lead to floating life forms, which environmentally might be limited in their intelligence.

If we examine spectra of various types of stars, we can learn something about their rotational velocities from the widths of their absorption lines: the broader the lines, the more rapid the velocity. What we find is that the distribution of velocities is not random. Hot, massive stars rotate very rapidly; Sun-type stars hardly at all. In fact, there is an abrupt decrease in velocity at about 1.5 $M_\odot$ (7000 K). In our Solar System, the planets contain most of the angular momentum while the Sun carries only a tiny portion. Since observational evidence is consistent with the conservation of angular momentum during stellar evolution, we can calculate the velocity at which the Sun would rotate if we could fuse it together with the planets. Its velocity would then be typical of stars more massive than 1.5 $M_\odot$. It is therefore probable that the Sun rotates so slowly because it has transferred its original angular momentum to the planets. One might reasonably conclude that all stars that are less massive than about 1.5 $M_\odot$ and also rotate slowly have shed their angular momentum into planetary systems. Thus solar systems would be abundant in the Galaxy.

We have so far considered only isolated stars, although there is in fact a high frequency of stars that occur in multiple systems, especially in pairs. R. S. Harrington (page 119 of this volume) discusses possible planetary orbits within such binary systems. Under suitable circumstances, there is no reason why life could not develop around such a system.

There is an increasing body of evidence that changes in solar behavior are correlated with climatic phenomena on Earth; for example, the Little Ice Age of the late 17th century occurred at a time when general solar activity was at a very low level. We are accustomed to thinking of solar activity as varying through an 11-year (strictly a 22-year) cycle. It appears, however, that there is no clear evidence, before 1700, for the existence of this regularity in sunspot frequency and other solar phenomena. This issue is controversial, but it shows how premature it would be to characterize the types of small-scale stellar variability that might be undesirable for the emergence of intelligent life. Only extreme flare activity, such as occurs during the early (pre-main-sequence) phases of stellar evolution, would be categorically inimical. There is no evidence for such flares among the restricted class of main-sequence stars with which we are now concerned.

The death of a sufficiently massive star is catastrophic: a supernova is created whose radiated luminosity may for a short period rival that of the entire galaxy in which it occurs. In our Galaxy we estimate roughly one such event each century. Their possible relevance to life is their high output of primary cosmic rays, which could have a significant influence on mutations. Short-lived organisms would double their mutation rate in response to a gross enhancement in background radiation, say 100- to 1000-fold. But long-lived forms would alter in response to only a fewfold increase. It has been suggested more than once that the extinction of the dinosaurs some hundred million years ago represented the response of rather specialized, long-lived creatures to a nearby supernova explosion. This is rampant speculation, but we may presume that a planet orbiting a star in a part of the Galaxy where the supernova rate was locally much higher than that in our vicinity (e.g., much closer to the nucleus of the Galaxy than we are) would not be a satisfactory abode for intelligent life.

To summarize in what will appear to be rather chauvinistic vein, intelligent life as we define it would be likely to arise on a planet orbiting a main-sequence star of mass between about 0.4 and 1.4 $M_\odot$ (between 3000 and 7000 K) in a part of the Galaxy roughly similar to our own neighborhood. Such a star is likely to have shed its angular momentum into planets, to have been stable in radiative output and temperature for at least 3 billion years, to warm a nearby planet without tidally locking the planet's rotation to its own, and not to be subject to frequent local supernova explosions.

Planetary Orbits in Multiple Star Systems

R. S. HARRINGTON

If processes occurring during the formation of multiple star systems are not prohibitory, it appears quite likely that planets can exist in stable, roughly circular orbits around one or two members of common types of multiple star systems.

The Universe is full of multiple star systems. The most conservative estimates suggest that at least one-third of all apparently stellar objects are multiple systems, with some estimates going all the way from more than half to practically all. These are not just binary stars, since at least a third of all multiple systems have three or more components (arranged in hierarchies of spacing such that any stable multiple system may be thought of as a set of binaries). When it comes to selecting objects that might be capable of supporting a planetary system with potential life-bearing planets, however, it has been traditional to rule out these multiple systems immediately. It has been argued that planetary orbits would be unstable and that planets would be ejected from the system, or at least that the perturbations on the planetary orbits would be so great that the constancy of conditions required to support life would be impossible. Thus multiple systems (binaries in particular) have not been considered as possible targets on many SETI lists. This question now needs to be examined more carefully.

Two questions must be considered. First, are planetary orbits in multiple systems sufficiently stable to allow life to develop? Second, could planets form in such systems in the first place? Considerable work has been done in the last few years on the first question. Before going further, we should clarify the definition of stability, since there are several in common use. Here we mean only what is usually referred to as orbital stability; that

119

is, an orbit is stable if there are no significant variations in its characteristic parameters (basically, the axes, or the major axis and eccentricity), and this must be true in particular for low-eccentricity orbits. Note that this is different from, and much weaker than, the more standard linear stability, but it is physically much more reasonable (after all, even the two-body problem is linearly unstable). Just what constitutes a significant variation is left somewhat vague, although in practice the transition from insignificance to significance is rather abrupt and easy to identify.

Szebehely (1977) and his students (Szebehely and Zare, 1977) examined the problem of stability from the point of view of the topology of Hill's zero-velocity surfaces. If a zero-velocity surface remains closed around the components of a binary system, a small mass cannot escape; if the surface remains closed around the entire body, an exterior body cannot get into the so-called region of interplay. If, however, the surface opens (Szebehely considers the first opening, the one at the neutral point between the components of the binary), then there is the possibility of exchange or escape, and the system is classified as unstable. Note that this is a sufficient condition for stability but only a necessary one for instability, in that a system need not actually be unstable just because a zero-velocity surface opens. Indeed, while many multiple stars are stable by this test, a planetary orbit in a binary (the case of a single small, but nonzero, mass) is unstable in general. This may be a bit too pessimistic.

An alternative approach is an experimental/statistical one. Here many multiple systems are integrated numerically on a computer, each system is evaluated for stability (usually qualitatively), and the pertinent parameters for stability are identified and quantified. The disadvantages of such an approach are that no system can be followed for anything like a cosmological length of time (because of error accumulation, if nothing else) and that, since we cannot consider all possible cases, the results are only empirical averages. Positive factors include a minimum of analytic assumptions, the fact that unstable systems do decay very quickly, and the fact that a statistical conclusion is all that is required now.

An extensive study of this nature was carried out by Harrington (1977), and the results will only be summarized here (see also Harrington and Harrington, 1978). Basically, a planet can stay in a stable orbit in a binary system if it stays close to one component (thus seeing the other star only as a distant perturbative influence) or well outside the system (thus seeing the two stars as approximately a single object). The first case would occur in the visual binary systems, the second in the spectroscopic systems. The actual limiting distance ratios depend on the eccentricity of the binary if the planet is close to one component, on the relative sense of revolution, and weakly on the mass ratio of the binary. A conservative all-inclusive statement is that a system will be stable if the distance ratios remain above 4:1.

The above limit establishes a rather extensive region in a binary (and, by extension, in any multiple system) in which planetary orbits would be quite stable. If the so-called habitable zone is also within this region, then the system is capable of supporting a life-bearing planet and is a possible SETI candidate. Note that all the problems associated with identifying the habitable zone (temperature, variability, evolution, etc.) still exist with multiple stars and that, if a system does not qualify for other reasons, the considerations here are irrelevant. However, if we take the simplest definition of the habitable zone — a region where the total energy received is equal to that received from a 1 $M_\odot$ star at 1 AU — then all of the popular nearby binaries qualify as possible SETI targets, with the exception of Sirius and possibly Procyon. Indeed, these are all cases in which the planet would have to orbit close to one component, which means that some of the nearby binaries might actually be potential multiple targets.

Perhaps more serious is the second question, that of whether planets could ever form in multiple star systems. The currently accepted mode of planet formation — condensation in the protostar nebula — is appealing in that it makes planetary formation a common phenomenon; it is therefore assumed that what could happen around single stars could also happen around components of a multiple system. However, when Heppenheimer (1974, 1978) studied the problem in detail, his conclusions were pessimistic. The basic mechanism assumed involves a cloud of particles orbiting a newly formed star and undergoing collisions reasonably often. If the relative collisional velocities are below a certain limiting value, the collision is inelastic and the particles stick together, eventually building up to planet-sized bodies. If the relative velocities are above the limit, the particles fragment, preventing formation by accretion.

For a single star, with all the planetesimals in essentially circular orbits, the mechanism works well. For a binary, the particles orbiting one component suffer perturbations by the other component. These perturbations will, in particular, produce secular increases in the eccentricities of the planetesimal orbits, which in turn produce increases in the velocity dispersion. With a set of reasonable assumptions about conditions within the cloud of particles and the nature of the particles themselves, this increased velocity dispersion may raise the relative collision velocities above the limiting velocity for accretion. Heppenheimer concludes that, since the semimajor axis of the binary needs to be greater than 50 AU, most nearby binaries are indeed unsuitable for planet formation. (Note that the case of a spectroscopic system with a planet orbiting the full system has not been treated.)

Several considerations may make the prospect seem less bleak. First, as Heppenheimer points out, the actual physical situation is more complicated. The protostar nebula is not a cloud of massless particles orbiting a point mass. The nebula has mass and its perturbations should be considered. More significantly, the nebula will produce aerodynamic drag on the particles, and this tends to decrease eccentricity, offsetting the effects of the other component. Unfortunately, the nebula has to be quite massive for either of these effects to be significant in the binaries under consideration, probably more massive than can be reasonably justified.

Another possibility is that the planets are created about the components before the final binaries are formed. It now seems quite reasonable that stars form out of fairly large clouds initially into loose clusters and associations. These unstable configurations undergo dynamical decay, leaving behind the observed distribution of single and multiple stars (Harrington, 1975). The escape energies for the decay are taken from the binding energies of the remaining multiple systems, so that these systems close up as the decay proceeds. If the nebulae around the stars contracted and formed planets before the decay of the system (Larson, 1978; Kobrick and Kaula, 1979), there would be no significant perturbations to increase the velocity dispersion within the nebula. The major drawback to this idea is that the time scales for decay (relaxation) of typical stellar systems are short compared to the time scales usually assumed for planet formation, making it difficult for planets to form before the system decays.

Finally, there is always the possibility that we do not yet fully understand the process (or processes) of planet formation. Indeed, this is the major area of uncertainty. If the Goldreich and Ward (1973) picture of the formation of planets is basically correct, and if the various assumptions concerning nebular conditions employed by Heppenheimer are correct, then it is difficult to reconcile the existence of planets in multiple stars. However, there are considerable uncertainties in these assumptions, which would make it premature to rule out multiple systems as possible locations of planets. Indeed, an observational determination of the relative numbers of planets around single and multiple stars would do much to place empirical limits on the possible mechanisms for planet formation. Multiple systems should remain possible targets for any SETI strategy, with the understanding that there are added uncertainties about the possibilities for such systems.

REFERENCES

Goldreich, P.; and Ward, W. R.: The Formation of Planetesimals. Astrophys. J., vol. 183, 1973, pp. 1051–1061.

Harrington, R. S.: Production of Triple Stars by Dynamical Decay of Small Stellar Systems. Astronom. J., vol. 80, 1975, pp. 1081–1086.

Harrington, R. S.: Planetary Orbits in Binary Stars. Astronom. J., vol. 82, 1977, pp. 753–756.

Harrington, R. S.; and Harrington, B. J.: Can We Find a Place to Live Near a Multiple Star. Mercury, vol. 7, 1978, pp. 34–37.

Heppenheimer, T. A.: Outline of a Theory of Planet Formation in Binary Systems. Icarus, vol. 22, 1974, pp. 436-447.

Heppenheimer, T. A.: On the Formation of Planets in Binary Star Systems. Astron. Astrophys., vol. 65, 1978, pp. 421–426.

Kobrick, M.; and Kaula, W. M.: Tidal Theory for the Origin of the Solar Nebula. Moon and Planets, vol. 20, 1979, pp. 61–101.

Larson, R. B.: Calculations of 3-Dimensional Collapse and Fragmentation. Monthly Notices Royal Astronom. Soc., vol. 184, 1978, pp. 69–85.

Szebehely, V.: Analytical Determination of the Measure of Stability of Triple Stellar Systems. Celestial Mech., vol. 15, 1977, pp. 107–110.

Szebehely, V.; and Zare, K.: Stability of Classical Triplets and of Their Hierarchy. Astron. Astrophys., vol. 58, 1977, pp. 145–152.

Constraints on Early Life by Earth's Accretional and Preaccretional Development

GUSTAF ARRHENIUS

The initial state of Earth set the conditions for incipient life, but its accretional history and early evolution remain hypothetical. Further space explorations should help immensely to clarify present major uncertainties.

The thermal state of the early Earth and the properties of the primordial ocean-atmosphere determine conditions for incipient life. In the pre-Apollo era, even some obvious accretional constraints were taken lightly, and it was common to assume an Earth created with internal volatiles. This primordial undifferentiated planet would then gradually or spasmodically yield up an atmosphere with composition similar to present-day volcanic gases, including water for an ocean.

The accretionary history and early evolution of Earth still remain hypothetical. However, exploration of our Moon, Jupiter, and the terrestrial planets, including the discovery of 3.7-billion-year-old Earth crust, and of seafloor subduction and reappearance have brought some earlier known limitations into focus and added a wealth of new information and suggestions with direct or indirect bearing on the problem.

Several different types of accretional hypotheses operate more or less within this new observational framework, each one leading to drastically different early states of Earth but all sharing common concepts such as evolution via planetesimals and with sterilizing and pyrolyzing runaway accretion at some stage. Some hypotheses rely on special assumptions of strong activity of the early Sun to remove early heavy atmospheres of the terrestrial planets. The recently established noble gas composition of the Venusian atmosphere places new boundary conditions on this aspect of the problem. Future space exploration plans include experiments which promise to further narrow the range of possible states for early Earth and other planets.

Cosmic Conclusions from Climatic Models: Can They Be Justified?

STEPHEN H. SCHNEIDER and STARLEY L. THOMPSON

Sweeping conclusions based on mathematical climatic models should be accompanied by a clear admission of the vast uncertainties in the climatic component of the argument, let alone other parts of the problem.

RECENT INTEREST IN CLIMATIC MODELING

Climatic modeling is coming of age. What was hardly a defined field only a decade ago (see SMIC, 1971) grew in scope very rapidly during the mid-1970s (see Schneider and Dickinson, 1974, or GARP, 1975) to the point that the field, going into the 1980s, can claim literally hundreds of adherents at dozens of institutions (GARP, 1979). Moreover, no longer are climatic models built primarily as tools to improve our understanding of climatic phenomena; they are increasingly being asked to shed light on two bolder fundamental questions: (1) What is the impact of human activities on climate? (2) How have climate and life coevolved on Earth? (And a corollary: how might climatic evolution on other planets offer conditions conducive to some forms of life?) Answers to these questions, particularly the second one, based on climatic modeling could well be termed, as we have, "cosmic conclusions."

How justified are conclusions derived from state-of-the-art climatic models? We will begin to address this question by briefly and selectively reviewing a few of the cosmic conclusions from early modeling efforts. We then turn to a review of the factors whose influences need to be modeled properly if conclusions from climatic models are to be reasonably credible.

127

We finish with some cosmic conclusions of our own on the justifications for using state-of-the-art (or likely near future) climatic models to study cosmic issues such as the coevolution of climate and life.

Modeling the Sensitivity of Earth's Climate

Ice catastrophe— A basic problem that has occupied most climate modelers is simply: what is the sensitivity of Earth's climate to perturbations in boundary conditions *external* to the atmosphere? These could include changes in the solar "constant" (S), atmospheric composition, the heat input from human activities, the land-sea distribution, orbital elements that govern the geographic and seasonal distribution of incoming solar radiation, or the surface properties of land, ice, or sea.

The simplest calculation we can perform with a model is to find the response, β, of the global average surface temperature, T, to a unit change in S. Budyko (1969) and Sellers (1969) independently published papers with the same dramatic conclusion: β is a very sensitive, nonlinear function of S. Whereas a continuous, but small, decrease in S (less than about 1.5%) would cause a continuous decrease in T, any minute decrease in S below a threshold of about 1.5–2% would lead to a discontinuous climatic response: an ice-covered Earth! Furthermore, as noted by North (1975) and Gal-Chen and Schneider (1975), Earth would remain glaciated unless S increased by tens of percent over its present value.

This "ice catastrophe" was the cosmic conclusion that gave the impetus for the dramatic growth during the 1970s in the number of climatic models and modelers. Later on we will digest the findings over the decade since the work of Budyko and Sellers (see, e.g., Thompson and Schneider, 1979, for an updated list of references) to reexamine the justification for the ice catastrophe, or any other cosmic conclusions that depend on it.

Faint early Sun paradox— It has been noted theoretically (e.g., Ulrich, 1975) that the Sun, like most main-sequence stars, gradually increases in luminosity with time, which would mean that S was some tens of percent less a few billion years ago than it is today. A paradox then arises: how could S have been that low and the Earth have escaped the ice catastrophe? Sagan and Mullen (1972) suggested that the answer might lie in the composition of the primordial atmosphere. Their hypothesis was that the added "greenhouse effect" of more infrared-opaque ammonia in the atmosphere could permit both a lower S and a nonglaciated Earth, even if the β of the Budyko or Sellers models were accurate. Henderson-Sellers and Meadows (1977) and Owen et al. (1979), for example, recently concurred with the idea that the infrared (IR) opacity of the primordial atmosphere was higher than that of the present atmosphere, but they argue that CO_2, rather than ammonia, was the "greenhouse gas."

Regardless of the outcome of the debates over the ice catastrophe results, the faint early Sun paradox, or the primordial atmospheric composition, one conclusion clearly emerges: climatic models must consider the nature of the climatic system and its boundary conditions (e.g., atmospheric composition and solar irradiance) for the time at which the model is applied. A model derived from modern conditions may not be applicable to past times when conditions were vastly different.

Modeling of Atmospheric and Climatic Evolution

A number of authors have recognized that the evolution of the composition of a planet's atmosphere has an important effect on the evolution of its climate. For example, Rasool and de Bergh (1970) investigated relationships among the planet-star distance, the evolution of planetary surface temperature, the IR opacity of the atmosphere, and the feedback effect of these factors on the evolution of atmospheric composition and climate. Hart (1978, 1979) and Zhe-Ming (1978) have made similar calculations. Although each of these studies use different modeling assumptions (e.g., in radiative-transfer calculations or cloudiness variations), they all consider the evolution of the composition of a planetary atmosphere as crucial to their results. This is, of course, necessary for credible conclusions. But is it sufficient?

In view of the uncertainty in the time evolution of *actual* planetary atmospheric compositions (compared to those assumed or computed in the models), there is reason to question specific results. And, perhaps even more importantly, atmospheric composition is not the only factor in the climatic system or its boundary conditions that could evolve in time. Surface composition and optical properties, cloudiness, orbital elements, orographic features, or solar irradiance could, singly or in combination, have varied significantly over geologic time. (See Pollack (1979) for a review of theories of climatic evolution on the terrestrial planets.)

Cosmic conclusions have been extracted from evolutionary climatic models that attempt to simulate changes on this billion-year time scale (e.g., Hart, 1978, 1979). How justified can the results of these pioneering studies be in light of (1) what is known (and unknown) about the properties of our present climate and (2) the likelihood that many of the factors that could contribute to vastly different climatic conditions are not included — or are included improperly — in climatic models?

Continuously Habitable Zones

Here we will explore in more detail the cosmic conclusions regarding the habitability of Earth for life. One may assume that there is a certain habitable zone around a main-sequence star of a given luminosity (Huang,

1959, 1960). For our purposes, the habitable zone may be defined as that region in which a planet can retain a significant amount of liquid water at its surface, assuming a suitable atmospheric pressure. The climatic extremes that would prevent a habitable Earth are then the cases where all the water has been evaporated from the surface (runaway greenhouse), or where Earth has become completely glaciated (ice catastrophe).

The evolution of solar luminosity has presumably caused the center of the habitable zone to move outward from the Sun. Thus the *continuously* habitable zone (CHZ) of Hart (1978, 1979) can be narrower than the habitable zone at a given time. It is doubtful that a unique CHZ can be defined for a given star since the CHZ width must depend on the characteristics of particular planetary climatic systems and boundary conditions. In addition, extrapolation of CHZ results from Earth models to other stellar systems will be credible only if it can be shown that a significant number of planets that form around main-sequence stars are similar to primitive Earth.

Rasool and de Bergh (1970) concluded that a runaway greenhouse would likely have occurred on Earth had the planet formed closer than about 0.9 AU to the Sun. Hart (1978), using a more physically comprehensive Earth model, estimated that the CHZ around a Sun-type star extends from roughly 0.95 to 1.01 AU. Hart (1979) applied his atmospheric evolution model to other main-sequence stars and found, as a consequence of the ice catastrophe, that no CHZ exists about most K or M stars (the Sun is a G2). Furthermore, stars greater than 1.2 solar masses induced a runaway greenhouse in the model planetary atmosphere. The results of these thought-provoking models are uncertain due to such factors as crude parameterizations, the possible omission of relevant processes, and the lack of good data for early Earth needed to verify the models' results.

In the remainder of this paper we will review results that help to bracket the uncertainties in some basic parameters determining the CHZ — uncertainties that arise from our incomplete knowledge of the climatic component. We begin by discussing Earth's present climatic sensitivity to solar constant changes as estimated from state-of-the-art climatic theory. Theory may be aided somewhat by observations of climatic "systems experiments." But even if the *present* climatic sensitivity were well determined, problems imposed by the evolution of the surface and atmospheric composition would remain formidable. As an example of a long-term complicating factor, we will discuss some possible climatic consequences of continental drift. We will conclude by suggesting that estimates of the CHZ should, at a minimum, bracket the extreme possibilities resulting from simulations using plausible extreme values of uncertain basic parameters in the models.

PRESENT CLIMATIC SENSITIVITY TO EXTERNAL FORCING

Radiative Effects

In order to determine a CHZ, we must be able to compute the complete time evolution of Earth's surface temperature. A less ambitious, but still relevant, preliminary task is to determine the sensitivity of Earth's present equilibrium climate to changes in the solar constant. As a first approximation, it is possible to consider only radiative effects. The simplest model of equilibrium surface temperature is the radiative balance:

$$F^{\uparrow}_{IR} = \epsilon \sigma T^4 = \frac{S}{4}(1 - \alpha_p) \tag{1}$$

where $F^{\uparrow}_{IR}$ is the outgoing terrestrial infrared irradiance, T is the planetary temperature, ϵ is a constant effective emissivity, σ is the Stefan-Boltzmann constant, S is the solar constant, and α_p is the planetary albedo. As defined in Schneider and Mass (1975), a global climatic sensitivity parameter is

$$\beta \equiv S_0 \frac{dT}{dS} \tag{2}$$

where S_0 is the present solar constant. From equation (2), for the present Earth case with no internal feedbacks (i.e., ϵ and α_p constant), β is about 70 K (i.e., a 1% change in S produces a 0.7 K change in T). If ϵ or α_p could change with T, these feedback effects could radically change β.

A very important positive temperature/radiation feedback results from the increase in atmospheric water vapor with increasing temperature. For the present Earth, this effect makes $F^{\uparrow}_{IR}$ more nearly a linear function of the *surface* temperature than equation (1) would indicate. Figure 1 shows the outgoing terrestrial infrared irradiance as measured by satellite plotted against observed surface temperature. (From here on, T refers to air temperature at the surface.) If we let $F^{\uparrow}_{IR}$ be a linear function of T, $F^{\uparrow}_{IR} = A + BT$, where A and B are constants, the climatic sensitivity is inversely proportional to B:

$$\beta = \frac{S_0}{4} \frac{(1 - \alpha_p)}{B} \tag{3}$$

(see Appendix of Schneider and Mass, 1975). The water vapor and other feedbacks implicit in the IR observations from satellites yield a B which, when used in the climate models of Schneider and Mass (1975) or Cess

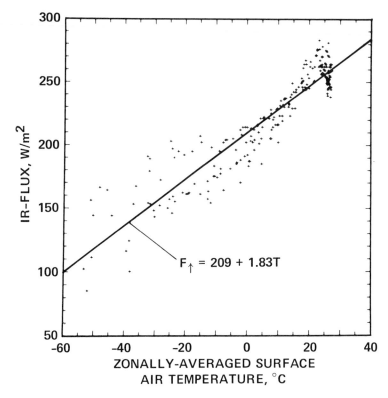

Figure 1. *Mean monthly values (12 months at 18 latitude zones) of zonally averaged out-going infrared irradiance* $F_{IR}^{\uparrow}$, *measured by satellites (Ellis and Vonder Haar, 1976) plotted against corresponding values of surface air temperature* T. *The straight line is a least-squares fit to the data. Note that physical processes in the real atmosphere lead to a fairly linear dependence of* $F_{IR}^{\uparrow}$ *on surface temperature, as opposed to the* T^4 *relationship of simple theory. (Source: Warren and Schneider, 1979)*

(1976) or Warren and Schneider (1979), increases β to about 150 K. But as the latter two authors point out, the estimate of B obtained from satellites varies considerably as a function of latitude and season, raising questions about the functional form of the parameterization.

When the effects of cloudiness are explicitly included, the climatic sensitivity becomes harder to determine. Clouds decrease $F_{IR}^{\uparrow}$ relative to a cloudless sky since their tops usually radiate at a lower temperature than the surface. Ramanathan (1977) found the effect of a given amount of cloud cover on $F_{IR}^{\uparrow}$ to be very nearly a linear function of the temperature difference between the cloudtop and Earth's surface. Thus, in a model calculation, β can vary by a factor of about 2, depending on the assumptions one makes concerning cloudtop temperature changes, let alone changes in cloud amount.

Except over highly reflective surfaces, clouds will increase the planetary albedo. Furthermore, the albedo of clouds appears to increase significantly with increasing solar zenith angle (Cess, 1976). Figure 2 shows the annual variation of α_p at several latitudes, as measured by satellite (Ellis and Vonder

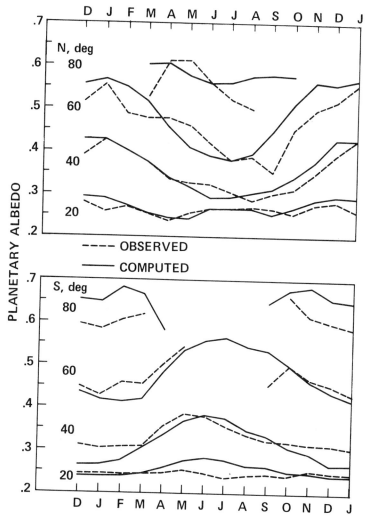

Figure 2. *Annual variation of Earth's zonally averaged planetary albedo as observed by satellites (Ellis and Vonder Haar, 1976) and as computed by a model accounting for variations in surface albedo, cloudiness, and solar zenith angle. Cloud albedo is assumed to obey the albedo/zenith-angle relationship suggested by Cess (1976). More than half the annual variation of the computed albedos is due to solar zenith angle changes. Most of the remaining computed variation is a result of surface albedo changes. (Source: Thompson, 1979)*

Haar, 1976) and calculated by Thompson (1979) using Cess's albedo/zenith-angle relationship. More than half the annual variation in α_p arising from this calculation is due to the solar zenith-angle change.

The net effect of a change in cloud amount on the global radiation balance remains controversial. Cess (1976) estimated no net effect, but the observations of Ellis (1978) indicate that the influence on planetary albedo dominates the influence on F_{IR} for the present climate. Cess's calculation, however, was for incremental changes in cloud amount away from present values, whereas Ellis's calculation was for a change in cloud amount from zero to total cloud cover. In any case, the uncertainties in basic parameters (such as cloud or surface albedos or cloudtop heights) are at the level of several percent or more, rendering any estimates of cloud amount/global radiation balance effects uncertain to a considerable extent.

Regardless of our ability to understand the direct radiative influences of prescribed changes in clouds, we must still predict such quantities as cloud amount and cloudtop temperature. There appears to be no obvious way to do this at present except to use highly resolved dynamical general circulation models (GCMs). Even in these models, the cloud parameterizations are at best highly empirical, limited largely by the coarse grid resolution (relative to the scale of real clouds) and lack of verifying observational data. Experiments with at least one GCM (Schneider et al., 1978) reveal that no simple parameterizations of the feedback processes between cloudiness and surface temperature are available yet for general use in lower resolution climate models (such as those used so far to estimate CHZs).

Yet, in his CHZ studies, Hart (1978, 1979) assumes arbitrarily that the fraction of the planet covered by clouds, f_c, is proportional to the total mass of water vapor in the atmosphere. In turn, the atmospheric water vapor partial pressure, given an available surface reservoir of H_2O, is approximately an exponential function of the surface temperature through the Clausius-Clapeyron relationship. The consequence is that f_c varies from about 0.25 at 280 K to 1.0 at 300 K in Hart's model. (This would certainly not produce a realistic cloudiness profile from equator to pole for the present Earth.) This strong $f_c(T)$ dependence is clearly highly debatable. The empirical evidence does not support such a parameterization (Cess, 1976); dynamical modeling studies (e.g., Roads, 1978; Schneider et al. 1978; Wetherald and Manabe, 1975) indicate a slight *decrease* of f_c with increasing global temperature. Furthermore, while Hart explicitly includes the effect of f_c on the planetary albedo, the influence of cloud amount on F_{IR} is incorporated only by tuning the infrared parameterization to the present conditions (i.e., $\partial F_{IR}/\partial f_c = 0$). This is an extreme assumption in view of the empirical evidence and theoretical reasoning mentioned earlier.

The net result of these parameterizations is the strong negative cloud amount/surface temperature feedback that stabilizes Hart's model, particularly against the ice catastrophe. Indeed, if more plausible assumptions of cloudiness change and radiative effects were made, the model would appear, presumably, to predict an ice-covered Earth at present. Thus, this potentially crucial factor in determining climatic sensitivity to external forcing is not based on verified assumptions.

In summary, state-of-the-art theory cannot yet resolve whether cloudiness changes could present positive, negative, or neutral feedback effects on surface temperature variations. Of course, our inability to confidently assess the influence of changes in cloudiness on climate does not require that the influence be overwhelming; it merely implies that it could be a serious deficiency of present models.

Another important feedback process is the coupling of surface temperature and surface albedo. This positive feedback is operative when snow and ice cover increase with decreasing surface temperature. Thompson (1979) calculates that somewhat less than half the annual variation in α_p shown in figure 2 is a result of seasonal snow cover and sea ice variations. Cess (1978) has proposed that major long-term (greater than decades) changes in vegetation cover associated with long-term temperature changes create an analogous positive temperature feedback in lower latitudes. This is based on field evidence that deserts were more extensive during ice ages. Simple climatic models may approximate these effects by making α_p a nonlinear function of the surface temperature.

Figure 3 illustrates the variation in global climatic sensitivity resulting from various choices of the coupling of temperature to $F_{IR}^{\uparrow}$ and α_p; B is as defined earlier for $F_{IR}^{\uparrow}$, and f is $-\partial\alpha_p/\partial T$ (for $T < 10°C$). The latitude of the edge of permanent ice, as deduced from a zonally averaged climate model, is plotted against the percentage change from the present value of the solar constant. Two types of climatic sensitivity can be seen. As defined by equation (2), β is approximately proportional to the slope of the curves at the point of no change in S. This is a *local stability* (more aptly, a local sensitivity) since it is only applicable to small deviations from the present climate. The *global stability* parameter is the decrease in S necessary to bring the ice line to the equator. The different values of the feedback coefficients (f and B) in figure 3 were chosen to indicate a plausible range of uncertainty in present theoretical estimates of β and global stability. The important thing to note is that, even for the present climate, including uncertainties only in radiative processes, estimates of global stability parameter can vary from 2 to 20% or more.

Meridional Heat Transport

The assumption of radiative balance at all latitudes cannot, of course, be justified in estimating the global climatic sensitivity. Nor can the effects of the atmospheric and oceanic circulations necessarily be averaged out by

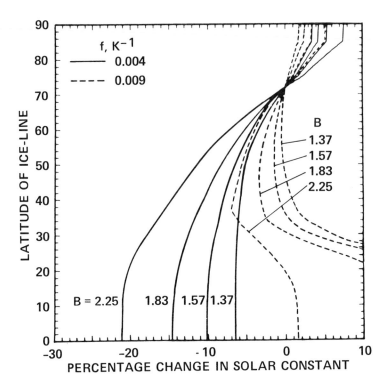

Figure 3. *Global mean climatic sensitivity and stability of a zonally averaged energy balance climate model as a function of the strength of two radiation-temperature feedbacks: (1) surface-temperature/planetary-albedo coupling and (2) surface-temperature/outgoing-infrared coupling. Global stability is defined as the percentage decrease in solar constant (from the present value) required to bring the edge of permanent ice to the equator. Local stability (or sensitivity) is proportional to the slope of the lines at the point of no change in solar constant (i.e., the global climatic sensitivity to small perturbations in solar constant). The values of B are plausible coefficients used in the empirical formula for outgoing infrared, $F_{IR} = A + BT$ (see fig. 1): f is the albedo-temperature coefficient for Sellers' (1969) planetary albedo parameterization. (Of the two values given, f = 0.004 is now thought to give a better simulation, although the validity of the parameterization for climatic change experiments is questionable.) Larger values of B (or f) indicate a stronger dependence of outgoing infrared irradiance (or planetary albedo) on surface temperature. Note that these plausible values of the parameters generate a wide range of climatic sensitivities. (Source: Warren and Schneider, 1979)*

considering only global means. It is well known that permanent ice on Earth would exist far equatorward of its present position were it not for the ameliorating poleward transports of heat by the atmosphere and oceans. While it is possible to estimate these meridional fluxes explicitly through numerical integration of the three-dimensional, time-dependent hydrodynamical equations for up to several simulated years, integration beyond decades exceeds the capability of present computers. For this reason the results of most climate models depend on the assumptions made in parameterizing the dynamical transports of heat in terms of surface temperature and its gradient. The scope of this paper does not permit an extended discourse on this important problem, but a few examples of circulation effects are in order. The reader is referred to Oort (1971) and Oort and Vonder Haar (1976) for a discussion of observations of the present heat transports. A review of the theory of the atmospheric general circulation is given in Lorenz (1967).

The poleward heat flux is considered to act usually as a negative feedback. For example, if the polar regions cool or the tropics get warmer, an increased poleward flux of heat is believed to arise, driving the system toward its previous state. It is not clear how strong this climatic restoring force is, although the increased poleward heat flux in winter relative to summer provides evidence that it does exist. It is not impossible, however, to imagine situations in which circulations could amplify temperature changes: standing planetary-scale waves in the atmospheric circulation may provide conditions favorable for the equatorward extension of snow and ice; a convectively stable atmosphere over extensive snow or ice fields may decrease the meridional heat flux by large atmospheric transient eddies, and so forth.

With regard to present climatic sensitivity, the effect of poleward heat transports is to reduce β by decreasing the extent of polar ice. The effect of transports on global stability is less clear. There is a possibility that the circulations may have a strong nonlinear influence on the extent of permanent ice. The plot of ice line vs change in S might level off a bit at some subtropical latitude if the ice edge advance were held back by intensified meridional heat transports out of tropical latitudes — at least until the tropics cooled sufficiently to permit ice cover. This "stability ledge" (Lindzen and Farrell, 1977) would increase the global stability even though its presence might not be felt during smaller climatic changes. On the other hand, reduced meridional transports could, under some circumstances, insulate the tropics from cooling felt in higher latitudes, thereby increasing global stability. State-of-the-art parameterizations of heat transports cannot yet be verified quantitatively over the wide range of variations for changes in S in figure 3 (i.e., tens of percent). The above uncertainties in the dynamical transport parameterizations of simple zonal climate models can only be compounded with the radiative ones indicated in figure 3, as discussed by Warren and Schneider (1979). Any cosmic conclusions drawn from climatic models will rest on these parameterizations.

SYSTEMS EXPERIMENTS

Forcing Responses and Transfer Functions

Earlier we discussed one way to estimate the climatic sensitivity parameter β, namely, simulation with models based on assumed (or semiempirical)
values of the parameters that determine the strength of the climatic feedback
processes (which, in turn, determine β). The integral effect of all such feedback processes still cannot be verified, given the uncertainties in current
climatic models and in the supporting observational data (see the discussion
in Schneider et al., 1978). This inability of theory alone to provide confident
(i.e., much better than order of magnitude) estimates of the sensitivity of the
present climate to unit external forcing leads us to search for an empirical
method of checking model estimates of β, what could be termed "climatic
systems experiments."

We define systems experiments as cases where large external forcings
and a statistically significant climatic response can both be documented
empirically. These "experiments" can, by analogy to systems analysis, help
to identify a "transfer function" that will characterize the climatic system
more adequately than do present theory and observation. (In linear theory
we might similarly look for the Greens function for the climate system.) A
transfer function can be estimated in practice by varying the parameters that
control climate sensitivity (e.g., α_p or $F_{IR}^{\uparrow}$) to reproduce in a model the
observed climatic response to a known forcing. For a zonal energy balance
model, we have

$$\frac{\partial T(\phi,t)}{\partial t} = Q(\phi,t)\frac{[1-\alpha_p]}{R} - \frac{[F_{IR}^{\uparrow}]}{R} - \frac{[\text{div } F]}{R} \qquad (4)$$

where div F represents net heat transport out of latitude zone ϕ, and R is the
thermal inertia of each zone (Thompson and Schneider, 1979). In this exam-
ple $Q(\phi,t)$ represents the seasonal solar forcing and $\partial T/\partial t$ the seasonal temperature response (both of which are well known). The transfer function is
made up of the parameters that relate the terms in each of the square brackets in equation (4) to the dependent (T) and independent (ϕ,t) variables. But
there are at least three separate feedback terms in the square brackets and,
moreover, R must be specified or calculated, so that one probably could
reproduce the response, $\partial T/\partial t$, to a known forcing, $Q(\phi,t)$, with a nonunique
set of plausible parameterizations comprising the transfer function. Systems
experiments are thus of limited usefulness in deriving or verifying *individual*
parameterizations that determine a model's sensitivity. Systems experiments

can be important for verifying the *overall* climatic sensitivity of models to a given external perturbation. Unless parameterizations are individually validated, however, a model may produce the right climatic sensitivity for certain kinds of external forcings, but the wrong sensitivity if those same parameterizations are used in experiments with other kinds of external forcings.

Examples of Systems Experiments

Seasonal cycle— We have already mentioned that the seasonal cycle is an excellent example of a clear external forcing and a statistically significant climatic response. For example, it is known that the range of the seasonal cycle of surface temperature in the northern hemisphere (NH) is about 14 K, whereas it is only 6 K in the southern hemisphere (SH). It is apparent from simple physical considerations — reconfirmed recently by modeling experiments (North and Coakley, 1979; Thompson and Schneider, 1979) — that the larger ratio of water to land in the SH relative to the NH leads to a larger value of thermal inertia in the SH relative to the NH. This larger R results in greater seasonal heat storage and a smaller seasonal temperature cycle amplitude. In this example, the amplitude of the seasonal cycle of temperature in a hemisphere could be simulated correctly if, say, thermal capacity were overestimated but the strength of temperature-albedo feedback processes were underestimated. The converse, or other combinations of compensating errors in estimates of transfer-function parameterizations, are also possible.

What is needed then is enough *independent* observational data to fix parameterizations for the terms in square brackets in equation (4) with fairly narrow ranges of uncertainty. Also, data for parameters like R are needed as well. Data such as those plotted on figures 1 and 2 can clearly help to narrow the range of uncertainty in the parameterizations of these feedback processes. After this is done, independent systems experiments are needed to help verify the overall climatic sensitivity of models.

Volcanic dust veils— It has long been suspected that the stratospheric dust veils that follow explosive volcanic eruptions affect climate by interfering with radiative transfer between Earth and space (see Mass and Schneider, 1977, for references). The existence of a potential volcanic signal in long-term climatic statistics has remained controversial because the magnitude of such hypothesized signals is comparable to the amplitude of the inherent interannual variability (or noise) of the climate. Even so, by composite techniques (i.e., superposed epoch analysis) it can be shown that a cooling of a few tenths of a degree Celsius can be detected in a few dozen long-term temperature records (Mass and Schneider, 1977). However, because of the weak signal/noise ratio from the volcanic events and the uncertainty in quantitative data on the radiative perturbations from historical dust veils, Mass and Schneider (1977) concluded cautiously that only order-of-magnitude insights for climatic sensitivity analyses could be extracted from these volcanic systems experiments. A few more major volcanic eruptions in which the radiation perturbations are well documented are needed before the observed climatic response can be usefully compared to global climate model calculations.

Orbital element variations– Orbital element variations cause a perturbation to the latitudinal and seasonal distribution of insolation but only a negligible change in global annual solar constant. That these perturbations could cause quarternary glaciations is commonly known as the "Milankovitch hypothesis." Spectral analyses of the time series of oxygen isotope ratios in two ocean sediment cores (Hays et al., 1976) show some power at frequencies near the three periodicities of Earth orbital element variations (i.e., 100,000, 40,000, and 22,000 years). Although this evidence is only statistical, it has motivated attempts to model physically possible connections between the insolation perturbations and the hypothesized glacial/interglacial response. This Milankovitch climatic change experiment has been performed with a variety of energy balance models (Suarez and Held, 1976; North and Coakley, 1979; Schneider and Thompson, 1979). In all cases similar results were obtained. The simulated temperature record in the NH led the observed record by some 5000 years, and the amplitude of simulated glacial/interglacial transitions was considerably less than the observed amplitude. The modelers have all speculated that the phase error could be attributed to the lack of explicit continental ice sheets in their models since the time scale to change appreciably the extent of continental glaciers in the NH is generally thousands of years. In fact, a simulation by Pollard (1978), which combined a zonal energy balance model with an ice sheet model, confirms that the phase error between simulation and observation can be reduced by inclusion of an interactive ice sheet model.

But the weak glacial/interglacial signal produced by the models is more difficult to rationalize than the phase error. For example, Cess (1978) has argued that slow vegetation changes occurring on time scales of centuries or more could change surface albedos enough to cause an underestimate of about a factor of 2 in the amplitude of the glacial/interglacial signal produced in simple Milankovitch forced energy balance models. Furthermore, the direct effects of radiation changes (from orbital element variations) on snow melt could alter albedo and thus temperatures on long time scales.

Therefore, a model that performs well in a seasonal simulation cannot necessarily be trusted to reliably estimate climatic sensitivity to forcings occurring over longer time scales, for example, centuries. Moreover, if the influence of deep-ocean heat storage or continental glacier effects on surface albedo, orography, or evapotranspiration are considered, it is quite possible (perhaps likely) that a β that agrees with short-term (i.e., up to a few years) systems experiments (or derived from short-term theory) could be an order of magnitude different from a β for forcings occurring over millennia. If even longer time scales are considered, then interactions among atmosphere, oceans, ice, and lithosphere could readjust the sensitivity estimates once again (Sergin and Sergin, 1976).

Climatic modeling is only beginning to assess quantitatively the potential importance of processes occurring over different time scales to the estimates of climatic sensitivity on these scales.

CONTINENTAL CONFIGURATION: A COMPLICATING FACTOR ON GEOLOGICAL TIME SCALES

In the last section we showed the large uncertainty in the estimates of global climatic stability without considering changes in surface conditions or atmospheric composition on geological time scales ($\geqslant 10^7$ yr). It is now accepted that continental shapes and locations have changed over geologic time. As one example of a long-term complicating factor, we will discuss some possible climatic effects of changes in continental configuration.

The grossest measure of land-sea distribution is the fractional area of the globe covered by land. This quantity has varied by 20% in the past 180 million years (Barron et al., 1980). A direct radiative effect exists since oceans generally have a lower albedo than land surfaces. An increase in total ocean area should thus decrease the planetary albedo. Using state-of-the-art estimates of β, one can determine that the direct radiative effect of a 20% change in land area is probably less than that of a 1% change in S.

A greater global influence might be the moderation of seasonal climatic extremes by the high thermal capacity of the oceans. (Recall that, at present, the range of hemispheric mean annual surface air temperature in the NH is about 14°C compared to 6°C for the more oceanic SH.) It is probably that a moderated annual cycle of temperature associated with an increase in ocean area would interact with variables nonlinear in T (e.g., ice cover, surface albedo, water vapor pressure) to create changes in the annual mean climate as well. A reduction in glaciation from much more moderate winters than today is a distinct possibility, for example.

We can say more if we consider the zonal distribution of land. A given change in land area at low latitudes, where the incident solar radiation is relatively large, will have a greater influence on the global absorbed radiation than the same change at high latitudes. Thus the distribution of land area with latitude is important for the planetary radiation balance. Polar icecaps could not form easily on continents if there were no land at high latitudes. In this case the snow/ice albedo-temperature feedback would be greatly weakened and global climatic stability would increase significantly. The absence of ice at high latitudes would imply a much reduced equator-to-pole temperature gradient and probably less vigorous atmospheric general circulation. Considering the complicated coupling of the general circulation with cloudiness and precipitation, it is difficult even to speculate on the magnitude, or direction, of possible feedbacks.

It is not unlikely that an adequate simulation of the climates in the geologic past will need to include the influence of the latitude/longitude distribution of land and the continental and ocean-bottom topography. These factors affect the circulation of the atmosphere and dominate that of the oceans. Ocean currents, which at some latitudes carry as much or more heat poleward as the atmosphere (Oort and Vonder Haar, 1976), are constrained by continental coastlines. Even a small land bridge can destroy a major circulation. From this point of view the establishment of the Antarctic circumpolar current about 30 million years ago and the subsequent development of Antarctic glaciation is an intriguing coincidence (NAS, 1975).

The continental configuration does not have as rigid an influence on the circulation of the atmosphere as it does on that of the oceans. Even so, the geographical distributions of topography and surface heat sources create a large stationary eddy component in the present climatological wind fields (Manabe and Terpstra, 1974). Furthermore, surface feedbacks induced by variations in the amplitude or phase of these stationary eddies could alter hemispheric annual temperatures by an amount equivalent to the effect of solar constant changes of several percent (Hartmann and Short, 1979).

The climatic history of the geologic past is not known in much detail, but more "recent" general trends are well documented. Figure 4 shows estimates of oceanic bottom water temperatures for the last 110 million years. Bottom water temperatures in the late Cretaceous were as much as 15°C warmer than at present. Since cold polar regions imply cold bottom water formation, it is likely that the poles were warm relative to today. This agrees with other geologic evidence that Earth was relatively icefree at that time (Hays, 1977). (Perhaps, as we speculated, this was related to warmer winters from the larger thermal capacity implied by submerged continents?) Assuming that the atmospheric composition 100 million years ago was not very different from today, and barring any unsuspected changes in solar output, the large temperature decline in figure 4 must be explained by other, probably internal, climatic mechanisms (i.e., allowing the continental configurations to be part of the "internal climatic system"). A credible quantitative explanation of this massive climatic change has yet to be given, although plausible speculations abound.

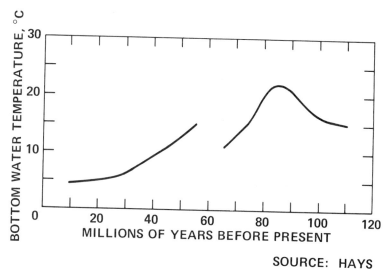

SOURCE: HAYS

Figure 4. *Bottom water temperatures during the last 10^8 years as estimated from the oxygen isotope ratio $^{18}O/^{16}O$ of the fossil shells of benthic foraminifera (after Hays, 1977).*

CONCLUSIONS

We have shown that, although climatic models can be used to estimate the sensitivity of the climate to external forcings, these estimates can vary by perhaps an order of magnitude for forcings occurring over different time scales. Furthermore, considerable uncertainty remains in the specific estimates for each time scale.

For the decadal time scale, we infer from the considerable simulation modeling and seasonal systems experiments that the climate sensitivity parameter, β, is accurate to within, perhaps, a factor of 2 (i.e., $\beta \approx 150 \pm 100$ K). If the upper limit is to be believed, and if one assumes that a change in CO_2 concentration is a similar external forcing to change in solar constant, then one major conclusion emerges: projected atmospheric CO_2 increases from human activities will cause significant climatic change by the end of this century (see Williams, 1978).

Truly cosmic conclusions depend on estimates of long-term climatic sensitivity to very large external forcings, estimates whose range makes the uncertainty in decadal β seem small by comparison. For example, we have seen that simulation in climatic models of the ice catastrophe critically depends on the parameterization of physical processes whose quantitative character leads to order-of-magnitude uncertainties. And these uncertainties increase as we consider climatic conditions for times further and further back from today. Not only does credible reconstruction of planetary climates become more difficult as we go back in time, but the likelihood increases that changes in atmospheric composition, continental locations, orography, solar irradiance, and even galactic dust (among other factors) could alter climatic sensitivity estimates based on today's climatic system and its boundary conditions. And at the billion-year time scale, precisely that needed for CHZ estimates, the latter uncertainties must render present climatic sensitivity estimates as "order-of-magnitude" at the very best!

If one estimates that continuously habitable zones exist in the "climatic space" between the ice catastrophe and the runaway greenhouse, both of these predicted by climatic models, then one should also point out the large range of uncertainty inherent in the present state of such modeling. At a minimum, the estimates given should attempt to bracket the extreme values of climatic sensitivity obtained by varying model parameters over their plausible limits.

None of this is meant to discourage further ingenious, or even speculative, use of climatic models on cosmic questions. But we conclude that cosmic conclusions from climatic models should be accompanied by clear admission of the vast uncertainties in the climatic component of the argument, let alone the other parts of the problem.

Acknowledgments

We thank Dr. C. Leovy, S. Warren, R. Cess, and V. Ramanathan for their criticisms of early drafts; C. Sagan for suggesting that we write the Icarus version (Schneider and Thompson, 1980); and H. Howard for typing the draft manuscript and its revisions.

REFERENCES

Barron, Eric J.; Sloan, J. L. II; and Harrison, C. G. A.: Potential Significance of Land-Sea Distribution and Surface Albedo Variations as a Climatic Forcing Factor; 180 m.y. to the Present. Palaeogeography, Palaeoclimatology, Palaeoecology, vol. 30, 1980, pp. 17–40.

Budyko, M. I.: The Effect of Solar Radiation Variations on the Climate of the Earth. Tellus, vol. 21, no. 5, 1969, pp. 611–619.

Cess, Robert D.: Climate Change: An Appraisal of Atmospheric Feedback Mechanisms Employing Zonal Climatology. J. Atmos. Sci., vol. 33, Oct. 1976, pp. 1831–1843.

Cess, Robert D.: Biosphere-Albedo Feedback and Climate Modeling. J. Atmos. Sci., vol. 35, Sept. 1978, pp. 1765–1768.

Ellis, J. S.: Cloudiness, the Planetary Radiation Budget, and Climate. Ph.D. Thesis, Colorado State University, Fort Collins, CO, 1978.

Ellis, J. S.; and Vonder Haar, T. H.: Zonal Average Earth Radiation Budget Measurements from Satellites for Climate Studies. Atmos. Sci. Paper 240, Colorado State University, Fort Collins, CO, 1976.

Gal-Chen, T.; and Schneider, S. H.: Energy Balance Climate Modeling: Comparison of Radiative and Dynamic Feedback Mechanisms. Tellus, vol. 28, no. 2, 1975, pp. 108–121.

Global Atmospheric Research Programme (GARP): The Physical Basis of Climate and Climate Modelling. GARP Publications Series No. 16, WMO-ICSU, Joint Organizing Committee, Geneva, 1975.

Global Atmospheric Research Programme (GARP): Joint Organizing Committee Study Conference on Climate Models. GARP Publications Series No. 16, 1979.

Hart, Michael H.: The Evolution of the Atmosphere of the Earth. Icarus, vol. 33, Jan. 1978, pp. 23–39.

Hart, Michael H.: Habitable Zones About Main Sequence Stars. Icarus, vol. 37, Jan. 1979, pp. 351–357.

Hartmann, Dennis L.; and Short, David A.: On the Role of Zonal Asymmetries in Climate Change. J. Atmos. Sci., vol. 36, March 1979, pp. 519–528.

Hays, J. D.: Climatic Change and the Possible Influence of Variations of Solar Output. In: The Solar Output and Its Variation, O. R. White, ed., Colorado Associated Univ. Press, Boulder, 1977, pp. 73–90.

Hays, J. D.; Imbrie, J.; and Shackleton, N. J.: Variations in the Earth's Orbit: Pacemaker of the Ice Ages. Science, vol. 194, Dec. 10, 1976, pp. 1121–1132.

Henderson-Sellers, Ann; and Meadows, A. J.: Surface Temperature of Early
    Earth. Nature, vol. 270, Dec. 1977, pp. 589–591.

Huang, Su-Shu: Occurrence of Life in the Universe. American Scientist,
    vol. 47, Sept. 1959, pp. 397–402.

Huang, Su-Shu: Life Outside the Solar System. Scientific American, vol. 202,
    no. 4, April 1960, pp. 55–63.

Lindzen, Richard S.; and Farrell, Brian: Some Realistic Modifications of
    Simple Climate Models. J. Atmos. Sci., vol. 34, Oct. 1977, pp. 1487–
    1501.

Lorenz, Edward N.: The Nature and Theory of the General Circulation of
    the Atmosphere. World Meteorological Organization Publ. No. 218,
    Geneva, 1967.

Manabe, Syukuro; and Terpstra, Theodore B.: The Effects of Mountains on
    the General Circulation of the Atmosphere as Identified by Numerical
    Experiments. J. Atmos. Sci., vol. 31, Jan. 1974, pp. 3–42.

Mass, Clifford; and Schneider, Stephen H.: Statistical Evidence on the Influ-
    ence of Sunspots and Volcanic Dust on Long-Term Temperature
    Records. J. Atmos. Sci., vol. 34, 1977, pp. 1995–2004.

North, Gerald R.: Theory of Energy-Balance Climate Models. J. Atmos. Sci.,
    vol. 32, Nov. 1975, pp. 2033–2043.

North, Gerald R.; and Coakley, J. A.: Differences Between Seasonal and
    Mean Annual Energy Balance Model Calculations of Climate and Cli-
    mate Sensitivity. J. Atmos. Sci., vol. 36, July 1979, pp. 1189–1204.

Oort, Abraham H.: The Observed Annual Cycle in the Meridional Transport
    of Atmospheric Energy. J. Atmos. Sci., vol. 28, April 1971, pp. 325–
    339.

Oort, Abraham H.; and Vonder Haar, Thomas H.: On the Observed Annual
    Cycle in the Ocean-Atmosphere Heat Balance Over the Northern
    Hemisphere. J. Phys. Oceanogr., vol. 6, Nov. 1976, pp. 781–800.

Owen, Tobias; Cess, Robert D.; and Ramanathan, V.: Enhanced CO_2 Green-
    house to Compensate for Reduced Solar Luminosity on Early Earth.
    Nature, vol. 277, Feb. 1979, pp. 640–641.

Pollack, J. B.: Climatic Change on the Terrestrial Planets. Icarus, vol. 37, 1979, pp. 479–553.

Pollard, David: An Investigation of the Astronomical Theory of the Ice Ages Using a Simple Climate-Ice Sheet Model. Nature, vol. 272, March 1978, pp. 233–235.

Ramanathan, V.: Interactions Between Ice-Albedo, Lapse-Rate and Cloud-Top Feedbacks: An Analysis of the Nonlinear Response of a GCM Climate Model. J. Atmos. Sci., vol. 34, Dec. 1977, pp. 1885–1897.

Rasool, S. I.; and de Bergh, C.: The Runaway Greenhouse and the Accumulation of CO_2 in the Venus Atmosphere. Nature, vol. 226, March 14, 1970, pp. 1037–1039.

Roads, John O.: Relationships Among Fractional Cloud Coverage, Relative Humidity and Condensation in a Simple Wave Model. J. Atmos. Sci., vol. 35, Aug. 1978, pp. 1450–1462.

Sagan, Carl; and Mullen, G.: Earth and Mars: Evolution of Atmospheres and Surface Temperatures. Science, vol. 177, July 7, 1972, pp. 52–56.

Schneider, Stephen H.; and Dickinson, R. E.: Climate Modeling. Rev. Geophys. Space Phys., vol. 12, Aug. 1974, pp. 447–493.

Schneider, Stephen H.; and Mass, C.: Volcanic Dust, Sunspots, and Temperature Trends. Science, vol. 190, Nov. 21, 1975, pp. 741–746.

Schneider, Stephen H.; Washington, W. M.; and Chervin, R. M.: Cloudiness as a Climatic Feedback Mechanism: Effects on Cloud Amounts of Prescribed Global and Regional Surface Temperature Changes in the NCAR GCM. J. Atmos. Sci., vol. 35, Dec. 1978, pp. 2207–2221.

Schneider, S. H.; and Thompson, S. L.: Ice Ages and Orbital Variations: Some Simple Theory and Modeling. Quaternary Res. (USA), vol. 12, no. 2, Sept. 1979, pp. 188–203.

Schneider, S. H.; and Thompson, S. L.: Cosmic Conclusions from Climatic Models: Can They Be Justified? Icarus, vol. 41, 1980, pp. 456–469.

Sellers, William D.: A Global Climatic Model Based on the Energy Balance of the Earth-Atmosphere System. J. Appl. Meteorology, vol. 8, June 1969, pp. 392–400.

Sergin, V. Y.; and Sergin, S. Y.: Systems Analysis of the Problem of Large-Scale Oscillations of the Climate and Glaciation of the Earth. Hydrometeorological Series. USSR Acad. Sci., 1976, pp. 5–51.

SMIC: Inadvertent Climate Modification. Report of the Study of Man's Impact on Climate. The MIT Press, Cambridge, Mass., 1971.

Suarez, Max J.; and Held, Isaac M.: Modelling Climatic Response to Orbital Parameter Variations. Nature, vol. 263, Sept. 2, 1976, pp. 46–47.

Thompson, S. L.: Development of a Seasonally-Verified Planetary Albedo Parameterization for Zonal Energy Balance Climate Models — Latitudinal Variation. In: WMO, Climate Models: Performance, Intercomparison and Sensitivity Studies, vol. 2, 1979, pp. 1002–1023.

Thompson, S. L.; and Schneider, S. H.: A Seasonal Zonal Energy Balance Climate Model with an Interactive Lower Layer. J. Geophys. Res., vol. 84, 1979, pp. 2401–2414.

Ulrich, Roger K.: Solar Neutrinos and Variations in the Solar Luminosity. Science, vol. 190, Nov. 14, 1975, pp. 619–624.

United States Committee for the Global Atmospheric Research Program. Understanding Climatic Change. NAS, Washington, D.C., 1975.

Warren, S. G.; and Schneider, S. H.: Seasonal Simulation as a Test for Uncertainties in the Parameterizations of a "Budyko-Sellers" Zonal Climate Model. J. Atmos. Sci., vol. 36, Aug. 1979, pp. 1377–1391.

Wetherald, Richard T.; and Manabe, S.: The Effects of Changing the Solar Constant on the Climate of a General Circulation Model. J. Atmos. Sci., vol. 32, Nov. 1975, pp. 2044–2059.

Williams, Jill, ed.: Carbon Dioxide, Climate and Society. Pergamon Press, Oxford, 1978.

Zhe-Ming, Chen: On the Evolution of Planetary Atmospheres and Their Response to External Sources. Intern. Conference on Evolution of Planetary Atmospheres and Climatology of the Earth, Nice, France, Oct. 16–20, 1978, pp. 45–47.

Evolution of Oxygen and Ozone in Earth's Atmosphere

J. F. KASTING and T. M. DONAHUE

It would appear from recent studies that the ozone screen was established prior to Silurian times and therefore was not directly linked with the spread of life onto land during that period.

One aspect of the early terrestrial environment that exerted a direct influence on the origin and evolution of life was the composition of the atmosphere. Earth's atmosphere is accepted to be of secondary origin, as evidenced by the observed large depletions of noble gases relative to their solar abundance. The bulk of our atmosphere was outgassed from the crust and mantle. The amount of time required to accumulate the current inventory of surface volatiles through the outgassing process is uncertain, but most authors agree that the initial degassing period lasted less than 1 billion years. Since that time, the total amount of surface volatiles has been maintained at a more or less constant level by dynamical exchange processes between the atmosphere, crust, and upper mantle. Figure 1 illustrates some of the more important processes (Walker, 1980). Volcanic gases, primarily H_2O and CO_2, are released into the atmosphere during eruptions. Water vapor precipitates into the oceans; carbon dioxide participates in weathering reactions that result in the deposition of carbonate sediments. When seafloor plates are subducted downward into the mantle, volatiles trapped in these sediments are released under the high pressure and are again recycled into the atmosphere through volcanic activity along the plate interface. Over the long term, these cycles are balanced so that the atmosphere and hydrosphere are maintained at approximately constant volume. This paper deals with the composition of the atmosphere after this steady-state situation was reached.

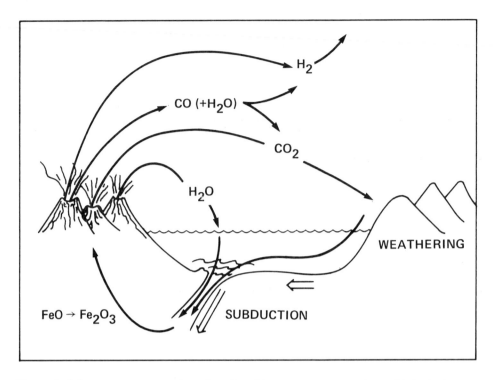

Figure 1. *Schematic representation of cycles of important gases in the prebiological atmosphere.*

From a biological standpoint, the most important aspect of atmospheric composition was the amount of oxygen present. Free oxygen would probably have been a poison to primitive life forms, which presumably lacked sophisticated mechanisms for dealing with it. This, coupled with geological evidence indicating an absence of oxidized sediments before about 2 billion years ago, leads us to believe that Earth's atmosphere was basically reducing during the first half of its history. The first part of this paper is concerned with predicting the ambient oxygen level in such an atmosphere.

The tool used to make such a prediction is a one-dimensional coupled flow-photochemical computer model, similar to those used to study ozone in Earth's current atmosphere. Eight long-lived chemical families are included in the calculation: $O_X \equiv O + O_3$, H, $H_XO_2 \equiv O_2 + HO_2 + H_2O_2$, $NO_X \equiv N + NO + NO_2 + HNO + HNO_2 + HNO_3$, H_2, H_2O, CO_2, and CO. The individual species within each family are calculated by assuming photochemical equilibrium at each height step. Other short-lived species included in the model are OH, $O(^1D)$, and $N(^2D)$. The numerical calculation is carried out on a variable-spaced grid, with step sizes ranging from 0.5 km at the ground to 5 km at 180 km. Present-day N_2 and CO_2 concentrations are used in the results presented here, following Walker's (1977) suggestion that the background levels of these constituents should not have varied greatly with time. Higher CO_2 levels may have been required to keep Earth's surface temperature above freezing in the face of decreased solar luminosity; but this possibility will not be considered here. The temperature and eddy diffusion profiles used in the model are also consistent with present-day values, with the exception of the removal of the temperature bulge due to ozone heating in today's stratosphere. More details on the model, including a list of chemical reactions and rate constants used, are given in Kasting et al. (1979).

Given these assumptions, the amount of free oxygen present in the atmosphere is determined by the balance between the sources and sinks for oxygen atoms. Since molecular oxygen is not released by volcanoes, the only obvious net source of oxygen atoms is photolysis of water vapor:

$$H_2O + h\nu \rightarrow H + OH$$

followed by the escape of hydrogen into space. It has been shown by Hunten (1973a,b), Liu and Donahue (1974a,b,c), Liu et al. (1976), and Hunten and Strobel (1974) that, under a wide variety of conditions, the escape of hydrogen from the terrestrial atmosphere is governed by the principle of limiting flux. Mathematically,

$$\phi_H \approx 2 \times 10^{13} \, f_T$$

where ϕ_H is the hydrogen escape flux, measured as the number of atoms escaping per square centimeter per second, and f_T is the total hydrogen mixing ratio in the stratosphere. Since the total hydrogen mixing ratio remains constant above the cold trap at the tropopause, that is, above the height at which water vapor condenses out of the atmosphere, the production rate of oxygen atoms may be evaluated if one knows the water vapor mixing ratio at the tropopause. In our model, which assumes present-day tropospheric temperatures, the 3.8 ppm of H_2O at 10 km yields a hydrogen escape flux of 1.47×10^8 H atoms/cm^2/sec, in agreement with Liu and Donahue. If no loss processes were operating, the oxygen left over as a result of this escape process would amount to 1 PAL (present atmospheric level) after about 3 billion years.

Two important loss processes for oxygen do operate: oxidation of reduced volcanic gases, primarily H_2 and CO; and oxidation of crustal materials at Earth's surface, which may be written schematically as

$$2FeO + O_2/2 \rightarrow Fe_2O_3$$

The crustal loss rate is difficult to evaluate. It turns out not to be necessary to do so, however, since the H_2 and CO outgassing from volcanoes appears to be more than sufficient to overwhelm the production of O_2 from H_2O photolysis followed by hydrogen escape. By assuming that volcanic outgassing must in the long run balance the loss of surface volatiles via subduction of sea-floor plates down into the mantle, Walker (1977) estimated the H_2O and CO_2 volcanic outgassing rates to be, respectively, 3×10^{14} and 1×10^{14} gm/yr. Holland (1962) has predicted thermodynamic equilibrium ratios of

$$P_{H_2}/P_{H_2O} = 1/105, \qquad P_{CO}/P_{CO_2} = 1/37$$

for present-day volcanic gases at 1225°C. These figures are in accord with averaged measured ratios of Hawaiian volcanic gases. Thus the estimated present-day outgassing rates for H_2 and CO are both on the order of 2×10^8 atoms/cm^2/sec, adding up to a combined reduced gas flux of about 4×10^8 atoms/cm^2/sec. This potential loss rate for oxygen atoms should be compared to a production rate of $1.47 \times 10^8/2$ O atoms/cm^2/sec from H_2O photolysis followed by hydrogen escape. The reduced gas flux is clearly sufficient to dominate oxygen production, and may have done so to an even greater extent if the crust was being recycled at a faster rate during earlier periods of Earth's history. The oxygen and hydrogen budgets become balanced when H_2 accumulates to a level such that the hydrogen escape rate balances the total ($H_2 + CO$) outgassing rate. For an outgassing rate of 4×10^8 atoms/cm^2/sec, the required H_2 mixing ratio is 3.8 ppm $\times$ ($8 \times 10^8/1.47 \times 10^8 - 1$) or about 17 ppm.

Once the hydrogen mixing ratio is known, the amount of oxygen present is determined from the photochemical model. The result for a solar zenith angle of 60° and the outgassing rate mentioned above is shown in figure 2. The O_2 number density profile exhibits a peak around 40 km due to the photolysis of CO_2 followed by

$$O + O + M \rightarrow O_2 + M$$

Below the peak, the O_2 density decreases rapidly toward the ground as a result of the three-body reaction with atomic hydrogen:

$$H + O_2 + M \rightarrow HO_2 + M$$

The O_2 mixing ratio at the ground amounts to less than 10^{-13} PAL, making oxygen a very rare constituent indeed. This result is in agreement with the predictions of low ground-level O_2 densities made by Walker (1977).

Proceeding in this same fashion, oxygen profiles may be calculated for a variety of different hydrogen outgassing rates. Figure 3 shows how the ground-level O_2 concentration varies as a function of total ($H_2 + CO$) outgas-

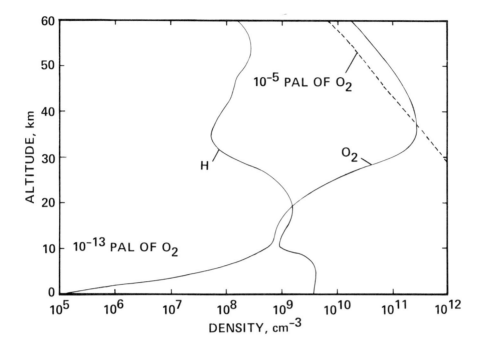

Figure 2. *Prebiological oxygen and atomic hydrogen profiles. Results shown are for 1 PAL (320 ppm) of CO_2 and a total hydrogen outgassing rate of 4×10^8 molecules/ cm^2/sec.*

sing rate. The vertical asymptote at an outgassing rate of 7.35×10^7 atoms/ cm^2/sec corresponds to the flux required to balance oxygen production exactly. If the outgassing rate should drop below this value, oxygen would begin to accumulate in the atmosphere, at least until crustal oxidation processes were able to offset the rate of oxygen production. For outgassing rates exceeding this critical flux, the O_2 density decreases toward a limiting value (horizontal asymptote) of 5.4×10^4 atoms/cm^3. The lower limit on oxygen is a result of the production of O_2 in lightning discharges, which can be estimated by assuming thermodynamic equilibrium at high temperatures. (Chameides et al., 1977; Chameides and Walker, unpublished data). The net result is that, for reasonable hydrogen outgassing rates, the ground-level oxygen concentration is rather closely contained between the values 5×10^4 and about 5×10^5 atoms/cm^2. At such a low density, free oxygen should have had little effect on any biological organisms that may have been evolving at the time.

The discussion thus far has neglected any influences early life may have exerted on its environment. During the earliest stages of biological evolution, these influences were probably small, so that such an omission is justifiable. At some point, however, perhaps as early as 3.3 billion years ago (Shidlowski

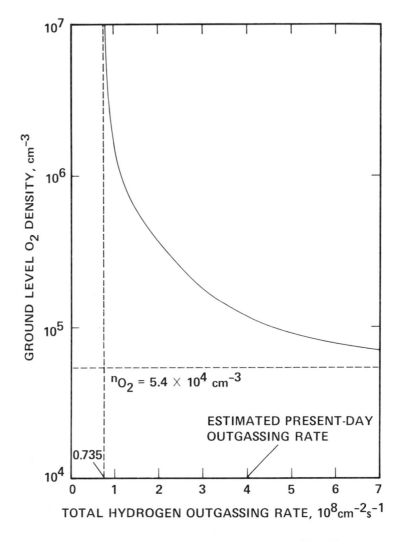

Figure 3. *Ground-level oxygen number density as a function of total hydrogen outgassing rate.*

et al., 1975), the process of oxygenic photosynthesis was invented. By enabling organisms to split water molecules into their constituent atoms, photosynthesis exerted a drastic influence on the oxidation state of the atmosphere. Burial of reduced carbon in organic material produced by photosynthesis was accompanied by the release of a stoichiometrically equivalent amount of free oxygen into the environment. After exhausting any leftover surface reservoirs of reduced material, particularly dissolved ferrous iron which may have been present in the early Archean oceans (Cloud, 1972), this oxygen eventually began to accumulate in the atmosphere.

We pick up the story again when the ground-level oxygen mixing ratio reached 10^{-5} PAL. Even at this level, oxygen was not yet well mixed in the lower atmosphere (see fig. 4). The oxygen bulge in the upper stratosphere due to CO_2 photolysis persists until the mixing ratio exceeds 10^{-4} PAL.

Accompanying the buildup of O_2 was a new atmospheric development that also had an important effect on the biosphere, namely, the emergence of an ozone layer. The presence of atmospheric ozone is essential to the existence of most land life since ozone is the only important absorber of solar near-ultraviolet radiation between 2000 and 3000 Å. This dependence led Berkner and Marshall (1964, 1965) and others to link the spread of life onto land in the late Silurian, about 420 million years ago, with the development of the ozone layer. An interesting and still unresolved question is whether this relationship was causal, with a rapid sequence of the two events, or whether the emergence of land life merely awaited the evolutionary advances necessary to make the transition well after the ozone screen had been established.

To examine the rise of atmospheric ozone, the computer model used to do the prebiological oxygen calculations is modified by the addition of nitrous oxide and methane, biogenic trace gases that are both important in

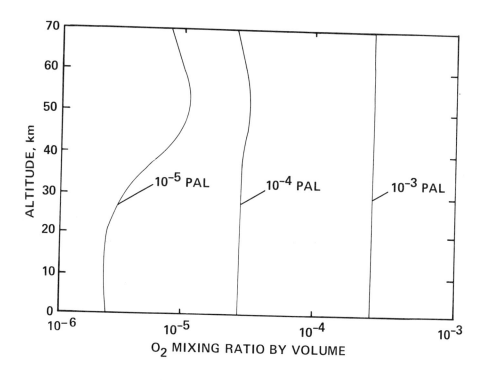

Figure 4. O_2 mixing ratio profiles at various oxygen levels.

influencing ozone concentrations in the present-day atmosphere. Nitrous oxide, which is produced by bacterial dentrification processes in anaerobic soils, reacts with atomic oxygen in the metastable ^1D state:

$$N_2O + O(^1D) \rightarrow 2NO$$

This reaction serves as an important source of odd nitrogen (NO_x) compounds in the stratosphere. Odd nitrogen provides a catalytic destruction pathway for ozone via the sequence:

$$NO + O_3 \rightarrow NO_2 + O_2$$
$$NO_2 + O \rightarrow NO + O_2$$
$$\overline{}$$
$$O + O_3 \rightarrow 2O_2$$

This destruction mechanism is the most important loss process for ozone in the present day middle stratosphere (Crutzen, 1970; Johnston, 1971).

Methane, produced primarily from fermentation and anaerobic decay processes in swamps and wetlands, is a source for ozone in today's troposphere (Fishman and Crutzen, 1977). Methane oxidation produces methyl peroxide and hydroperoxyl radicals which react with nitric oxide:

$$NO + CH_3O_2 \rightarrow NO_2 + CH_3O$$
$$NO + HO_2 \rightarrow NO_2 + OH$$

When these reactions are followed by photolysis of NO_2, ozone is formed:

$$NO_2 + h\nu \rightarrow NO + O$$
$$O + O_2 + M \rightarrow O_3 + M$$

The NO in our model troposphere is produced in lightning discharges, according to the predictions of Chameides et al. (1977).

Including these species in our model necessitates extrapolation of their production rates back to times when the atmosphere contained much less oxygen than at present. The rate of NO production in lightning is scaled by assuming thermodynamic equilibrium at 2300 K in the immediate vicinity of the lightning discharge. This yields the following scaling factors relative to today's production rate:

PAL of O_2	NO scaling factor
10^{-1}	0.316
10^{-2}	.1
10^{-3}	.04

The rates of biological generation of N_2O and CH_4 at lower oxygen levels are much more difficult to assess. The choice has been to hold the production rates constant by assuming a constant upward flux of each species at the ground. The decreasing validity of this assumption at the lowest O_2 levels is compensated by the fact that both N_2O and CH_4 become less effective in influencing ozone densities when oxygen is less abundant in the atmosphere. Increased tropospheric OH densities and rapid photolysis lead to decreased methane and nitrous oxide concentrations by the following reactions:

$$CH_4 + OH \rightarrow CH_3 + H_2O$$
$$N_2O + h\nu \rightarrow N_2 + O$$

Making use of the above assumptions, model experiments have been carried out for oxygen levels ranging from 10^{-5} to 1 PAL of O_2. The resulting ozone profiles are shown in figure 5. A solar zenith angle of 45° was used, and the resulting photolysis rates were multiplied by a factor of 0.5 to account for the diurnal variation. The total ozone column depth for the 1 PAL case is 1.12×10^{19} molecules/cm^2, which is somewhat higher than

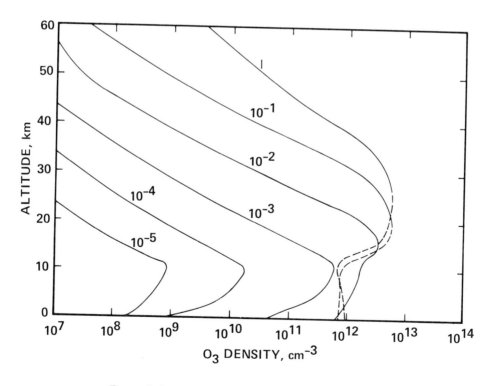

Figure 5. *Ozone profiles for various oxygen levels.*

the mean global value of 8.6×10^{18} molecules/cm^2 reported by McClatchey et al. (1971). Experiments with higher solar-zenith angles yield smaller O$_3$ column depths, so that these results represent a crude upper limit on the available ozone. The temperature profile used is "primordial" (i.e., we assume an isothermal stratosphere) for 10^{-1} PAL of O$_2$ and below and present-day for 1 PAL. Thus the model includes a first-order approximation to the coupling between stratospheric temperature and ozone abundance.

The ozone density for the present atmosphere peaks at a height of 24 km, where the number density is 5.37×10^{12} molecules/cm^3. At lower oxygen levels the ozone peak moves downward due to the increased depth of penetration of solar ultraviolet radiation. At 10^{-3} PAL of O$_2$ and below, the O$_3$ peak is found at 10 km, the height of the tropopause in our model. Below 10 km, photolysis of water vapor leads to large concentrations of odd hydrogen (HO$_x$ = H + OH + HO$_2$) radicals, which destroy ozone via a number of catalytic reactions.

The characteristic of these ozone profiles which is important from a biological standpoint is the total column depth since that is what determines how much ultraviolet radiation may leak through. The original estimate by Berkner and Marshall (1964, 1965) was that an effective ultraviolet shield would be provided by a column depth of 0.2 atm cm, or 5.4×10^{18} molecules/cm^2 (1 atm cm = 2.687×10^{19} molecules/cm^2). This column depth is sufficient to reduce the ultraviolet flux at the surface of Earth to less than 1 erg/cm^2/sec at 50 Å. Ratner and Walker (1972) argue that this flux is still unacceptably high; hence they adopt a somewhat more stringent lower limit on ozone column depth of 7×10^{18} molecules/cm^2.

Figure 6 shows the ozone column depths calculated by our model for various oxygen levels. Also shown are the original results of Berkner and Marshall (1964, 1965) and those of Levine (1977), who performed a similar set of calculations with a one-dimensional model less sophisticated than ours. The differences between our results and those of Levine are mostly due to the fact that he modeled O$_3$ separately rather than including it as part of odd oxygen (O$_x$ = O + O$_3$), as is done in current stratospheric ozone models. Also, Levine did not include H$_2$O and CO$_2$ in attenuating the solar ultraviolet flux, which caused his results to be defective at the lower oxygen levels.

The ozone column depth curve derived from our model crosses Ratner and Walker's "critical level" at an O$_2$ mixing ratio of about 0.05 PAL. This again is for an assumed solar-zenith angle of 45°, so that the predicted ozone column depth represents an upper limit. For a solar-zenith angle of 57.3°, corresponding to that used by Ratner and Walker (1972) and Levine (1977), the critical level is not passed until the O$_2$ mixing ratio reaches about 0.1 PAL — the value predicted originally by Berkner and Marshall, which is considerably higher than the 10^{-2} PAL found by Levine or the 10^{-3} PAL estimated by Ratner and Walker.

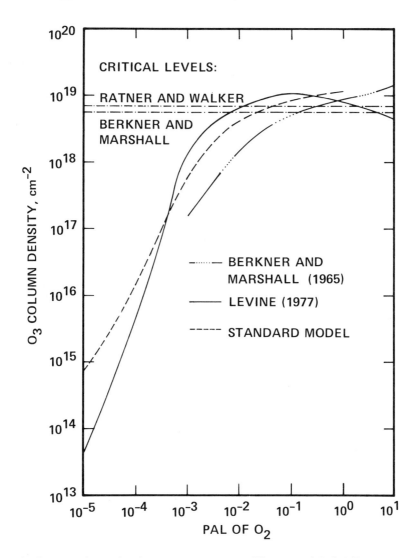

Figure 6. *Ozone column depth vs oxygen content. The curve labeled "standard model" corresponds to the ozone profiles shown in figure 5.*

The significance of this result may be ascertained by comparing our critical O_2 level with the estimate by Rhoads and Morse (1971) for the minimum atmospheric oxygen content during the Cambrian period (600 million years ago). By studying modern anaerobic marine basins, Rhoads and Morse determined that calciferous fauna require a dissolved oxygen concentration of at least 1 mℓ/liter, compared to the 4–9 mℓ/liter that would be in equilibrium with the present atmosphere. Since the Cambrian period was marked by the sudden appearance of abundant shelled organisms, they conclude that the atmospheric oxygen content must have been at least as high as 0.1 PAL during that time. Their lower bound on Cambrian O_2 is the same as our estimate for the amount of oxygen necessary to produce a biologically effective ultraviolet shield. Thus it would appear that the ozone screen was established before the Silurian and therefore was not directly linked with the spread of life onto land during that period. However, the uncertainty inherent in both calculations leaves open the possibility that a causal relationship between the evolution of atmospheric ozone and the appearance of land life did indeed exist.

REFERENCES

Berkner, L. V.; and Marshall, L. C.: The History of Oxygenic Concentration in the Earth's Atmosphere. Disc. Faraday Soc., vol. 37, 1964, pp. 122–141.

Berkner, L. V.; and Marshall, L. C.: On the Origin and Rise of Oxygen Concentration in the Earth's Atmosphere. J. Atmos. Sci., vol. 22, May 1965, pp. 225–261.

Chameides, W. L.; Stedman, D. H.; Dickerson, R. R.; Rusch, D. W.; and Cicerone, R. J.: NO_x Production in Lightning. J. Atmos. Sci., vol. 34, Jan. 1977, pp. 143–149.

Cloud, P.: A Working Model of the Primitive Earth. Amer. J. Sci., vol. 272, June 1972, pp. 537–548.

Crutzen, P. J.: The Influence of Nitrogen Oxides on the Atmospheric Ozone Content. Quart. J. R. Meteor. Soc., vol. 96, Apr. 1970, pp. 320–325.

Fishman, J.; and Crutzen, P.: A Numerical Study of Tropospheric Photochemistry Using a One-Dimensional Model. JGR, vol. 82, Dec. 1977, pp. 5897–5906.

Holland, H. D.: Model of the Evolution of the Earth's Atmosphere. In: Petrologic Studies: A Volume to Honor A. F. Buddington, Geological Soc. of Amer., New York, 1962, pp. 447–477.

Hunten, D. M.: The Escape of H_2 from Titan. J. Atmos. Sci., vol. 30, May 1973a, pp. 726–732.

Hunten, D. M.: The Escape of Light Gases from Planetary Atmospheres. J. Atmos. Sci., vol. 30, Nov. 1973b, pp. 1481–1494.

Hunten, D. M.; and Strobel, D. F.: Production and Escape of Terrestrial Hydrogen. J. Atmos. Sci., vol. 31, Mar. 1974, pp. 305–317.

Johnston, H. S.: Reduction of Stratospheric Ozone by Nitrogen Oxide Catalyst from SST Exhaust. Science, vol. 173, Aug. 1971, pp. 517–522.

Kasting, J. F.; Liu, S. C.; and Donahue, T. M.: Oxygen Levels in the Prebiological Atmosphere. JGR, vol. 84, June 1979, pp. 3097–3107.

Levine, J. S.: The Evolution of Stratospheric Ozone. Ph.D. dissertation, University of Michigan, 1977.

Liu, S. C.; and Donahue, T. M.: The Aeronomy of Hydrogen in the Atmosphere of the Earth. J. Atmos. Sci., vol. 31, May 1974a, pp. 1118–1136.

Liu, S. C.; and Donahue, T. M.: Mesospheric Hydrogen Related to Exospheric Escape Mechanisms. J. Atmos. Sci., vol. 31, July 1974b, pp. 1466–1470.

Liu, S. C.; and Donahue, T. M.: Realistic Model of Hydrogen Constituents in the Lower Atmosphere and Escape Flux from the Upper Atmosphere. J. Atmos. Sci., vol. 31, Nov. 1974c, pp. 2238–2242.

Liu, S. C.; Donahue, T. M.; Cicerone, R. J.; and Chameides, W. L.: Effect of Water Vapor on the Destruction of Ozone in the Stratosphere Perturbed by CL_X or NO_X Pollutants. J. Geophys. Res., vol. 81, June 1976, pp. 3111–3118.

McClatchey, R. A.; Fenn, R. W.; Selby, J. E. A.; Volz, F. E.; and Garing, J. S.: Optical Properties of the Atmosphere. Tech. Rept., AFCRL-71-0279, Air Force Cambridge Res. Labs, Bedford, MA, 1971.

Ratner, M. I.; and Walker, J. C. G.: Atmospheric Ozone and the History of Life. J. Atmos. Sci., vol. 29, July 1972, pp. 803–808.

Rhoads, D. L.; and Morse, J. W.: Evolutionary and Ecologic Significance of Oxygen-deficient Marine Basins. Lethaia, vol. 4, 1971, pp. 413–428.

Schidlowski, M.; Eichmann, R.; and Junge, C. E.: Precambrian Sedimentary Carbonates: Carbon and Oxygen Isotope Geochemistry and Implications for the Terrestrial Oxygen Budget. Precambrian Res., vol. 2, 1975, pp. 1-69.

Walker, J. C. G.: Evolution of the Atmosphere. New York: Macmillan, 1977.

Walker, J. C. G.: Atmospheric Constraints on the Evolution of Metabolism. In: Origins of Life, vol. 10, June 1980, pp. 93–104.

Prospects for Detecting Other Planetary Systems

A comprehensive search for other planetary systems must use ground- and space-based instruments. Indirect search can be done from the ground, but direct detection must be tried from space. Prospects for the latter are not clear at this time.

Whether life can originate and evolve in the absence of a planet is an open question, but the available evidence suggests that planets may be necessary to provide the conditions that are conducive to the formation and evolution of life. Thus we might regard planets as "cosmic petri dishes."

The role of planets in the origin and evolution of life in the Universe is sufficient rationale for considering the detection of other planetary systems. However, the rationale for undertaking such a search extends beyond the question of life in the Universe. Perhaps the strongest reason is that the results of the search, be they positive or negative, are *essential* if we are ever to understand the origin of the Solar System: only data concerning the frequency of occurrence of planetary systems and their general properties will allow us to test empirically the many working hypotheses on the origin of the Solar System. A comprehensive search may also yield valuable insight into the process of star formation, since the prevailing view on this process suggests that planetary systems should be a frequent by-product. Data from a search will either affirm this view or indicate the need for a new view of star formation.

A reasonable question to pose is whether any other planetary systems have been discovered. The answer is that *there is no unambiguous evidence for the existence of other planetary systems.* This answer may come as a surprise, given the tremendous increase in sophistication of astronomical observations. To be sure, advances in technology have had a significant impact on astronomical instrumentation. Until very recently, however, none of these advances has taken a form that pertains directly to a search for other planetary systems.

There have been observations that have been interpreted as giving evidence for the existence of other planetary systems — the best-known example concerns a small and cool M-dwarf star known as Barnard's star. An extensive discussion of the controversy regarding the existence of companions to Barnard's star is beyond the scope of this paper; however, the essential elements are summarized below, and more detail can be found in Black (1980a).

Van de Kamp and coworkers at Sproul Observatory have conducted a prolonged study of Barnard's star and believe they have found some indication of a dark companion. The observational technique employed by the Sproul group was that of astrometry — the determination of the apparent position of a star as a function of time. Van de Kamp found that his data were consistent with either a single dark companion of mass comparable to that of Jupiter, revolving about Bernard's star in an eccentric orbit with a period of 24 years (van de Kamp, 1963), or two dark companions with masses comparable to that of Jupiter, moving in coplanar circular orbits with periods of 12 and 24 years, respectively (van de Kamp, 1969).

Hershey (1973) and Gatewood and Eichhorn (1973) have cast doubt on the validity of the earlier Sproul data. Hershey found evidence for small "sudden" changes in the apparent right ascension position of stars in a star field unrelated to the position of Barnard's star. The timing of these major discontinuities (1949 and 1957) coincide with times when there were adjustments and/or alterations to the Sproul telescope. They also correspond in both time and amplitude to the dominant apparent perturbation in the position of Barnard's star. Gatewood and Eichhorn analyzed data on Barnard's star taken with two other astrometric telescopes, one at Van Vleck Observatory and one at Allegheny Observatory. They were unable to confirm the perturbation that appears in the Sproul data. It appears likely that instrumental effects in the Sproul telescope gave rise to spurious data. It should be emphasized that these effects are at a very low level and do not vitiate other research done on the Sproul telescope. Further, these results on Barnard's star should not be taken as evidence that it does not possess a planetary system. Continued studies of Barnard's star, with increased accuracy, are currently underway at Sproul, Allegheny, and the U.S. Naval Observatories. However, the Barnard's star story emphasizes the difficulty of detecting other planetary systems. The expected magnitude of signals due to another planetary system is very small, generally at or well below the accuracy obtainable with existing telescopes.

The first detailed examination of techniques for detecting other planetary systems was a 1976 Workshop study sponsored by NASA-Ames

Research Center (Morrison et al., 1977). A second NASA-sponsored work-shop on ground-based techniques for detecting other planetary systems was held in 1979 (Black and Brunk, 1980a,b). During the summer of 1976, a systems design study was sponsored jointly by NASA-Ames Research Center and Stanford. That study — Project Orion — centered on the design of a novel ground-based astrometric telescope that would be 30 to 50 times more accurate than existing systems (Black, 1980b). The material summarized here derives mainly from the findings of these three studies.

The review is divided into four sections. Some background on the problem of planetary detection is given in the next section. Ground-based techniques for detecting other planetary systems are then discussed, followed by space-based techniques. The paper concludes with a summary of the prospects. Of necessity, we will not discuss details of the many techniques or of the possible instruments that could be used in a search for other planetary systems. Readers interested in such detail can refer to Black (1980a) and references cited therein.

DETECTION PROBLEM

Before embarking on a discussion of the prospects for detecting other planetary systems, it will be useful to examine the nature of the detection problem itself. What are the likely observable manifestations of another planetary system?

To appreciate the possible range of observables that might arise from a planetary system, consider a model of a simple system (fig. 1). The model

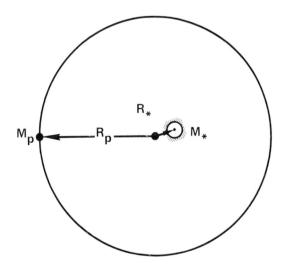

Figure 1. *Schematic representation of a simple planetary system.*

system is comprised of a planet of mass M_p revolving about a star of mass M_* in a circular orbit of radius R_p. The instantaneous center of mass of the system is located on a line between the star and the planet and is depicted in figure 1 as lying a distance R_* from the star. In addition to this geometric characterization, we must also characterize the radiative properties of each member of the system. An important parameter in this regard is the temperature of each body (T_p and T_* for the planet and star, respectively). Also important is size, and we designate by d_p and d_* the diameters of the respective bodies (taken to be spheres for simplicity). More complicated characteristics are needed to completely specify radiative properties. However, these additional characteristics (such as albedo and magnetic field strength) are not possible to specify a priori in any reasonable way, so we will begin by specifying only a limited number of parameters.

How might this simple planetary system be detected? If there were no planets, the center of mass of the star would coincide with the center of mass of the total system. The presence of a planetary companion has the effect of displacing the star with respect to the center of mass of the system (barycenter). Further, because the planet revolves about the barycenter, the star must also revolve about the barycenter. It is essentially a celestial teeter-totter; the star constantly adjusts its position to "balance" the effect of the planet. For this simple system, the displacement of the star is simply

$$R_* = \left(\frac{M_p}{M_*}\right) R_p \tag{1}$$

If the planet is revolving with an orbital period of P, the star must also revolve with that period. If this orbital motion is superposed on a general linear motion of the entire planetary system, a distant observer will see that the star does not move at a constant rate in a straight line (the barycenter does); rather it will appear to "wobble" as it moves across the sky. The magnitude of the true wobble is given by equation (1), but the magnitude of the *apparent* wobble depends on the distance between the observer and the planetary system. If we denote that separation distance by D, the apparent angular extent of the wobble, θ_*, is given by

$$\theta_* = \arctan(R_*/D)$$

Because R_* is always much smaller than D, we can use the approximation

$$\theta_* \approx R_*/D \text{ radian} \tag{2}$$

Combining equations (1), (2), and Kepler's Third Law gives

$$\theta_* = 9.8 \times 10^{-4} \, K_p M_p \text{ arcsec} \tag{3}$$

where

$$K_p = \frac{1}{D} \left(\frac{P}{M_*}\right)^{2/3}$$

The units of P, M_p, M_*, and D are, respectively, years, Jovian masses ($\sim 1.9 \times 10^{30}$ gm), solar masses ($\sim 2 \times 10^{33}$ gm), and parsecs ($\sim 3 \times 10^{18}$ cm). This grouping of parameters may seem strange, but note that K_p involves distance, stellar mass, and orbital period. Each parameter is determined observationally, as is θ_*; M_* is determined independently, based on the type of star. This isolates M_p, the planet's mass, as the principal unknown. We will consider some numerical examples later in this section. The measurement of the angular displacement, θ_*, is accomplished through astrometric observations (such as those by van de Kamp).

In addition to causing an angular displacement in the position of a star, the presence of a planet causes the star to orbit the barycenter of the system. The orbital speed, v_*, is given by

$$v_* \approx 29.8 \, K_2 M_p \text{ m/sec} \tag{4}$$

where $K_2 \equiv (PM_*^2)^{-1/3}$. Now, when a body emitting or absorbing certain spectral features moves with respect to an observer, the wavelength λ of that emission or absorption changes relative to the wavelength λ_0 it would have if the source were not moving. The change in wavelength, $\Delta\lambda \approx \lambda - \lambda_0$, due to this Doppler shift is given by

$$\Delta\lambda/\lambda_0 = v_{LOS}/c \tag{5}$$

where $c = 3 \times 10^8$ m/sec is the speed of light and v_{LOS} is the speed of the star toward (or away from) the observer, that is, along the line of sight. For our model planetary system, we have

$$\Delta\lambda/\lambda_0 = (v_*/c)\cos i \tag{6}$$

where i is the inclination angle between an observer's line of sight and the orbital plane of the planetary system. Note that, if $i = \pi/2$ (observer looking normal to the orbital plane), then $\delta\lambda = 0$ and is *independent* of the value of v_*. The quantity that is measured is essentially $\Delta\lambda$, which can then be related to the product $v_* \cos i$, which is proportional to $M_p \cos i$. Instruments that measure these wavelength shifts are called radial velocity meters. As we see from a comparison of equations (2) and (6), the amplitude of the apparent astrometric observable is smaller for more distant stars than it is for nearby stars (for identical R_* values). This is not the case for the radial velocity; the amplitude of $\Delta\lambda$ is independent of the distance D. On the other hand, a small θ_* almost always implies that a low-mass companion is present, whereas a small $\Delta\lambda$ does not necessarily imply that a small companion is present (because of the $\cos i$ factor).

Another possible observable manifestation of a planetary system involves the transit of a star by a planetary companion. During a transit, the planet passes between the star and an observer and blocks out part of the starlight. The associated apparent dimming of the star could, in principle, be detected with very accurate photometric observations. Because this technique for detecting other planetary systems is somewhat limited, we will not discuss it further here.

Astrometry and radial-velocity studies involve observations of the stellar member of a planetary system; the presence of planetary companions is *inferred* from some observable effect that those companions have on the star. For that reason we refer to astrometry and radial velocity as *indirect detection techniques.*

Both the planet and the star depicted in figure 1 would be sources of radiation. Generally, the total radiation flux $\phi(\nu)$ from the planet will consist of a sum of three components: thermal, reflected, and nonthermal (reflected radiation is nonthermal, but it is convenient to treat it separately); that is,

$$\phi(\nu) = \phi_{th}(\nu) + \phi_{ref}(\nu) + \phi_{nt}(\nu) \tag{7}$$

where ν is the frequency of the radiation. Although there are many known examples of nonthermal planetary radiation (e.g., Jupiter's radio bursts), it is virtually impossible to describe in a general way the range of phenomena that could give rise to nonthermal emission from a planet. In this paper we will ignore this contribution to $\phi(\nu)$ and concentrate on the thermal and reflected components.

If we assume that the intrinsic thermal flux from a planet is that of a black body of temperature T_p, then

$$\phi_{th}(\nu) = \frac{2\pi^2 d_p^{\ 2}}{c^2} \frac{h\nu^3}{\exp(h\nu/kT_p) - 1} \tag{8}$$

where $\phi_{th}(\nu)$ is the power per unit frequency interval, and the constants c, h, and k are the speed of light, Planck's constant, and Boltzmann's constant, respectively. The reflected component is given by

$$\phi_{ref}(\nu) = \eta f \phi_*(\nu) \tag{9}$$

where f denotes the fraction of the total stellar radiation flux, $\phi_*(\nu)$, that is incident upon a planet, and the parameter η denotes the fraction of the incident flux that is reflected. In the context of figure 1,

$$f \approx (d_p/4R_p)^2 \tag{10}$$

The parameter η is generally a function of frequency and of the composition of the reflecting medium; it also depends on the position of an observer relative to the star-planet pair. Clearly, the radiated flux from a planet depends on a number of factors, and we consider next a specific example to obtain a better feeling not only for the magnitude of the problem, but also for the frequencies that are best from a detection standpoint.

Figure 2 shows $\phi(\nu)$ as a function of frequency for the Sun and for Jupiter. In deriving these curves, it was assumed that the Sun radiates as a black body and hence its radiation is described by equation (8) with the solar diameter and temperature replacing d_p and T_p, respectively. The Jupiter curve was obtained by taking $\eta f \approx 2 \times 10^{-9}$. The contrast ratio between

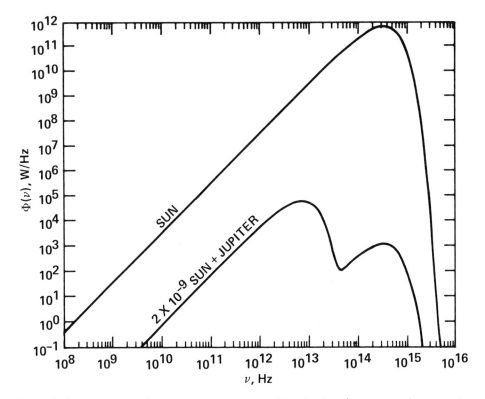

Figure 2. *Power per unit frequency interval emitted by the Sun (upper curve), assumed to radiate as a black body, and by Jupiter. Jupiter's radiation is a sum of thermal and reflected components.*

the Sun and Jupiter is lowest at low frequencies, approaching a limiting value of $d_\odot{}^2 T_\odot/d_{2\downharpoonleft}{}^2 T_{2\downharpoonleft} \approx 4500$. The contrast ratio at the frequency of peak emission by Jupiter is somewhat larger, about 10^4. By comparison, the ratio in the visual part of the spectrum, where Jupiter is simply reflecting sunlight, is about 5×10^8.

The contrast ratio between star and planet is only part of the detection problem. Another major aspect of the problem is the angular separation between the star and planet. Pursuing the model of a Jupiter-Sun pair, we note that at a distance of 10 parsecs (32.6 light years), the linear separation between Jupiter and the Sun corresponds to only 0.5 arcsec angular separation. It is this combination of a relatively bright object located very close (in angular measure) to a relatively dim object that makes direct detection of planets revolving around other stars so difficult.

We next examine the indirect detection of a model Jupiter-Sun system as viewed from a distance of 10 parsecs. The maximum angular extent of the wobble in the motion of the Sun due to Jupiter alone is given by equation (3). In this case, $M_p = 1$, $M_* = 1$, $D = 10$, and $P = 11.8$, giving $\theta \approx 5\times10^{-4}$ arcsec. Some appreciation for the smallness of θ_* can be had by realizing that 5×10^{-4} arcsec corresponds to only a few angstroms ($1\ \text{Å} = 10^{-8}$ cm) as viewed from a distance of 15 inches. Thus our planetary detection problem is akin to measuring the displacement of atoms in this page of paper due to thermally induced vibrations! If we were to consider the Earth-Sun pair alone, we would find $\theta_* \approx 3\times10^{-7}$ arcsec (again as viewed from a distance of 10 parsecs), smaller than the Jupiter-Sun example by a factor of nearly 1600. What are the expected radial velocity effects for our model planetary system? Using equation (4) with $M_p = 1$, $M_* = 1$, $P = 11.8$ (for Jupiter), we find that $v_* \approx 13$ m/sec. By way of comparison, the Earth-Sun system would lead to $v_* \approx 0.09$ m/sec.

These numerical examples are not meant to represent the expected run of observable parameters for nearby planetary systems (if any exist), but they should give the reader a feeling for the magnitude of the detection problem. Most nearby stars are less massive and fainter than the Sun. The former characteristic is generally helpful from the standpoint of indirect detection techniques, while the latter is either helpful or not, depending on the wavelength used, for direct detection techniques. Against this backdrop of the obvious difficulties inherent in trying to detect other planetary systems, we turn next to a brief review of the capability of existing instruments.

GROUND–BASED TECHNIQUES

Earth's atmosphere presents a number of difficulties that effectively preclude the use of direct detection techniques from the ground. In contrast, although the atmosphere does present problems in varying degree for indirect detection techniques, these problems can be overcome to a great extent. In discussing the various ground-based techniques, we first briefly review the current status of a technique and then the prospects for improvement.

Astrometry

Astrometry is the technique most widely used in searches for dark companions to stars. The accuracy of a single observation, using existing telescopes and detectors (i.e., photographic plates) is typically around 0.02–0.03 arcsec. By observing a given star on many nights during a year, the accuracy of a so-called yearly normal point can be increased to 0.005 arcsec or better. The maximum expected value of θ_* for the Sun-Jupiter system as seen from 10 parsecs is 0.0005 arcsec, an order of magnitude below the accuracy limit of typical astrometric systems.

What are the prospects for increased accuracy in astrometric systems? In the near term (the next 4 years), it appears that studies either currently underway or soon to be started will provide a thorough examination of three general ways to make high-accuracy astrometric observations. Each of these would employ photoelectric detectors in lieu of photographic plates, thereby minimizing a major current limitation to accuracy. Two of the three techniques require the construction of new telescopes; the third does not. Finally, the limitations on accuracy arising from Earth's atmosphere appear to be no worse than about 10^{-4} arcsec, and in some cases as low as 10^{-5} arcsec.

The three general techniques to be explored are (1) classical narrow-field astrometry, (2) single-aperture interferometry, and (3) two-aperture interferometry. The first adopts the technique currently used in astrometry but introduces modern technology in both the detector and the telescope so as to reduce the magnitude of the errors they contribute. The second technique uses a single-aperture telescope (no new telescope is needed) either as a speckle interferometer (see Worden, this volume) or as an amplitude interferometer. Astrometric studies done with this technique differ from those done with the first technique in that only close binary systems are observed. One does not look for a wobble in the motion of a star as measured against a reference frame defined by other stars; one looks instead for changes in the *separation* of the binary pair. Use of an internal reference frame like this means that one cannot tell which of the stars has a planetary companion (if one is detected). The third technique uses two relatively widely separated ($\gtrsim$ 10 m) apertures. The basic observing scheme is identical to that of the classical method (i.e., measurements of the position of a star relative to a reference frame defined by other stars), but one works with fringe patterns rather than the usual Airy disk type of image.

Single-aperture interferometry involves a rather modest expenditure of money and will almost certainly be developed for use in planetary searches. This technique should routinely yield observations accurate to 10^{-4} arcsec and, in special cases, to 10^{-5} arcsec. The two-aperture interferometry technique is currently under study by Shao at MIT. The theoretical accuracy of such a system depends on the properties (e.g., aperture size and separation) of the telescope; the currently envisioned system should have errors no larger than 10^{-4} arcsec. However, the hardware and data acquisition aspects of two-aperture systems are complex, and it will take 2 or 3 years before we know whether an operating system can be constructed that will yield the predicted level of accuracy. By way of contrast, the classical technique involves relatively simple hardware and data acquisition; the principal question is whether the accuracy of a "modern" system is significantly better than that of current systems. Preliminary tests indicate that the gain in accuracy will be significant (i.e., more than an order of magnitude).

The long-term prospects for ground-based astrometry in a search for other planetary systems are excellent. One or more of the techniques discussed above are likely to be developed to the point where a routine observational program will be conducted. It is too early to tell which technique will prove best suited for the problem, but the near-term activity should provide an indication in that regard. It is clear that the critical assessment of the capability of ground-based astrometry which has occurred as a consequence of the interest in detecting other planetary systems will give rise to a new generation of astrometric systems — systems of vastly improved accuracy for use in a wide range of astronomical studies.

Radial Velocity

Until recently, radial-velocity studies of stars yielded measurements accurate to about 1 km/sec. The introduction of new techniques, based on ideas first advanced in the early 1950s, has led to stellar radial-velocity observations accurate to a few hundred meters per second. Much higher accuracies (on the order of meters per second) have been obtained from radial-velocity studies of planets in the Solar System, but to date no stellar velocity system routinely yields observations sufficiently accurate for planetary detection (i.e., $\lesssim$ 10–20 m/sec).

The near-term prospects for obtaining the required high-accuracy radial-velocity measurements are good. A major source of error in stellar radial-velocity studies has been the nonuniformity of the stellar image itself. Recent work has shown that the use of high-quality optical fibers eliminates this source of error. The basic idea is to image a star on an optical fiber (100–200 μm in diameter), and to use the output from the fiber as the input to the radial-velocity meter. The effect of the fiber is to scramble the input image so as to yield a uniform, steady output. The fiber also makes it possible to physically decouple a radial-velocity system from a telescope. The system can sit on the observatory floor and be fed a signal through a flexible fiber. This eliminates errors due to flexure arising from changes in the orientation of a telescope with respect to Earth's gravity field.

One of the principal uncertainties with this technique is that we do not yet have any data at the expected level of accuracy for the intrinsic radial-velocity variations of stars. A significant step in this area can be made by studying the radial-velocity variations of the nearest and brightest star, the Sun. Most astronomers feel that virtually all main-sequence stars will undergo intrinsic velocity variations at the 10 to 20 m/sec level, but data are needed. The major concern is that such variations may be quasi-periodic and could thereby give a spurious indication for the presence of a planetary companion.

The projected near-term activities in the area of radial-velocity systems should provide a clear indication of which system(s) are best suited for the search for other planetary systems, and they should also provide valuable data concerning intrinsic radial-velocity variation for main-sequence stars. The knowledge gained from these activities will be used to guide a long-term, radial-velocity program. There is little question that stellar radial-velocity observations can be conducted that will have errors on the order of 1 m/sec, sufficiently small to permit discovery of other planetary systems.

SPACE–BASED TECHNIQUES

All known methods for direct detection of planetary systems involve space-based systems. The indirect detection methods discussed in the previous section can also be employed with space-based systems. As Earth's atmosphere does not appear to be a limiting factor in the obtainable accuracy of radial-velocity systems, there is little point in conducting a radial-velocity search from space. The accuracy of astrometric methods is limited ultimately by Earth's atmosphere (particularly the narrow-field astrometric method), so one might consider doing astrometry from space.

There are no current space-based systems that could be employed in a search for other planetary systems. NASA is planning to launch two systems in the 1980s that might be used for such a search. The first of these, the infrared astronomical satellite (IRAS), is a joint venture between the Netherlands and the United States. The IRAS is basically an infrared survey satellite that will scan the sky using a variety of instruments and will operate for approximately 1 year. Although the IRAS will not be used specifically to look for other planetary systems, the significantly increased sensitivity and wavelength coverage of the IRAS over that of existing infrared systems might provide unexpected results. The second NASA space-based telescope is the space telescope (ST). As with the IRAS, the ST was not designed for the purpose of searching for other planetary systems. Although the imaging capability of the ST will far exceed that of ground-based telescopes, it appears unlikely that it will be able to cope with the severe imaging problems discussed above. The ST will also have subsystems with which astrometric observations can be made. Preliminary estimates indicate that the accuracy of the ST's astrometric systems will be about 10^{-3} arcsec, better than that of *current* ground-based astrometric work but one or more orders of magnitude worse than the accuracy of ground-based systems that will be operating when the ST is launched. The European Space Agency (ESA) is currently considering construction of a special-purpose astrometric space system. The primary function of the ESA satellite would be to obtain parallax (distance) measurements on a large number of stars. Estimates indicate that the accuracy of this system, for the type of observation needed in a search for other planetary systems, is comparable to that of the ST's astrometric systems ($\sim 10^{-3}$ arcsec). Thus, the prospects for meaningful contributions to a search for other planetary systems with currently planned or considered space-based systems are not good.

   This bleak outlook derives mainly from the need for special instrumentation, and there is no reason why that instrumentation could not be realized. We now know enough about the sources of error in astrometric telescopes to construct a space-based device capable of conducting an astrometric search for other planetary systems with an accuracy approaching 10^{-6} arcsec. The prospects for direct detection systems are less clear. Studies indicate that a system to search for other planetary systems at visual wavelengths would push technology, particularly in the area of mirror construction, to or beyond its current capability. Direct detection at infrared wavelengths looks more promising, mainly because the brightness ratio between star and planet is more favorable in the infrared than in the visual part of the spectrum. The IRAS may provide evidence for other planetary systems, but its short lifetime (about 1 year) coupled with its modest spatial resolution suggest that we can only hope for tentative results. A special-purpose infrared system, perhaps an interferometer, will be required. Although there would be major technological problems with such a space-based infrared system, they do not appear to be as severe as those facing a visual direct detection system.

SUMMARY

   Observational evidence concerning the existence of other planetary systems, their frequency of occurrence, and general properties of the planetary members of those systems is an important ingredient in considerations of life in the Universe. Planets apparently are the cosmic petri dishes of nature's experiments with life, but *to date there is no unequivocal evidence for the existence of any planetary system other than our own.*

We have briefly reviewed the nature of the planetary detection problem, some of the techniques that might be used in a search for other planetary systems, and the prospects for success. Almost certainly, the initial stages of any comprehensive program to search for other planetary systems will involve ground-based instrumentation. The prospects for vastly improved accuracy in both astrometric and radial-velocity observations are good; measurements accurate to 10^{-4} arcsec and 1 m/sec, respectively, are feasible. Observations at this level of accuracy are sufficient to conduct a meaningful search.

For a search program to be fully comprehensive, it must involve both ground-based and space-based instruments. Techniques for the indirect detection of other planetary systems can be conducted from the ground, while direct detection techniques require space-based instruments. The prospects for a space-based search are less clear than are those for a ground-based search. None of the existing or currently planned space systems has the capability to mount a meaningful search for other planetary systems. A special-purpose telescope, but one with the capability to perform other useful observations, will be required.

A search for other planetary systems is, in my view, one of the most important and timely problems for astronomy in the coming decade. Knowledge about planetary systems is essential not only to the subject of this conference, but also to understanding the origin of the Solar System. Further, and no less importantly, whether we find that planetary systems are ubiquitous or rare, that finding will have a significant philosophical impact on all mankind. That we will find an answer is certain, for we are looking for physical effects, the absence of which is as telling as their presence. We are limited in our quest only by our willingness to invest time, thought, and money.

REFERENCES

Black, D. C.: In Search of Other Planetary Systems. Space Sci. Rev., vol. 25, Jan. 1980a, pp. 35–81.

Black, David C., ed.: Project Orion — a Design Study of a System for Detecting Extrasolar Planets. NASA SP-436, 1980b.

Black, David C.; and Brunk, William E., eds.: An Assessment of Ground-Based Techniques for Detecting Other Planetary Systems, Vol. I: An Overview. NASA CP-2124, 1980a.

Black, David C.; and Brunk, William E., eds.: An Assessment of Ground-Based Techniques for Detecting Other Planetary Systems, Vol. II, Position Papers. NASA CP-2124, 1980b.

Gatewood, George; and Eichhorn, Heinrich: An Unsuccessful Search for a Planetary Companion of Barnard's Star (BD + 4°3561). Astronom. J., vol. 78, no. 8, Oct. 1973, pp. 769–776.

Hershey, J.: Astrometric Analysis of the Field of AC + 65°6955 from Plates Taken with the Sproul 24-Inch Refractor. Astronom. J., vol. 78, no. 5, June 1973, pp. 421–425.

Morrison, Philip; Billingham, John; and Wolfe, John, eds.: The Search for Extraterrestrial Intelligence — SETI. NASA SP-419, 1977.

van de Kamp, Peter: Astrometric Study of Barnard's Star from Plates Taken with the 24-Inch Sproul Refractor. Astronom. J., vol. 68, no. 7, Sept. 1963, pp. 515–521.

van de Kamp, P.: Alternate Dynamical Analysis of Barnard's Star. Astron. J., vol. 74, Aug. 1969, pp. 757–759.

Detecting Planets in Binary Systems with Speckle Interferometry

SIMON P. WORDEN

Speckle interferometry is particularly suited for work with close-spaced binaries, where conventional astrometry has difficulties. The technique does not require special telescopes to study the 188 known close binary pairs nearer the Sun than 65 light years.

Several methods have been suggested for the detection of low-mass, nonluminous companions of nearby stars. Among the most promising are indirect methods, that is, methods for detecting the effects of an invisible object on its visible companion. Spectroscopic methods can be used to observe radial-velocity fluctuations in the visible star caused by orbital reflex motion relative to the invisible companion. Astrometric detection involves observing the positional perturbation caused by the orbit of the unseen object. Astrometry therefore requires establishing the position of the center of light in a stellar image and referencing this center to a fixed frame.

Astrometry is limited by our ability to establish the center of a star image and by uncertainties in the reference frame. Star images, perturbed by atmospheric seeing, are roughly gaussian in shape. Our ability to find the center of an image obviously depends on the stability of the gaussian profile. Long exposures are used to collect enough photons to define the center of each star image and to average out seeing for image motion. The reference frame of these measurements can in principle be established by the centers of three or more other star images. Since the proper motions of these stars can influence the coordinate frame, the reference stars are chosen to be distant background stars. Earth's atmosphere can also distort the reference frame as a function of atmospheric conditions and the colors of the reference stars. These uncertainties can be reduced by using a relatively large number of reference stars in conjunction with sophisticated mathematical error analysis schemes. Despite such difficulties, conventional astrometry

177

now yields positional accuracies of ±0.05 arcsec for a single exposure; this may be improved to better than ±0.001 arcsec (Gatewood, 1976) for the results of a year's observation.

Astrometric accuracy could be improved by getting smaller star images. Clearly, the smaller the image, the easier it is to find an image center. If images were improved to the diffraction limit, the increase in accuracy would be substantial: finding the center of a diffraction-limited spot of 0.05 arcsec from a 2-m telescope can be done with much higher precision than finding the center of a 1-arcsec seeing disk with the same number of photons. We can also increase accuracy by removing the atmospheric effects on the reference frame. Both improvements are possible using some form of interferometry.

Stellar interferometry, first demonstrated by Michelson (1920), makes it possible to approach the full theoretical (diffraction-limited) resolving power of large optical systems. Instruments with baselines up to hundreds of meters long have been demonstrated (Hanbury Brown, 1968). Smaller-scale instruments have been perfected to adapt existing telescopes into diffraction-limited systems using techniques known as amplitude interferometry (Currie, 1968a,b; Currie et al., 1974) and speckle interferometry (Gezari et al., 1972). These techniques have been extensively reviewed by Dainty (1975), Worden (1977), and Labeyrie (1978). Long-baseline interferometers have been proposed which would convert the image scale to the full diffraction limit and remove atmospheric effects on the reference frame. Such systems are ideal for detecting small-scale astrometric perturbations due to objects of planetary mass. However, these instruments are costly and have yet to be completely demonstrated in the field. On the other hand, amplitude and speckle interferometry have already demonstrated the ability to extend conventional telescopes to the diffraction limit.

Binary stars are suitable candidates for planetary searches. McAlister (1978) has used speckle interferometry with relatively simple detectors and calibration methods in a systematic program to determine binary separations to accuracies of a few thousandths of an arcsecond. If a planet orbits one component of a binary system, then the orbit of that component about the other star will "wobble." For the nearby stars and Jupiter-sized planets, this may be a modulation of up to 0.01 arcsec. Harrington (1977) has shown that most binary systems have potentially stable planetary orbits. Interferometric searches for planetary perturbations of binary star orbits are therefore important. Indeed, interferometry may be the only way to search small-separation binaries. Binary star interferometry is therefore important to assure a complete search for low-mass stellar companions through all nearby stars and all types of stellar systems.

The speckle interferometric method to search binary star systems for planets was presented in some detail by McAlister (1977). In this paper, I will discuss only briefly the physics behind atmospheric degradation of images and the speckle interferometric technique to remove this degradation. I will restrict myself to speckle systems, although much of what follows is also true of amplitude interferometers. A more detailed discussion of both speckle and amplitude interferometric searches for planets has been given by Currie et al. (1980).

ATMOSPHERIC IMAGE DEGRADATION

Small-scale temperature inhomogeneities in Earth's atmosphere produce changes in the atmosphere's index of refraction. These changes cause phase delays along any incoming plane wave, such as light from a stellar point source (fig. 1). Without phase errors, optical systems would produce the image in figure 1(a), which is said to be "diffraction-limited"; here the image of a point source is the classical Airy disk for a circular telescope aperture. The size of this image is inversely proportional to the telescope diameter. With phase errors, telescope resolution is degraded to that of an optical system only as large as the scale over which some phase coherence exists (i.e., over which the phase is the same). Since the atmosphere breaks an incoming plane wave into fragments of about 10 cm, all telescopes produce images with resolution no better than a 10-cm telescope, namely, 1 arcsec (fig. 1(b)).

Labeyrie (1970) proposed a method to recover information down to the diffraction limit of a large telescope. He pointed out that short-exposure photographs (~0.01 sec) "freeze" the turbulence in the atmosphere. Although the phase coherence size in this frozen system is still only 10 cm, there will be some 10-cm patches scattered over the entire aperture which are at the same phase. These portions act in concert as a form of multiple-aperture interferometer that provides some information down to the diffraction limit of the *entire* telescope aperture. As shown in figure 1(c), the image of a point source seen through a multiple-aperture interferometer is a series of nearly diffraction-limited images modulated by a 1-arcsec seeing disk. This process is known as speckle interferometry since the short-exposure photographs (fig. 2) look like laser speckle photographs.

SPECKLE INTERFEROMETRIC INSTRUMENTS AND DATA REDUCTION

A diagram of the Kitt Peak photographic speckle interferometer is shown in figure 3. There are six similar systems in use at the present time. The Kitt Peak camera was designed by Lynds (Lynds et al., 1976;

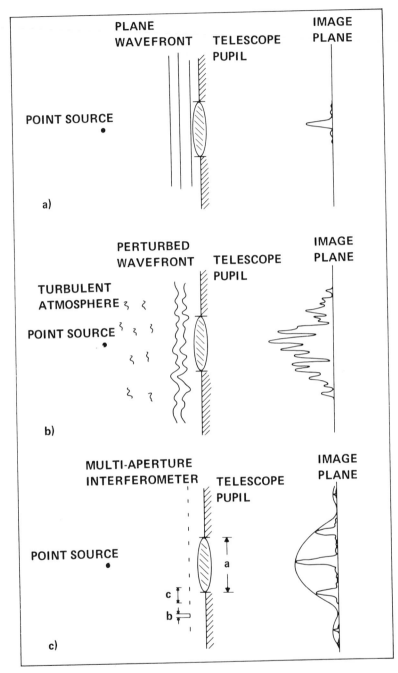

Figure 1. *Schematic diagram of image formation through a turbulent atmosphere.*
(a) Image formation outside the atmosphere: diffraction-limited image. (b) Image forma-
tion through Earth's atmosphere; image is badly degraded. (c) Image formation using
speckle techniques; envelope results from atmosphere seeing while the high-resolution
features result from the resolution of the full aperture.

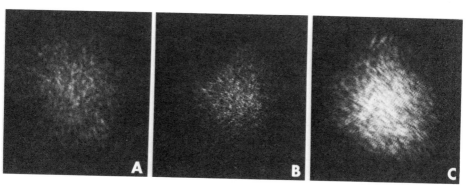

Figure 2. *Speckle photographs for three stars from the Kitt Peak 4-m telescope; note the different characters of the three objects. (a) Resolved supergiant star α Orionis (Betelgeuse), (b) Point source star γ Orionis (Bellatrix), (c) Close double star; separation, 0.05 arcsec, α Auriga (Capella).*

Breckinridge et al., 1979). As shown in the figure, light from the telescope passes through a shutter and focuses at the telescope image plane. The shutter is necessary to ensure exposures shorter than the atmospheric change time, typically 20 msec. The telescope image is relayed and magnified by a microscope objective. The magnification is set to provide a pixel resolution oversampling the telescope diffraction spot size by at least a factor of 4. For the Kitt Peak 4-m telescope, this provides a final image scale of 0.2 arcsec/mm. Atmospheric dispersion blurs speckle image patterns in the sense that the "red" portion of an image focuses at a slightly different place than the "blue" portion. Since this may be significant for even 200-Å band-pass photographs, a set of rotating atmospheric compensating prisms are included to counteract the dispersion. Since there are about 20 orders of optical interference across even a narrow band ($\Delta\lambda \approx 200$ Å), an interference filter is used to preserve coherence across the entire speckle photograph. If this were not included, the "speckles" near the edge of the photographs would be elongated. A three-stage image tube intensifies the image enough to allow photographic data recording. A transfer lens relays the intensified image to a data recording system, in this case a 35-mm film camera.

The speckle photographs in figure 2 were taken with the Kitt Peak system. The different character of these photographs is readily apparent. This is understandable from the analogy to a multiple-aperture interferometer. Each speckle should be a diffraction-limited image of the object. Indeed, the binary-star (α Auriga) speckles are double, the point-source speckles are roughly diffraction spots, and the resolved-star (α Orionis) speckles are somewhat larger. This aspect led Lynds et al. (1976) to a direct speckle-image reconstruction scheme whereby individual speckles were identified and

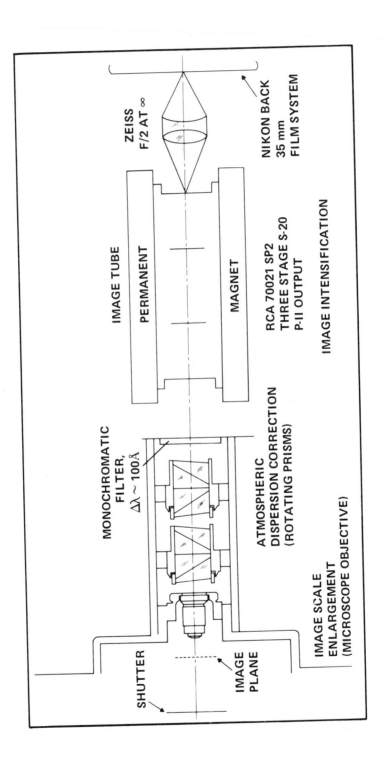

Figure 3. *Diagram of the Kitt Peak photographic speckle interferometry camera.*

co-added to produce a nearly diffraction-limited image for the special case of stars like α Orionis.

A number of methods can be used to reduce speckle interferometry data. Labeyrie's original method is widely used — particularly for measurements of binary stars. Individual speckle photographs are Fourier-transformed either optically or digitally, and the Fourier modulus is computed. If the speckle image is represented in one dimension as $i(x)$ and its transform as $I(s)$, this process is mathematically represented by

$$I(s) = \int_{-\infty}^{\infty} i(x)e^{-2\pi i x s} \, dx \tag{1}$$

The modulus or power spectrum, $|I(s)|^2$, of this transform contains the diffraction-limited information in an easily extractable form. Examples of mean power spectra for several binary systems are shown in figure 4. The power spectra for such systems show banding that represents the separation of the elements: the farther apart the bands, the closer the elements. The orientation of these bands represents the position angle of the binary system. Superimposed on this signal is a background attributable to the residual effects of seeing. For stars brighter than about +7 visual magnitudes, about 50 individual speckle snapshots are transformed to produce a mean power spectrum. A least-square fit to the spacing and orientation angle of the bands in this power spectrum yields the binary separation and position angle — all accomplished from less than 1 sec actual exposure time at the telescope!

The residual effects of seeing must be removed to achieve maximum accuracy. Even though the bands (fringes) are readily visible in raw speckle power spectra, their spacing is affected by the residual seeing effects. Labeyrie's method uses observation of point-source stars to determine and remove these seeing effects. If $p_j(x)$ are point-source speckle photographs with a mean power spectrum $\langle |P(s)|^2 \rangle$, and $\langle |I(s)|^2 \rangle$ is the mean power spectrum of the object speckle photographs $i_j(x)$, the diffraction-limited power spectrum of the object is

$$|O(s)|^2 = \frac{\langle |I(s)|^2 \rangle}{\langle |P(s)|^2 \rangle} \tag{2}$$

Point-source data are usually derived from speckle observations of point-source stars situated near on the sky to the program objects. Since these point-source objects are not generally observed within the same isoplanatic angle or at the same time, their power spectrum can only represent the residual seeing effects in a statistical sense. Worden et al. (1977) have developed a method to calibrate for residual seeing effects using the same set of speckle photographs as were used to study the program objects.

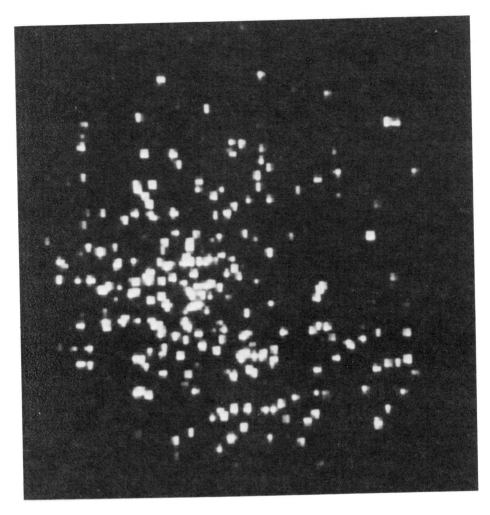

Figure 4. *Mean speckle power spectra for two binary stars. Larger separated fringes are for* i *Serpentis (separation, 0.1 arcsec), the smaller for* β *Cephei (separation, 0.25 arcsec).*

Current photographic speckle cameras are generally limited to objects brighter than +7. The photographic recording systems are therefore being replaced with high-quantum-efficiency, digital recording systems that record individual photon events. The University of Arizona's speckle camera uses a CID (charge-injected device) television system to record photon arrivals. This system simply replaces the photographic emulsion, and it can record data for objects faint enough so that only a few photons arrive in a 20-msec exposure. Figure 5 shows data from this system for Saturn's moon Rhea, which is a 10th magnitude object. For faint objects like this, only the few hundred photon locations are recorded rather than the entire frame. This allows such systems to run at the maximum speckle data rate of one speckle

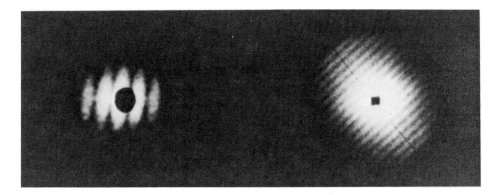

Figure 5. *Speckle data showing individual photons for Saturn's moon Rhea taken with the University of Arizona's CID speckle camera.*

frame every 20 msec. This form of data is ideal for fast computer reduction. Data reduction is simple enough that a direct computer interface can compute the results in real time at the telescope. The limiting requirement for the method is that at least two photons arrive in a 20-msec exposure. This translates to about a +16 stellar magnitude limit. Although angular diameters are more difficult to derive than binary separations, we have used this system to derive angular diameters for 13th magnitude stars accurate to ±5% with less than 5 min total observing time.

ACCURACY

McAlister has initiated a substantial speckle program to derive binary star parameters at Kitt Peak. The first results (McAlister, 1978) give a basis for estimating the precision possible with speckle interferometry.

Internal errors in speckle interferometry include the basic uncertainty in the data itself and the error due to uncertain calibration of the image scale. McAlister has computed errors based on 46 pairs of observations for 5 binary systems, with each pair separated in time by one day to one month. An observation is defined as the result from a single 50-frame set of speckle photographs. For these data (with binary separations of 0.2 to 3.25 arcsec), McAlister concludes that the error due to basic uncertainty in the data is ±0.3% in separation and ±0.2° in position angle for each 50-frame data set. If the calibration errors are included, the angular separation error is reduced to ±0.6%.

Calibrations of image scale and position angle are made by placing a double slit with known slit separation over the telescope aperture. Since the telescope is then effectively a two-slit interferometer, the fringe spacing and position angle in the power spectra of data taken through this slit provide accurate calibrations of angular separations and position angles. Calibrations are generally done only a few times each night. If a set of built-in double slits were used to calibrate each star after every observation, calibration errors could be reduced to much less than the inherent error in the data. For binaries with separations less than 1 arcsec, accuracies of ±0.002 arcsec are already obtainable, and accuracies of ±0.1 arcsec are obtainable for stars of 5 arcsec separation, in single observations.

McAlister has computed possible external errors in his results by comparing binary orbits derived from speckle interferometry with high-quality published orbits. He concludes that speckle orbits match the published orbits to within the accuracies of these orbits; this result precludes large systematic errors in speckle measurements of binary stars.

The above analysis for speckle interferometry is based on photographic data-recording systems. Advanced photoelectric data acquisition systems have several advantages. Since the new systems run at essentially television rates (60 frames/sec), a single 50-frame sample takes less than 1 sec to obtain! McAlister observes about 150 stars per night, spending several minutes on each star. We might expect that 50 observations of 5-min duration would be possible in an observing session with a dedicated telescope. If we assume that errors are reduced as the square root of observing time, then the over 10^4 50-frame data sets obtained per year would refine the accuracies by a factor of 10^2 over the McAlister values. This corresponds to 2×10^{-5} arcsec/yr on binaries with separations smaller than 1 arcsec, and 10^{-4} arcsec/yr on a binary with 5-arcsec separation. The higher quantum efficiency and linearity of the digital system indicate that these numbers should apply to stars brighter than about +9, compared to McAlister's limit of +7. The accuracies on binaries near the faint limit at +14 would probably be a factor of 10 worse for the same observing time.

Another limitation is the isoplanatic angle — the maximum binary separation at which speckle interferometry will work. Conventional wisdom, not based on any real observations, places the isoplanatic angle at about 3 arcsec, meaning that interferometry of binary stars with separations much larger than this would be impossible. Recent measurements by Hubbard et al. (1979) of binary stars with larger separations indicate that this angle is closer to 6 arcsec and may be as large as 10 arcsec. It may therefore be possible to use as a reference star an unrelated background star rather than the other binary component. This may extend interferometric position determinations to wider binaries and some single stars.

Photographic speckle systems have been limited to binary stars in which the components are within 5 magnitudes of each other. Photoelectric systems may extend this limit to 7 or 8 magnitudes. However, the requirement of two photons in each exposure practically limits us to systems in which both stars are brighter than +16 magnitudes.

SUITABLE TARGET BINARY SYSTEMS

This section will examine a set of possible program binary systems and discuss detection probabilities. As a data source I have used Gliese's *Catalogue of Nearby Stars* (1969), which includes all stars with known parallaxes equal to or greater than 0.045 arcsec, plus borderline cases.

There is some conjecture that the formation of binary stars inhibits planet formation. Since definitive models for planet formation are not available, however, there is absolutely no reason to dismiss binary systems a priori as possible planetary systems. Knowledge of the frequency of binaries and the mass distribution of companions would further illuminate the physical processes of star and planetary-system formation. The extensive satellite systems of Jupiter and Saturn point strongly to the hierarchical formation of such systems. There are, however, dynamical constraints on possible planetary orbits in binary systems. Harrington (1977) examined the dynamical stability of a planetary body in a binary system in terms of the restricted three-body problem. He concluded that stable planetary orbits are possible in two classes of binary systems: those in which the planetary orbit is large compared to the binary orbit and those in which it is small. In both cases, the planetary orbit must be a factor of 3 to 4 larger (or smaller) than the maximum (or minimum) binary separation. Since there are no detectable effects of a planet in the case where the binary separation is small compared to the planetary orbit, we restrict our discussion to the opposite case. If we use Jupiter's orbit at approximately 5 AU radius as a benchmark, we can examine which binaries may have a stable Jovian orbit. Table 1 lists the separations for a stable Jovian orbit as a function of parallax. Table 2 shows the effects of a Jovian planet ($m = 10^{-3}$ $M_\odot$) in a Jovian orbit (5 AU) and a large terrestrial planet ($m = 10^{-5}$ $M_\odot = 3$ $M_\oplus$) at 1 AU. These effects are shown as a function of the reflex motion on a Sun-type primary (1 $M_\odot$) and a late type-M dwarf (0.15 $M_\odot$). Based on our previous calculations of 10^{-4} arcsec

TABLE 1.– ANGULAR SIZE OF BINARY
ORBITS FOR A STABLE JOVIAN ORBIT
(in arcsec)

Parallax	Size of Jovian orbit	Minimum binary orbit for stable Jovian orbit
0.2	1.000	4.000
.1	.500	2.00
.075	.375	1.500
.050	.250	1.00

TABLE 2.– EFFECTS OF PLANETS ON PRIMARY-STAR
ORBITAL AMPLITUDE AS A FUNCTION OF DISTANCE
(in arcsec)

Parallax	$10^{-3}M_\odot$ effect on $1M_\odot$	$10^{-3}M_\odot$ effect on $0.15M_\odot$	$10^{-5}M_\odot$ effect on $1M_\odot$	$10^{-5}M_\odot$ effect on $0.15M_\odot$
0.2	2×10^{-3}	1.3×10^{-2}	2×10^{-6}	2.6×10^{-5}
.1	1×10^{-3}	6.7×10^{-3}	1×10^{-6}	1.3×10^{-5}
.075	7.7×10^{-4}	5×10^{-3}	7.5×10^{-7}	9.7×10^{-6}
.050	5×10^{-4}	3.3×10^{-3}	5×10^{-7}	6.5×10^{-6}

accuracy, we see that Jovian planets are detectable for all binary separations out to 20 parsecs. In the special case of nearby M dwarf stars, it may even be possible to detect large terrestrial planets.

In the Gliese catalog there are 248 star systems with binary separations between 0.2 and 15.0 arcsec. Of these 248 systems, 188 lie north of –30°. Table 3 lists the separation distributions for these 188 star systems; while table 4 lists their parallax distributions. Using Harrington's stability criterion,

TABLE 3.– BINARY SEPARATIONS
OF STARS NORTH OF –30° IN
THE GLIESE CATALOG

Separation, arcsec	Number of systems
0.2–0.5	23
.5–2.0	64
2.0–6.0	64
6.0–10.0	19
10.0–15.0	18

TABLE 4.– PARALLAXES OF
NORTHERN BINARY STARS
WITH SEPARATIONS LESS
THAN 15 ARCSEC IN THE
GLIESE CATALOG

Parallax, arcsec	Number of systems
>0.200	9
0.200–0.100	24
0.100–0.075	21
0.075–0.050	76
<0.050	58

we find that roughly half the 188 systems could have a stable Jovian orbit, while almost all could have a stable terrestrial orbit. The sample therefore permits large numbers of planetary orbits at distances similar to those in the Solar System.

For the 182 systems that list magnitudes for both components, we found the following magnitude differences between the two components (see table 5). Almost half the systems have nearly equal magnitudes, while 82% have less than the 5th magnitude difference needed for the present photographic system. These magnitude differences should be about 1 magnitude less if observations were made around 8000 Å since the secondary is almost invariably redder than the primary. The magnitudes for 375 of the component stars are shown in table 6. Of these, 50% are brighter than +9, the magnitude limit for the maximum accuracy. The spectral classification of 251 of these stars is given in table 7. As might be expected, the listings, which are generally for the primary component only, are weighted heavily toward late spectral types. The secondary components should be weighted even more heavily toward later spectral types. Almost all of these stars are main-sequence (luminosity class V), but 16 are subgiants (luminosity class IV) and 2 are giants (luminosity class III).

TABLE 5.— MAGNITUDE DIFFER-
ENCES FOR 182 NORTHERN
BINARY SYSTEMS IN THE GLIESE
CATALOG

Magnitude difference	Number of systems
0–1	74
1–3	48
3–5	28
>5	32

TABLE 6.— MAGNITUDE OF COMPO-
NENT STARS IN NORTHERN GLIESE
CATALOG BINARY STARS

Apparent visual magnitude	Number of stars
0–5	35
5–7	79
7–9	68
9–11	90
11–13	74
>13	29

TABLE 7.– SPECTRAL
TYPES OF NORTHERN
GLIESE CATALOG BINARY
STARS

Spectral type	Number of stars
A	10
F	40
G	43
K	59
M	99

I conclude that there is a sizable sample of binary candidates for planetary search. Even if we apply the restrictive requirements that one component be brighter than +9, that the system have a stable Jovian orbit, that the magnitude difference be less than 5, and that the separation be less than 6 arcsec, we have nearly 40 candidate systems.

The results of systematic long-term searches of these systems would be extremely valuable. The binary orbits that would be a by-product of a planetary search would allow a very accurate calibration of lower main-sequence masses and solar neighborhood distance scales. The problem should also turn up large numbers of low-mass, but not planetary, stellar components. These data will be essential for calibrating binary star mass functions.

SUMMARY

A small-aperture speckle interferometer can obtain accuracies of better than 10^{-4} arcsec/yr on binary star orbits. There are 188 accessible binary systems within 20 parsecs of the Sun. About half should have stable orbits for a Jovian planet, which would be easily detectable within the accuracies possible.

I conclude that stellar speckle interferometry is a viable option for detecting planets, and it is an option that could be easily implemented. In fact, interferometry is ideal for precisely those systems in which conventional astrometry has some difficulty, namely, the close binary systems.

REFERENCES

Breckinridge, J. B.; McAlister, H. A.; and Robinson, W. G.: Kitt Peak Speckle Camera. Applied Optics, vol. 18, April 1979, pp. 1034–1041.

Currie, D. G.: On a Detection Scheme for an Amplitude Interferometer. Woods Hole Summer Study on Synthetic Aperture Optics. NAS-NRC, Washington, D.C., 1968a.

Currie, D. G.: On the Atmospheric Properties Affecting an Amplitude Interferometer. Woods Hole Summer Study on Synthetic Aperture Optics, NAS-NRC, Washington, D.C., 1968b.

Currie, D. G.; Knapp, S. L.; and Liewer, K. M.: Four Stellar-Diameter Measurements by a New Technique — Amplitude Interferometry. Astrophys. J., vol. 187, Jan. 1974, pp. 131–134.

Currie, Douglas G.; McAlister, Harold A.; Schneeberger, Timothy, J.; and Worden, Simon P.: Using Small Aperture Interferometry to Detect Planets in Nearby Binary Star Systems. In: An Assessment of Ground-Based Techniques for Detecting Other Planetary Systems, Volume II: Position Papers. NASA CP-2124, 1980.

Dainty, J. C., ed.: Laser Speckle and Related Phenomena. Springer-Verlag, Berlin, 1975.

Gatewood, G.: On the Astrometric Detection of Neighboring Planetary Systems. Icarus, vol. 27, Jan. 1976, pp. 1–12.

Gezari, D. Y.; Labeyrie, A.; and Stachnik, R. V.: Speckle Interferometry: Diffraction-Limited Measurements of Nine Stars with the 200-Inch Telescope. Astrophys. J., vol. 173, April 1972, pp. L1–L5.

Gliese, Wilhelm: Catalog of Nearby Stars, vol. 22. Veroff des Astron. Rechen-Institut, Heidelberg, 1969.

Hanbury Brown, R.: Measurement of Stellar Diameters. Ann. Rev. Astron. Astrophys., vol. 6, 1968, pp. 13–38.

Harrington, R. S.: Planetary Orbits in Binary Stars. Astrophys. J., vol. 82, Sept. 1977, pp. 753–756.

Hubbard, G.; Hege, K.; Reed, M. A.; Strittmatter, P. A.; Woolf, N. J.; and Worden, S. P.: Speckle Interferometry, I — The Steward Observatory Speckle Camera. Astronom. J., vol. 84, Sept. 1979, pp. 1437–1442.

Labeyrie, A.: Attainment of Diffraction Limited Resolution in Large Telescopes by Fourier Analysing Speckle Patterns in Star Images. Astron. Astrophys., vol. 6, May 1970, pp. 85–87.

Labeyrie, A.: Stellar Interferometry Methods. Ann. Rev. Astron. Astrophys., vol. 16, 1978, pp. 77–102.

Lynds, C. R.; Worden, S. P.; and Harvey, J. W.: Digital Image Reconstruction Applied to Alpha Orionis. Astrophys. J., vol. 207, 1976, pp. 174–180.

McAlister, H. A.: Speckle Interferometry as a Method for Detecting Nearby Extrasolar Planets. Icarus, vol. 30, 1977, pp. 789–792.

McAlister, H. A.: In: Modern Astrometry, F. V. Prochazka and R. H. Tucker, eds., Proceedings of IAU Colloquium 48, Vienna, Austria, Sept. 1978.

Michelson, A. A.: On the Applications of Interference Methods to Astronomical Measurements. Astrophys. J., vol. 51, 1920, p. 257.

Worden, S. P.: Astronomical Image Reconstruction — Speckle Interferometry. In: Vistas in Astronomy, vol. 20, pt. 3, 1977, pp. 301–311 and 313–318.

Worden, S. P.; Stein, M. K.; Schmidt, G. D.; and Angel, J. R. P.: The Angular Diameter of Vesta from Speckle Interferometry. Icarus, vol. 32, Dec. 1977, pp. 450–457.

Panel Discussion

Panel Discussion

SUMMARIZED BY M. A. STULL

Panel Members:

Philip Morrison, Chairman   Massachusetts Institute of Technology
Helmut Abt   Kitt Peak National Observatory
A. G. W. Cameron   Center for Astrophysics
Harold P. Klein   NASA Ames Research Center
Harold Masursky   U.S. Geological Survey
John Oro   University of Houston
James Pollack*   NASA Ames Research Center
James Walker   Arecibo Observatory

*James Pollack served on the panel in place of Tobias Owen (from State Univ. of New York at Stony Brook) who was unable to attend the Conference.

The panel was introduced by its Chairman, Philip Morrison. He proposed that there should be an opportunity for members of the audience to question the panelists but that, first, the panelists should have the opportunity to make a few individual remarks.

Harold Klein opened by saying that he was a sessile invertebrate. From this viewpoint he had been impressed by the extent of our ignorance, and thought that we should expect the unexpected. He then gave two examples to illustrate this degree of ignorance. In his first example, he recalled that once, when he was young, he had asked Fritz Lipman why, in the proteins of all living things, the amino acids were all L (left-handed) as opposed to D. Lipman answered that he had thought about this for a long time and believed that L-acid forms of life had "defeated" D-acid forms in a tremendous struggle for survival between competing metabolisms during the first 2 billion years of life's Earthly history. What we see today is but the remnant of a great complexity of organisms no longer extant. Thus the apparent simplicity of this remnant may obscure the existence of many alternate evolutionary paths that might have been viable, given other conditions. In Klein's second example, he recalled that a few years ago a Space Science Board Panel had assessed the probability of life on Mars and designed the life-detection instrumentation for Viking. They decided that if the Viking biology instruments gave positive results, but the gas-chromatography/mass-spectroscopy (GCMS) results were negative, it would mean that there had been an instrumental failure. It was inconceivable, at the time, that the biology experiments could give positive results in the absence of sufficient numbers of living organisms for the GCMS to detect the presence of complex organic molecules. This combination of negative and positive results did in fact occur, but it was not due to instrument failure. Rather it was caused by the existence of strange and unexpected surface chemistry. We should draw a conclusion from these examples: as scientists, we are often successful at explaining what has been observed, but we are notoriously poor at anticipating the existence of phenomena before empirical evidence has been discovered.

James Walker was next to speak. He said that ignorance notwithstanding, we are making progress. He proposed to illustrate this by talking about atmospheric modeling for the purpose of explaining the densities of trace constituents. The great virtue of theoretical models, he claimed, is that they are essentially logical devices which show you that particular results must inexorably follow once specific, well-defined assumptions have been made. If the results of a given model are seen to be in conflict with observation, one then knows with certainty that at least one of the model's assumptions is wrong and must be relaxed. This approach, applied to the environment of primitive Earth, results in significant insights into physical processes affecting biogenesis and could eventually guide us to an understanding of how environment and environmental change affect the evolution of metabolic processes. Walker proposed to illustrate this with three examples.

In the first place, he said, consider the sort of calculation Kasting talked about. Here, if one specifies the abundances of nitrogen, carbon dioxide, and water, as well as the temperature structure, one can calculate precisely the oxygen concentration in the atmosphere. If one is concerned with the origin of life, this claim of precision is not contradicted by the fact that there may be several orders of magnitude uncertainty in the oxygen concentrations. From a biological viewpoint, it does not matter whether one has 10^5 or 10^{10} molecules of O_2 per cubic centimeter; there is effectively no oxygen in that atmosphere. If, as Towe maintains, there must have been O_2 in the atmosphere before the origin of life, Kasting's calculations tell you what assumptions must be relaxed. One must change either the water vapor mixing ratio in the stratosphere or the rate of release of hydrogen by volcanoes. Thus one has placed constraints on the physical environment.

For his second example, Walker took the possible role of lightning as an energy source for early chemical evolution. He noted that the study of the production of trace atmospheric constituents by lightning has received significant attention in the past few years. It is straightforward to take any assumed composition for a primitive atmosphere and calculate the rate of synthesis of organic compounds in that atmosphere by lightning, provided one also assumes a lightning rate. These assumptions are clearly defined: the lightning rate and the chemical composition. When one considers a wide range of possible compositions, one arrives at the conclusion that lightning, at its presently observed rate, was not adequate to produce significantly large abiotic synthesis of organic compounds. This implies either that there was a lot more lightning in the primitive atmosphere or that lightning was not responsible for the synthesis of organics. And we can conclude this with certainty.

As his third example, Walker said he would like to pose an unsolved problem — the problem of methane consumption by primitive organisms. In the primitive anaerobic ecosystems of early Earth, photosynthetic organisms produced organic material that was, in turn, consumed by fermenting organisms. But the latter must have produced methane. There are no anaerobic organisms that oxidize methane and, in fact, it appears that methane cannot be oxidized or consumed by organisms in an anaerobic environment. Thus the problem is to find a nonbiological process that cycles methane by converting it into consumable form. One can attack this problem by modeling. One can make various sets of assumptions, develop a model based on each, and rule out those that conflict with empirical data. Such modeling has the potential for showing that certain environmental characteristics (here the physical conditions that made possible the consumption of methane) must have been closely linked to the early evolution of life.

Walker was followed by James Pollack. Pollack argued that the important environmental questions concerning the origin and evolution of life are not restricted to the nature of physical processes in the atmosphere of early Earth, but also include details of the evolution of the Sun and even of the evolution of the Universe. In particular, consideration of the processes of star and planetary-system formation and of stellar evolution is essential to determining the size of the continuously habitable zone (ecozone), in which temperatures on a planet are neither too high nor too low for living organisms to thrive. It is not clear whether the situation on Earth is the norm or a fortuitous and unlikely happenstance. One problem is that we cannot yet

define the temperature range outside of which life cannot exist. Nevertheless, Venus and Mars today seem clearly to lie outside the ecozone. However,
early in the history of Mars, conditions were sufficiently clement for there to
be liquid water on the surface, and this period may have lasted for a long
time. Moreover, although evidence one way or the other is presently lacking,
Venus may have experienced a temperate climate during the early history of
the Solar System, when the Sun's luminosity is believed to have been some
tens of percent lower than at present.

Thus the question that must really be addressed is — what is necessary
for a planet to sustain clement conditions over at least part of its surface for
a period on the order of 5 billion years? It is possible that Mars may have
experienced a self-destructive situation in which liquid water dissolved carbon dioxide out of the atmosphere and deposited it in rocks, disastrously
reducing the greenhouse effect. The same process may have been balanced
fortuitously on Earth by the rising solar luminosity since, due to its initially
higher temperature, Earth may not have been as vulnerable as Mars to a
lessened greenhouse effect. A second possibility is that the evolution of the
Sun may have been different than what we believe because of changes in the
physical constants due to cosmological evolution. Yet another factor is the
mass of the planet, which affects the length of time over which the lithosphere remains thin enough so that atmospheric gases that are lost to rocks
(e.g., CO_2 to carbonates) can be eventually recycled due to subduction and
metamorphism. Earth is massive enough so that this condition is met, while
the opposite is true for Mars. He closed by emphasizing that input from a
wide variety of disciplines is crucial if we are ever to answer the question of
the size of the ecozone, as well as many other questions concerning the
origin and evolution of life.

Al Cameron followed by saying he would like to try to make some
people angry by describing, in two lessons, how one becomes a SETI enthusiast. First of all, he summarized the previous two speakers by saying that
on the one hand it is very easy to calculate everything about a complicated
system provided you know all the rate constants and what assumptions to
make, while on the other hand the history of the Solar System is really terribly complicated. This leads to the two lessons: First, since you must estimate the likelihood that extraterrestrial life exists, you write down a string
of probabilities — which, since you are an enthusiast, you have set equal to
one — and then you multiply them all together. Second, you assume that all
those guys out there that are smarter than you are will maintain a beacon
specifically designed to be easy for you to detect. Cameron suggested that
this notion about beacons may be a form of cargo cultism. He concluded
that a strategy based on eavesdropping may be the only viable approach to
SETI, although it may be more difficult since it would allow less specific
assumptions about transmitter characteristics.

Morrison asked if anybody had been provoked.

Walker replied that he had been provoked. He said that we do know all the rate coefficients.

Morrison then asked whether Walker's models didn't also depend on the nature of the radiation field, that is, on whether one has a blackbody spectrum at ultraviolet or x-ray frequencies. Walker replied that yes, one must say what the radiation spectrum is, but, given that, we can do the calculations. Morrison asked about particles. Walker said they were not likely to be important. Morrison replied that that was the usual attitude of theorists. Where this kind of argument goes wrong is that the solution to a particular problem often turns out *not* to be a matter of choosing between inputs you know about; what hurts you is the input you didn't know was there. It may do little good to make alternate assumptions to see what conclusions they lead to, when one is unaware of factors that render invalid the rationale for choosing those assumptions in the first place. Walker saw Morrison's point, but maintained that his approach was nonetheless useful. He suggested that one can gradually establish constraints on a situation. By eliminating various possibilities, one can tell if one is heading toward a satisfactory solution.

J. William Schopf said he would like to make one observation, and this speaks to the strength of space exploration and the sort of things NASA has done. There is only one court of last resort for scientists — reality. With respect to the origin and evolution of life, reality is ascertained by looking at the rock record. NASA has provided data on the rock record, and the rock record has supplied us with constraints. For example, the rock record has limestones 3.8 billion years old. This implies that there must have been carbon dioxide in the atmosphere. It used to be claimed that there wasn't any carbon dioxide that long ago, but that claim is no longer tenable. Moreover, there must have been liquid water 3.8 billion years ago because rocks of that age have detrital pebbles in them.

Harold Masursky spoke next. He noted that there are some things that can be tested immediately. For example, Mars has very red soils, worse even than those of eastern Brazil which are highly anorganic. But there is also a great canyon on Mars with 7 km of rocks, and we can see that these are of various kinds: light, dark, etc. The opportunity is there to look at the Grand Canyon of Mars and examine the record of past Martian environments. In particular, we can hope to find rocks dating from the eras in which the channels were formed. Much could be learned should this prove to be the case. We already know from crater counts that volcanoes have been active at epochs in early, middle, and recent Martian history, and that the (H_2O?) channels have also been produced throughout these periods. Studies of samples from the canyon could allow channel episodes to be correlated with the past climate of Mars, since channel episodes could be dated radiologically from volcanic deposits above and below them. We might find times when the

Martian soil was not anorganic. Moreover, Martian chronology could be compared to Earth's. Among other things this could tell us how long liquid water
must be present before substantial organic developments take place. Other
solar-system bodies also offer opportunity for study. Venus has a great
plateau 3 to 5 km high. This, too, must give us a rock record and could show
whether the Venusian climate was more clement early in its history. Io has
real gases and hot springs; these could be favorable environments for life.
Lightning in Jupiter's atmosphere could allow us to study Miller-Urey processes there. If we follow Schopf's advice and seek to ascertain reality
through data, the Solar System offers us an enormous abundance of it.

The next panel member to speak was John Oro. He said that there are
three major obstacles to be surmounted in our quest toward the origin of
life. In the first place, we know *nothing* about the first 750 million years of
Earth's history, and it is essential that we obtain this knowledge, although it
is not clear how we can do it. In the second place, we do not know anything
about other planetary systems, not even if any exist. NASA has an obligation
to make special efforts to detect other planetary systems. Third, regardless
of terrestrial history, the fundamental principles of organic life are based on
the chemistry of carbon, as well as that of hydrogen, oxygen, nitrogen, sulfur, and phosphorus (the six organogenic elements). We have not yet been
able to produce a self-replicating system comparable (but not necessarily
identical) to the living organisms of today. This question must be addressed
primarily by organic chemists and biochemists. Oro then said he thought
there ought to be some questions from the audience.

Kenneth Towe noted that certain meteorites (carbonaceous chondrites)
contain amino acids and other organic compounds. He asked to what extent
seeding by organic materials in the solar nebula or from space instead of
synthesis of organics in Earth's atmosphere could have been responsible for
the origin of life on Earth.

Oro replied that there are at least three phases in the formation of
organic molecules: a cosmic phase, involving formation in stars and the
interstellar medium; a solar nebula phase; and a phase of formation on primitive Earth. Only the last phase really matters, provided the organogenic elements were available on early Earth and conditions were favorable for them
to react. If those were the circumstances, then whether additional organic
compounds arrived on Earth due to meteorite bombardment is irrelevant; to
this extent the answer to Towe's question is negative. Four essential ingredients are required for life: (1) a membrane of some sort to surround the
organic material, (2) an informational or replicative material, (3) a catalytic
molecule, and (4) a molecule that can translate from informational molecules to catalytic molecules. Once you have these four things you have life;
meteorites per se do not give you these.

Morrison asked if Sherwood Chang had anything to add. Chang said that one can take Walker's approach and make assumptions about how many meteorites fell into Earth during the relevant period of Earth's history. Then one can estimate how many of these meteorites would have been carbonaceous, how many would have survived passage to the surface, how efficiently could the amino acids be extracted from a meteorite, and how large were the bodies of water into which the meteorites fell. If one does all these things, he gets an answer. Chang said he would not give exact numbers, but if the estimates required are very conservative, the result is a 10^{-6} to 10^{-7} molar solution. This is very dilute, although it is comparable to what many believe would result from a Miller-Urey type of synthesis. Therefore, the important question may not be how you get organic material, but how you concentrate it once it has been produced and transformed from simple molecules to complex ones. He thought that Oro was correct. It does not matter if organic material was brought in by meteorites; they should be regarded as just one of many sources.

Towe disagreed. He said that it is important to distinguish among sources of organic materials. This is because the atmosphere is supposed to have contained no oxygen since organics could not have been synthesized had oxygen been present. But if organic matter originated in meteorites, then there is no need to postulate an oxygen-free atmosphere. If the atmosphere contained oxygen, an ozone shield could have permitted nucleic acids to form since ozone absorbs ultraviolet radiation at the same frequencies as nucleic acids.

Oro said that Whipple had recently calculated that 10^{25} to 10^{26} gm of carbon-containing material from comets or comet-like bodies could have accumulated on Earth. Apparently this assumed a dense atmosphere to aid capture. Oro's own work indicated that 10^{22} to 10^{23} gm was the appropriate number. He noted that one of the main Apollo results was that 3.9 to 4.1 billion years ago there was a big peak in meteorite impacts; however, it has been suggested that this was not really a peak at all, but just the tail end of many more collisions of loose small bodies left over from the formation of the Solar System, because remains of previous impacts would have been obliterated by the final showers of 3.9 to 4.1 billion years ago. This implies that 10^{22} to 10^{23} gm of carbon-containing material could well have been acquired by Earth. The present biosphere contains only 10^{18} gm, so even an amount two orders of magnitude smaller would have been sufficient to account for life.

Another member of the audience said he thought we faced a different dilemma. Irrespective of the origin of organic molecules, he thought there was an unresolved time constraint. Earth was formed 4.5 billion years ago, but the earliest fossil life dates from 3.0 to 3.5 billion years ago. This implies that it took only about 1 billion years to evolve primitive bacteria from simple organic molecules. He felt that this tremendous quantum leap, in so short a time, was not understood.

Oro replied that 1 billion years was a more than ample amount of time. He noted that Gustav Arrhenius had said earlier in the day that Earth could have cooled in as little as 10 million years. Beyond this, Oro stated that chemical reactions are easy; they take only minutes to produce something as complex as a self-replicating molecule. Granted, this is still a long way from having an *E. coli*, but given the speed of chemical processes, there appears to be no difficulty in understanding the observed time scale.

Schopf noted that George Gaylord Simpson had said in his 1949 book that the oldest trilobites were 500 million years old (this turns out not to be correct), and that these must have been preceded by protozoans, back to 1 billion years. Simpson thought that this scenario made good sense because it seemed it should be a larger step from the origin of life to an amoeba-like organism than from an amoeba to anything subsequent. Schopf said that such statements seemed to be nonstatements and nonquestions. Rather evolution proceeds at some rate and we do not yet understand what determines that rate. Biologists do not appreciate the magnitude of geologic time. Seven hundred million years, the time from which the environment became clement to the first fossil record, is a long time — longer than the time from before the first trilobite to man! It is long enough for lots of things to happen. Since we do not understand what determines the rate of evolution, there is no basis at all for saying there is a problem just because bacteria are complex. That complexity could have been produced rapidly; moreover, there was a tremendous amount of time available.

David Usher asked if he could add one point. He said that his favorite example of the time that has been available for evolution is the Grand Canyon. If you just stand on the edge and look down and reflect that *it* formed in 3 to 5 million years, you acquire some perspective. The Grand Canyon is no small accomplishment, but evolution has had a thousand times longer to achieve its result.

A member (unknown) of the audience responded that he was not challenging this dogma. Rather he merely meant that, given the complexity of the simplest microbial organism, it could be a mistake merely to assume that lots of time was available; evolution of that degree of complexity is not necessarily something that happens easily or quickly. Moreover, he wondered why we did not see a variety of trials and errors in the record. He felt that there was indeed a potential problem with trying to fit everything, from the primeval solar nebula to the appearance of simple organisms, into the time available. He said that it remains to be seen whether we can explain this.

Oro replied that they had not meant to imply that the evolution of early life need not be explained in detail, but only that they thought there was a good detailed explanation, and that the processes involved were not terribly difficult to set in motion. He said that the audience member was simply expressing our lack of knowledge, but that Usher was right to emphasize that quite a long period of time was available. Oro said that what he would really like would be for Al Cameron or Gustav Arrhenius to say when Earth first became cool enough for processes leading to life to take place.

Karl Turekian asked whether Helmut Abt had something to say about stellar phenomena that might be relevant to the discussion.

Abt said that a key datum is the fraction of stars that have planets. It is clear that a very large fraction (between 60 and 100%) have companions of some sort. If one considers the stars of a given spectral class (e.g., that of the Sun, class G), about 18% have companions approximately the same mass. Another 14% have companions approximately half their mass, 10% have companions 1/4 their mass, 7% have companions 1/8 their mass, and 5% have companions 1/16 their mass. But there is where the observations stop, because we cannot yet detect companions less massive than about 1/16 the mass of the primary. Therefore, all we can say is that approximately 54% of stars have companions with masses greater than 1/16 the mass of the primary. Now it so happens that 1/16 the mass of the Sun ($M_\odot$) is the mass below which nuclear burning stops, although bodies more massive than 0.01 $M_\odot$ (10 times the mass of Jupiter) can radiate at stellar luminosities for up to a billion years as a result of energy released by gravitational contraction. If we extrapolate the observed mass distribution, we conclude that 78% of stars have companions more massive than 0.01 $M_\odot$. However, we do not really know whether this distribution continues. If we really want to tread on thin ice, we can push the extrapolation to planetary masses and conclude that at least some stars must have companions of planetary mass. Abt felt uneasy about extrapolating so far into a region for which we have no data.

Indeed, Abt felt that there were reasons to believe that the observed mass distribution did not extend to masses much smaller than the presently observed limit of 1/16 $M_\odot$. And even if it did, he continued, the existence of companions of planetary mass would not imply the existence of solar systems. Rather there are fundamental differences between known multiple star systems and our Solar System, and these raise the question of whether there are two distinct star system formation mechanisms, one leading to stars with close companions, the masses of which are described by the observed distribution, while the other results in solar systems in some degree similar to our own. He noted that Cameron had long been proposing that the Solar System was created from a disk around a single star with relatively low angular momentum, whereas protostars with significantly higher angular momenta undergo fission to give rise to double- or multiple-star systems. The latter mechanism could account for the formation of all close pairs of stars. Abt said that he had in mind two fundamental differences between our Solar

System and multiple-star systems. In the first place, all solar-system objects lie in a plane, but multiple-star systems are spherical (i.e., in systems of three or more stars there is no correlation between the planes of the individual orbits). Second, the orbital periods of solar-system objects obey a Bode's law relationship, such that the periods get longer by a factor of approximately 1.5 as one goes to each more distant planet. In multiple-star systems, though, the factors are much larger. Indeed there is no known triple-star system in which the longer period is significantly less than 10 times greater than the shorter period, and sometimes the difference is as much as a factor of 100.

Robert Harrington said he agreed with most of Abt's remarks, but not with the idea that there are only two classes of systems of primary star plus companions. He said that the double systems Abt was talking about were the very close, spectroscopic binaries. These may well have been formed by a bifurcation mechanism, but there are also wide binaries with separations on the order of tens to hundreds of astronomical units. In this third class of star system, the possibility exists that each of the individual stars might be accompanied by a planetary system. Given such large separations, the systems would be stable once formed. But there remained a question concerning whether planets could form in wide binaries in the first place. Harrington said that Heppenheimer had been arguing that they could not. Did that argument have any validity?

Cameron responded that he disagreed with most of what everybody had said all day about how planets are made. However, he emphasized that it doesn't really matter; we just do not understand. We cannot estimate formation times or cooling times, even to within orders of magnitude. His own belief was that planet formation occurs via massive gravitational instabilities in primitive protostellar nebulae; these instabilities he called "giant gaseous protoplanets." If they are formed in the inner Solar System, they cannot grow very large before the growing central star tidally strips away their envelopes. In the outer Solar System, though, they can contract, lose only a small amount of mass through a planetary wind mechanism, form cores, and become giant planets. The open question (and obvious evidence points to an answer) is whether the giant gaseous protoplanets in the inner Solar System manage to precipitate condensed matter into a central core before they get tidally stripped. If they can, then the cores can become 80% of the mass of terrestrial planets. By this mechanism it may be possible to make the bulk of terrestrial planets in a very short time ($\lesssim 10^5$ yr), with a final sweep-up of the debris remaining in space adding a final layer, perhaps over a period of several hundred million years. The early Solar System was probably very complicated, and any other planetary system is likely to have been equally complicated and quite different from our own. Some spectacular discoveries may await us; we probably have no real appreciation yet for what is out there — for the complexity of nature.

Cameron then said he thought attention should be focused on one more question: To what degree does the evolution of life respond to environmental change? He thought it clear that evolution, at least to some degree, is driven by environmental change. But what happens if environmental change is speeded up? Does the rate of evolution also speed up, or does it slow down? There may be some rate of environmental change that drives evolution fastest. Cameron thought it very important to try to determine this optimum rate of environmental change and suggested that Earth might turn out to be an example of it.

Abt said he wished to say something more about double stars, in response to Harrington. Harrington was indeed correct to point out the distinction between close and wide binaries, but he did not go far enough. There can be all kinds of separations, and in *none* of these systems are companions of planetary mass precluded. There could be distant planets orbiting both stars of a close system, and there could be nearby planets orbiting one or more of the individual stars of a wide system. However, Abt again emphasized that having one or two companions of planetary mass is not the same as having a solar system, and it is possible that one of these two possibilities will turn out to be the rule, and the other the exception. He then noted that a popular astrophysical topic of recent years has been accretion disks, particularly in close binaries. Theoretical studies have indicated that disks tend to be thin; moreover, infrared observations have discovered disks around stars. Is it possible, he asked, that solar systems originate due to the occurrence of disk phenomena?

Cameron responded that some of the disks probably behave much as one would expect, and that planetary objects probably form from many of these disks. He pointed out that a large number of the observed disks are associated with massive stars that probably do not live long enough for life to evolve.

Cameron then returned to Harrington's question concerning planetary formation in multiple-star systems. He said there was a distinction to be made between an object like Jupiter and the members of observed multiple-star systems, which are all of stellar mass ($\gtrsim 0.05\ M_\odot$). The latter tend to be essentially of solar composition, but Jupiter, in bulk, is not. To the best of our knowledge, Jupiter is enriched in heavier elements, most of which are probably condensed into a core. Moreover, there is a difference between the character of Jupiter's orbit and that of a typical binary companion. Although the latter might have a semimajor axis of 5 AU (the same as Jupiter's), its orbit would likely be highly eccentric. The orbits of Jupiter and the other planets are nearly circular. In a triple-star system, the orbits do not even share the same plane; in the Solar System they do. Jupiter shows evidence of a formation more complicated than just a gravitational condensation of solar material. It is very possible, therefore, that planets are made by a different mechanism than binary companions in eccentric orbits. We do not understand these mechanisms. It follows that it is not possible to say whether planets do or do not form in multiple-star systems. But it is clearly dangerous to put much confidence in the extrapolation of the observed distribution of the masses of companions that Abt discussed.

Oro asked whether there was any estimate of the probable abundance of wandering planets, that is, small, nonluminous objects without a companion star. These might generate enough energy for life. Are they plausible?

Morrison said it was a very speculative proposition since we do not understand how stars form. Of course, there could be all sorts of ejections, and therefore there must be at least a limited number of dark wanderers. If they exist, this would support the notion that life occurs elsewhere, but not very strongly.

Morrison said that time had run out, and the discussion must be ended. It seemed quite clear to him that we are dealing in matters about which we know only a little, and that we must make great efforts to narrow the enormous range of possibilities, both by theory and by experiment. We see a landscape through mist and fog, but the features are real. We see a structure, but we cannot yet tell whether it is two separate towers or only one, joined at the base. This is the job for the next generation.

III — Evolution of Complex Life in the Galaxy

The Darwin-Wallace theory of biological evolution through natural selection makes the adaptability of life forms to their environment the factor most important to their survival and success. It follows that environmental change ought to result in faunal and floral change, and that a particular environment ought to favor certain faunal and floral characteristics over others. On a superficial level, the truth of these propositions is obvious to any careful observer of Earthly life; indeed Darwin was led to his theory when he drew exactly these conclusions from his empirical studies. But on a more fundamental level, the relationship between life's environment and biological evolution, even after a century of success in post-Darwinian paleontology and genetics, remains obscure. We cannot predict, even qualitatively, the rate and direction of evolution from a knowledge of environmental parameters and their history.

Our lack of understanding can best be illustrated by posing some questions that we cannot answer: what is the effect on the rate and direction of evolution of gross planetary characteristics such as orientation of spin-axis, rotation rate, atmospheric composition, incident stellar energy flux, and size of contiguous land masses? How do changes in these characteristics influence evolution? Can cataclysmic cosmic phenomena (e.g., nearby supernovae, asteroid impacts) cause mass extinction events such as those of the Permo-Triassic and Cretaceous-Tertiary boundaries? Do mass extinctions change the rate of gross evolutionary trends such as the development of an increasingly sophisticated central nervous system? Only when such cause-and-effect relationships are understood will it be possible to develop a quantitative model that can predict the gross evolutionary pattern once the environmental history has been specified.

Research aimed ultimately at developing such a model is important today for more reasons than mere scientific curiosity. At stake is nothing less than our ability to understand the nature and distribution of life in the Universe about us. It will soon be possible to discover other planetary systems and to measure the gross characteristics of their components. But we must understand what an environment portends for evolution if we are to be able to determine from such measurements whether a given planet is likely to support life and, if so, what form that life is likely to take. And it is of special interest to identify systems with planets on which intelligent life may have evolved since we have the ability to detect manifestations of technological activity, such as radio signals, if only we knew in which direction to look.

It would seem that today we have for the first time capabilities that should make discernible the causal relations between life's environment and evolution. Our abilities to study the environments of Earth's sister planets at first hand and to date accurately the rock record on these bodies, as well as on Earth, ought to allow us to determine whether events in the history of life on Earth were related to environmental changes, such as an increase in solar luminosity or a decrease in Earth's rotation rate. Improved knowledge of biochemistry and genetics should be able to tell us what possibilities are open to evolution and what paths are barred, both very generally and under specific environmental constraints (e.g., temperature, moisture availability, surface gravity). Modern techniques for studying ecological systems in the field can highlight the ways in which the environment and changes therein influence survival prospects. And modern data-processing capability makes possible the modeling of complex interactions, such as energy flows, in ecological systems.

These approaches are not likely to be exclusive. The papers in this session present the thoughts of some of our leading scientists on various aspects of the relation of environment to evolution. It is hoped that their efforts will not only bear fruit, but will also stimulate the research of others.

Dr. Mark A. Stull
Extraterrestrial Research Division
Ames Research Center
(now practicing law in Maryland)

Transfer RNA and the Origin of Protein Synthesis

ALEXANDER RICH

Knowledge is accumulating rapidly about the structural details of protein synthesis in contemporary systems. This remarkable body of knowledge may eventually shed light on how this process evolved in prebiotic times.

Exobiological research is oriented toward finding those conditions which led to the establishment of life on this planet. In this work one starts with an environment comparable to that believed to have existed in prebiotic times, and using natural processes we attempt to create those chemicals which have given rise to living systems. Research in this field may be used to interpret the evolutionary state of planets to be explored in the Solar System, especially in assessing the molecular complexity found on planets and its relevance to a particular stage of biological evolution.

A number of workers in exobiological research have made significant advances. Using a reducing atmosphere, Miller, Urey, and many other investigators have established conditions that give rise to the production of amino acids, nucleotides, sugars, and other biological molecules. We now have a number of plausible prebiological chemical models that not only aid our understanding of the development of these monomeric constituents of living systems, but even lend some insight into the production of polynucleotides, polysaccharides, and polypeptides. Work on polymerization has proceeded in several directions. Most interesting in the polynucleotide field is the development of complementary polynucleotides that use the hydrogen bonding found in the double-helical nucleic acids of contemporary biological systems (Lohrmann et al., 1980). This system contains its own molecular constraints, which allow a single polynucleotide strand to aid in selecting individual

211

monomeric nucleotides to build up a double helix. Strand separation and repetition of this process lead to a plausible prebiological model for nucleic acid replication, and this reproduces an important stage in the early evolution of life on this planet.

There are many ways to polymerize amino acids, including the use of heat alone. Although such prebiotic polymers may have had some role in the early establishment of living systems, they are not biological products because they do not yield multiple copies of molecules with the same amino acid sequence. The central event of contemporary biological systems is the polynucleotide-directed synthesis of polypeptides. In this system, information found in nucleotide sequences directs the sequential assembly of amino acids. The next step in the development of a prebiological model of the origin of life is the elucidation of this system. There is today a great chasm in our understanding. Although we have a general comprehension of the role of messenger RNA codons in bringing about the sequential assembly of amino acids in ribosomes, we do not know the detailed nature of the interactions that made this possible. The contemporary system for protein synthesis is outlined in figure 1. In attempting to create a model of amino acid polymerization in protein synthesis, it is not at all clear in which direction efforts should be expanded. One of the central research goals is the definition of the physical nature of the interactions which occur between messenger RNA (mRNA) and transfer RNA (tRNA). We need to understand the important constraints found in contemporary biological systems to gain insight into the nature of the models needed to develop prebiological research. We need to know how information coded in the randomly polymerized nucleic acids can be translated into polypeptides in the absence of ribosomes, or at least their protein constituents. If we can create a prebiotic, protein-free system of polynucleotide-directed polypeptide synthesis, we will in principle have reached a stage sensitive to the selection pressures that are the basis of biological evolution.

Transfer RNA is the central actor in protein synthesis today. It is also an ancient component of biological systems. It is likely that it arose some 4 billion years ago in the early prebiotic period, at which time the increasing complexity of organic molecules led to the creation of polynucleotide chains that had the capacity to carry out molecular self-replication. Although the nucleic acids are effectively selected for carrying genetic information and carrying out self-replication, they cannot express genetic information. The major mode for expressing genetic information is the synthesis of proteins, which, through their varied amino acid side chains, are able to provide the large variety of complex chemical environments needed to create the catalytic activities and structural assemblies that are the molecular basis of living systems. The tRNA molecule acts at the crossroads between the information-containing polynucleotide chains and the proteins that express genetic information. At one end of the tRNA molecule, the three anticodon bases

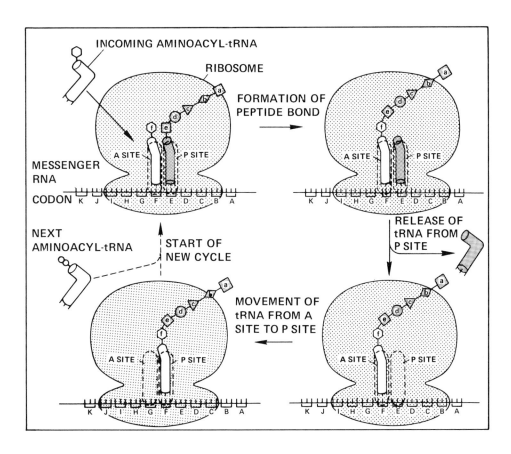

Figure 1. *The function of transfer RNA (tRNA) in the synthesis of a protein molecule is to make a chain of amino acids that reflects the nucleotide sequence of the template represented by messenger RNA (mRNA). The aminoacyl-tRNA synthetase first joins a specific RNA molecule to its corresponding amino acid with a covalent bond. In the diagram, amino acids are represented by squares, hexagons, etc. The tRNA with the amino acid attached to it binds at the A site to the ribosome. This interaction requires specific hydrogen bonding between the three codon bases on the mRNA strand that specify the amino acid and the three anticodon bases of the tRNA. From the tRNA molecule in the adjacent P site there is transferred the growing polypeptide chain to the tRNA in the A site. The "empty" tRNA leaves the P site, and the ribosome moves along the mRNA a distance of one codon so that the tRNA carrying the polypeptide chain is shifted from the A site to the P site. Then the cycle can begin anew. It takes about 5 msec for one cycle in procaryotes.*

interact with mRNA; at the other end, the molecule is attached to the growing polypeptide chain during protein synthesis (fig. 1). This molecule is
probably as ancient a component of biological systems as the system of
expressing genetic information through the polymerization of amino acids.
Here we consider the two types of tRNA molecules found in protein-
synthetic systems and speculate on how the system may have started.

The tRNA molecules are active in protein synthesis where they partici-
pate in two major activities: aminoacylation, which takes place on the surface of the aminoacyl synthetases, and protein synthesis, which takes place
inside ribosomes. In a sense, there is a paradox in trying to understand this
central biochemical function of tRNA. On the one hand, the tRNA molecules must be sufficiently unique to be discriminated by different aminoacyl
synthetases. On the other hand, they must be sufficiently identical that they
can all go through the same ribosomal apparatus during protein synthesis.
We would like to understand the manner in which this twofold chore is
carried out in an effective and relatively error-free mode. Further, a distinction has to be made between the initiator tRNA, which lays down the first
amino acid, and chain-elongation tRNAs, which add subsequent residues.
Initiator tRNAs go to the ribosomal P site (fig. 1), while chain-elongation
tRNAs go to the ribosomal A site.

Holley et al. (1965) reported the first nucleotide sequence of a tRNA
molecule. They noted that there were segments of the polynucleotide chain
that appeared complementary, which suggested that the chain folded back
on itself. One of these foldings has given rise to the cloverleaf diagram in
which sequences are represented as a series of stems and loops, the stems
largely composed of complementary bases with Watson-Crick hydrogen
bonding much as in the DNA double helix. As additional sequences were
reported, it became apparent that this cloverleaf folding was expressing
something of a fundamental nature concerning the molecule. At present,
over 100 different tRNA molecules have been sequenced; figure 2 shows the
cloverleaf sequence for yeast phenylalanine tRNA (yeast tRNA^{Phe}) (Sprinzl
et al., 1980). There are a number of invariant and semi-invariant nucleotides
in all tRNA sequences. The reasons behind this large number of conserved
residues remained unknown until the elucidation of the three-dimensional
structure of yeast tRNA^{Phe} (Rich and Rajbhandary, 1976).

The cloverleaf arrangement plus the large number of invariant positions
imply considerable constraints on the three-dimensional form of the molecule. Prior to the elucidation of the three-dimensional structure of yeast
tRNA^{Phe}, a number of attempts were made to anticipate the conformation
by building models of tRNA folding. These models usually maintained the
stem regions as double helices, while additional tertiary base-base interactions were used to stabilize the three-dimensional conformation. None of the

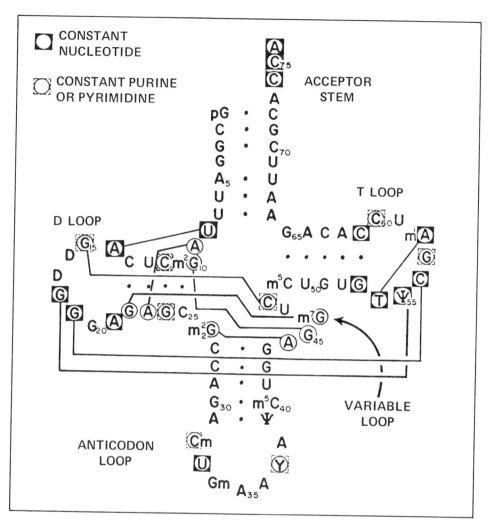

Figure 2. *Cloverleaf diagram of the nucleotide sequence of yeast tRNA^{Phe}. Tertiary base-base hydrogen-bonding interactions are shown by solid lines, which indicate one, two, or three hydrogen bonds. The invariant and semi-invariant positions are indicated by solid and dashed lines around the bases; Y_{37} is a hypermodified purine.*

proposed models was correct. The principal reason was that they all empha-sized Watson-Crick hydrogen bonding in the tertiary interactions between residues that were not in stems. There are 9 base-base tertiary hydrogen-bonding interactions in the three-dimensional structure of yeast tRNA^{Phe}, as indicated by the solid lines in figure 2. Only one of these is a Watson-Crick interaction; the other 8 involve alternative types of hydrogen bonding. Watson-Crick hydrogen bonds are of great utility in building a regular helical structure, but the hydrogen-bonding potential of bases is much more varied, and it is utilized in many different ways in building the highly irregular and nonrepeating interactions found in the globular coiling of the tRNA polynucleotide chain.

CRYSTALS AND STRUCTURE

Transfer RNA molecules were first crystallized in 1968 independently in several different laboratories. A large number of species were found to form crystals, and this gave rise to optimism that the three-dimensional structure would be known in a relatively short time. However, it was not generally appreciated that all the crystal forms had one defect in common: they were all disordered. In protein crystallography, it is recognized that one must have an x-ray diffraction pattern with a resolution of between 2 and 3 Å to trace the polypeptide chain. To fix the position of the bases, sugars, and phosphate groups, a comparable resolution is required.

The origin of the disorder in tRNA crystals is not readily apparent. A large part of it must result from the polyelectrolytic nature of the molecule. These molecules have from 75 to 90 negative charges, and the exact ordering of the tRNA molecules in the crystal lattice is very sensitive to the nature and number of cations found in the crystallizing mixture. It is likely that the polyelectrolytic nature of the molecule gives rise to frequent mistakes in the building of the lattice. After much effort, we reported in 1971 that if one added the oligo cation spermine to yeast tRNA[Phe], an orthorhombic crystal could be formed which yielded an x-ray diffraction pattern with a resolution of nearly 2 Å (Kim et al., 1971).

In early 1973, the folding of the polynucleotide chain was traced from an electron density map at 4-Å resolution (Kim et al., 1973). The tracing was possible because the electron-dense phosphate groups could be seen even at this low resolution. The folding of the chain (fig. 3) was highly unusual and had not been anticipated by any of the model builders. The stem regions, which had been assumed to be in the form of RNA double helices from the sequence data, were indeed in that form; but they had an unusual organization. The molecule was L-shaped, with the acceptor stem and T stem forming one limb of the L while the D stem and anticodon stem formed the other limb. The 3' terminal adenosine to which the amino acid was attached was at one end of the L while the anticodon was at the other end, some 76 Å away. The molecule is fairly flat, about 20 to 25 Å thick. Although the coiling of the chain could be seen at 4-Å resolution, the details of the tertiary interactions awaited the results of a higher resolution analysis.

In 1974 the analysis was extended to 3 Å and revealed a number of tertiary interactions (figs. 2 and 3). In particular, it showed the detailed

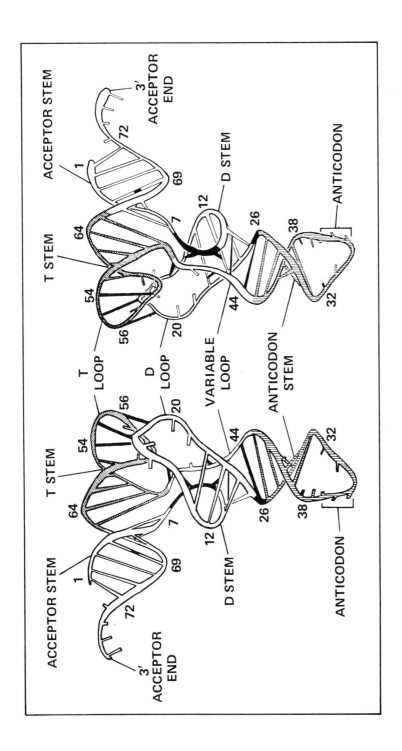

Figure 3. *A schematic diagram showing two side views of yeast tRNA^Phe^. The ribosephosphate backbone is depicted as a coiled tube, and the numbers refer to nucleotide residues in the sequence. Shading is different in different parts of the molecules, with residues 8 and 9 in black. Hydrogen-bonding interactions between bases are shown as crossrungs. Tertiary interactions between bases are shown as solid black rungs, which indicate either one, two, or three hydrogen bonds between them (as described in text). Those bases not involved in hydrogen bonding to other bases are shown as shortened rods attached to the coiled backbone.*

manner in which the T and D loops interact to stabilize the corner of the molecule (Kim et al., 1974a). Robertus et al. (1974) also presented 3-Å results for the same spermine-stabilized yeast tRNA^Phe in the monoclinic crystal lattice. This showed a virtually identical folding of the molecule. It can be seen (fig. 3) that many of the base-base tertiary hydrogen-bonding interactions involve the invariant or semi-invariant nucleotides. This suggests that the structure of yeast tRNA^Phe may be a generalized model for understanding the structure of all tRNAs (Kim et al., 1974b).

The analysis has since been extended to 2.5-Å resolution, and crystallographic refinement calculations have been carried to the point where many details concerning the tertiary interactions in the molecule can be observed (Quigley and Rich, 1976).

The major structural unit in yeast tRNA^Phe is the RNA double helix that forms the cloverleaf stems. These are very close to an RNA helix with approximately 11 residues per turn. This conformation is maintained due to the presence of the ribose 2' OH group in the backbone. DNA lacks this group, and its double helix is substantially different.

The presence of a 2' hydroxyl group in the ribose phosphate backbone contributes to the stabilization of the standard ribose 3' *endo* conformation generally found in RNA molecules. Its absence in the DNA molecule leads to a deoxyribose 2' *endo* conformation. This is responsible for the difference in the overall shape and form of the RNA double helix compared to the DNA double helix. The RNA double helix differs from the familiar DNA double helix in that the base pairs in the double helix are not perpendicular to the helix axis, but rather are tilted about 15°. Furthermore, they are not found on the axis of the helix, but rather are displaced away from the center. A hole approximately 6 Å in diameter extends through the center of the RNA double helix, so that the molecule looks more like a band wrapped around an imaginary cylinder than a double-helical twisted molecule that fills its axis. Consequently, the deep groove of the RNA double helix is extremely deep, whereas the narrow groove is extremely shallow.

The helix joining the acceptor stem and T stem seems almost uninterrupted. The sequence of yeast tRNA^Phe shows a G-U pair in the acceptor stem. This introduces only a slight perturbation in the helix. Examination of the electron density map at 2.5 Å shows that these bases are held together by two hydrogen bonds in a typical "wobble" pairing (Crick, 1966).

HYDROGEN BONDING AND BASE STACKING

One of the outstanding features of the yeast tRNA^Phe molecule is the fact that most of the bases are involved in base-stacking as well as hydrogen-bonding interactions. That this involves the bases in the double-helical stem regions was not a great surprise, but the extent to which these same interactions also involve the bases in the nonhelical loop regions was of considerable interest. Virtually all the bases in the molecule are organized into two large base-stacking domains along each of the two limbs of the L-shaped molecule. The horizontal stacking domain in figure 3 includes bases of the acceptor and T loop and some from the D loop. Most of the remaining bases are involved in the vertical stacking domain, which extends down to and includes bases in the anticodon loop. Only 5 of the 76 bases are not involved in stacking interactions. The two dihydrouracil residues, D16 and 17 and G20, are not stacked, nor is U47 from the variable loop. These all come from regions with variable numbers of nucleotides, as can be seen in a survey of different tRNA sequences (Sprinzl et al., 1980). In addition, the 3' terminal adenosine A76 is unstacked. It is interesting that most of the unstacked bases are found in the segments of the polynucleotide chain that do not have constant numbers of nucleotides in all tRNA molecules. This suggests a structural explanation for this variability. These bases are found in regions in which there is a bulging or arching of the polynucleotide chain backbone. This arching makes it possible to accommodate variable numbers of nucleotides in different tRNA molecules; a larger arch is likely to be found in molecules containing larger numbers of nucleotides. The structure thus provides information relative to the question of whether yeast tRNA^{Phe} can serve as a model for all tRNAs. In this case, something special in the structure is found in regions of nucleotide variability.

DOUBLE HELIX TURNS A CORNER

Two regions in the yeast tRNA^{Phe} molecule are marked by moderate structural complexity. One is the corner of the molecule where the T and D loops interact to stabilize the L-shaped conformation. The other is in the core of the molecule near the D stem, where there is considerable additional hydrogen bonding.

The T loop is organized in such a manner as to stabilize its interaction with the D loop and thereby maintain the two limbs of the molecule at approximately right angles to each other. This is accomplished using both detailed hydrogen-bonding and base-stacking interactions. All bases in the T loop are stacked parallel to the bases in the T stem, except U59 and C60, which are excluded from the stacking interactions . Bases U59 and C60 are oriented at right angles to the rest of the bases in the loop; they serve to nucleate the stacking interactions on which the vertical stacking domain in figure 3 is built.

In the corner of the molecule where the T and D loops come together, we see many features that have the overall effect of bringing about a rather intricate or detailed fitting together of different components to stabilize the unusual conformation. Broadly, the tRNA molecule has the appearance of a molecule in which the double helix has been engineered to make a right angle turn. The corner of the molecule has several features that appear to be designed to stabilize this conformation.

The core of the molecule immediately beneath the T stem includes some complex hydrogen-bonding interactions. Four segments of polynucleotide chains come into this region; two of the chains are part of the D stem, and the others are from the variable loop and the section of chain joining the acceptor stem to the D stem (residues 8 and 9). These chains are hydrogen-bonded in several ways, including groups of three bases hydrogen-bonded in triplets.

The anticodon stem is an RNA double helix with about 11 residues per turn. The conformation of the bases in the anticodon loop is somewhat similar to that suggested by Fuller and Hodgson (1967) in that the three anticodon bases are at the end of a stacked series of bases at the 3' end of the loop. The constant residue U33 plays an interesting role in maintaining the conformation of the loop since it is hydrogen-bonded through N3 to the phosphate group of P36. Indeed, there is great similarity between the conformation of the polynucleotide chain in the region of the T loop and the conformation in the anticodon loop (Quigley and Rich, 1976). In both places the polynucleotide chain makes a sharp bend and a uridine residue (U33 or U55) plays a key role in stabilizing this sharp turn through the formation of a hydrogen bond to a phosphate residue on the other side of the loop.

The configuration of the anticodon bases is shown in figure 4 as viewed from the bottom of the molecule. The three anticodon bases have the form of a right-handed helix with approximately 8 residues per turn. They are in a conformation such that they can form hydrogen-bonding interactions with the codon. At present it is not clear whether the detailed anticodon conformation seen in the crystal is maintained when it interacts with mRNA.

TRANSFER RNA IN SOLUTION

The orthorhombic crystal of yeast tRNA^{Phe} contains approximately 75% water. This suggests that the transformation from the crystalline state to a completely aqueous solution is not likely to have enormous structural consequences. The available evidence strongly supports this interpretation. A variety of experimental techniques have been used to correlate the structure observed in the crystal with that found in solution. All of these studies reach the same general conclusion: the structure appears to be the same in solution as it is in the crystalline state (Rich and Rajbhandary, 1976).

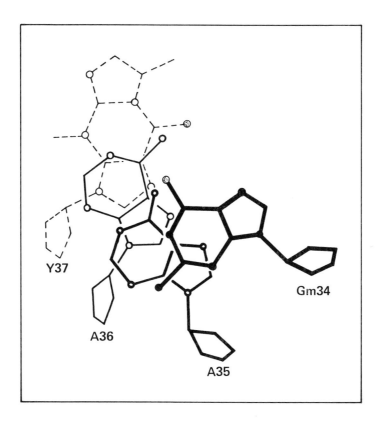

Figure 4. *Diagram illustrating the stacking of yeast tRNA^{Phe} anticodon bases 34 to 36 on the hypermodified Y base, as viewed from the exterior of the molecule. Ribose rings are shown as irregular pentagons, oxygen atoms are stippled circles, and nitrogen atoms are unstippled circles. Not all of the side chain of the Y base is included. It can be seen that the anticodon takes the form of a right-handed helix.*

There is good reason to believe that the molecular structure of yeast tRNA^{Phe} can be used as a model for understanding the molecular structure of all tRNAs (Kim et al., 1974b). Differences in the number of nucleotides in different tRNA sequences can be accommodated by variably sized arches looping out of the molecule. However, there is uncertainty about the conformation of those tRNAs that have very large variable loops containing 13 to 21 nucleotides. These undoubtedly form a stem and loop structure that is likely to project away from the molecule, but further work will be necessary before the details of their conformation are known.

It is likely that some altered forms of hydrogen bonding will be found in other sequences. Basically, those hydrogen bonds in yeast tRNA^{Phe} which do not involve invariant or semi-invariant positions are probably altered in other tRNA molecules. In several cases it is possible to guess the detailed nature of these modifications, but many will need to be determined by further experimentation.

INITIATOR AND CHAIN-ELONGATION TRANSFER RNAs

As mentioned above, the initiator tRNA is special in that it goes to the P site of the ribosome, whereas all other chain-elongation tRNAs go to the ribosomal A site. The initiator tRNA is specifically designed to recognize the initiator codon AUG and to insert a methionine in the N terminus of the polypeptide chain of all proteins. Initiator tRNAs have several features in common. Although their nucleotide sequences differ from species to species, they can nonetheless be substituted for each other in *in vitro* protein-synthetic systems. Furthermore, when cleaved by S1 nuclease, which cuts single strands and loops of nucleic acids, a common cleavage pattern is found for all initiator tRNAs (Wrede et al., 1979). This pattern differs from the cleavage pattern that is common for all chain-elongation tRNAs.

The structural basis for this modification in cleavage pattern has recently been revealed in the three-dimensional structure of the *E. coli* tRNA$_f^{Met}$, which has been solved to a resolution of 3.5 Å (Woo et al., 1980). A comparison of the folding patterns of the chain-elongation yeast tRNA^{Phe} and the initiator *E. coli* tRNA$_f^{Met}$ is shown in figure 5. In the side views, it can be seen that the general folding patterns are very similar, except a difference in the conformation at the 3' acceptor end of *E. coli* tRNA$_f^{Met}$. The 3' OH end is folded over, and this may be associated with a modification of the hydrogen-bonding pattern in the acceptor stem of that molecule. An interesting and apparently functional change is found in the folding of the anticodon loop. The three anticodon bases have a similar stacking, as seen in the upper part of figure 5; but there are differences, which can be seen in the lower part of figure 5. The tracing of the anticodon loop in yeast tRNA^{Phe} is rather rounded. It can be seen that residue 33, which is emphasized in the figure, has a conformation such that the uracil residue is hydrogen-bonded to phosphate 36. In contrast, the nucleotide U33 has a different orientation in *E. coli* tRNA$_f^{Met}$. The nucleotide is rotated relative to its position in yeast tRNA^{Phe}. Instead of having U33 hydrogen bonding to P36, the altered conformation has the 2'-hydroxyl group of the ribose in a position where it may form this hydrogen bond. This change in the conformation of uridine is associated with a radically different folding of the anticodon loop. This in turn is no doubt responsible for the differences observed in the S1 nuclease cleavage pattern (Wrede et al., 1979).

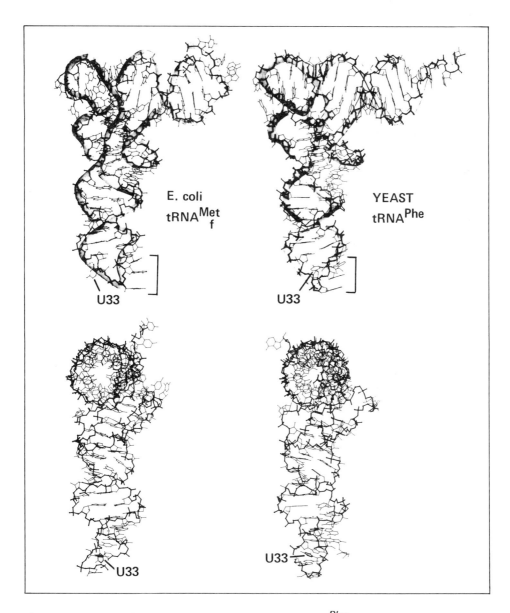

Figure 5. *Views of the chain elongation yeast tRNA^{Phe} and the initiator E. coli tRNA$_f^{Met}$. The computer drawing emphasizes the backbone with a thicker line. In the upper side views, the stippled ribbon is added to facilitate tracing of the backbone. The brackets show the position of the anticodon base triplet. The major difference shown in this view is at the acceptor end, where the 3' OH end of E. coli tRNA$_f^{Met}$ is folded back on the acceptor stem; in yeast tRNA^{Phe} it continues to form part of a single-stranded helix. The lower end views are made looking down the acceptor–T stem helices, with a hole in their center. It can be seen that the anticodon loop folding is different in the two molecules. U33 is crosshatched in the bottom views. In the upper side views, the U33 can be seen to have different positions in the two molecules.*

Initiator tRNAs go into the ribosomal P site, while chain-elongation tRNAs go into the ribosomal A site. Thus it is reasonable to suggest that these two conformations may be similar to the conformations seen in these ribosomal sites. If this is true, then the tRNA may undergo a conformational change in going from the A site to the P site in the ribosome. This process may be facilitated by the formation of a ternary complex involving the two tRNAs and the mRNA.

Knowledge of the three-dimensional structure of the tRNA-mRNA complex is essential to understand the manner in which these molecules may come together in a prebiotic system. Protein synthesis had to be initiated in the absence of a ribosome, although not necessarily in the absence of ribosomal RNA. Observing how this is done in contemporary biochemical systems may provide clues that allow us to carry out this reaction in a manner that seems plausible for a prebiotic environment.

TRANSFER RNA AND PROTEIN SYNTHESIS

Now we return to the paradox described at the beginning of this paper. What is the mechanism whereby nature differentiates between different tRNA species during aminoacylation, and at the same time allows all these molecules to pass through the same ribosomal apparatus? In short, where are the components of uniqueness and the components of commonality?

There has been considerable effort toward understanding the mechanism of aminoacylation and finding the regions of the tRNA molecule that may be recognized by synthetase enzymes. Different workers have suggested the acceptor stem, the D stem, or the anticodon (Kisselev and Favorova, 1974). Much experimental work on synthetase-tRNA interactions is in accord with the suggestion that synthetases recognize varying aspects of the tRNA structure along the diagonal side of the molecule (Rich and Schimmel, 1977). It is clear that the synthetases are probably all different, and there is not likely to be a common recognition system even if they all approach the same side of the tRNA molecule. It is likely that tRNA-synthetase recognition takes two parts. One is a recognition of the ribose-phosphate chain, which would be sensitive to the folding of the tRNA molecules. Second, there must be a recognition by the protein of specific bases in the double-helical stems or among the unpaired segments of the molecules. There are several ways in which proteins can recognize nucleic acid sequences (Seeman et al., 1976). It is likely that the basis of specificity resides in this detailed sequence recognition rather than in any conformational differences between tRNA molecules. Furthermore, there is an adequate basis for specificity in these sequences. The number of nucleotides or base pairs recognized by the enzyme need not be very large to obtain the requisite specificity.

Far more puzzling is the question of what goes on in the ribosome. Our information in this field is scanty. As mentioned above, there are two sites within the ribosome, one occupied by the peptidyl-tRNA and the other by the aminoacyl-tRNA. It is likely that both sites are occupied at the same time, and for this reason the tRNA molecule may have been designed to have the form of a double helix that turns a corner. The L shape of the molecule may make it possible for two adjacent tRNAs to come close together at one end near their anticodons where they are interacting with adjacent codons of the mRNA. At the same time, the CCA acceptor ends may be able to come close together because of the L shape. These two CCA ends must come close to allow the ribosomal peptidyl transferase to transfer the peptide chain from one tRNA molecule to the other. It is not at all clear how this is accomplished, although it has been suggested that the two codons become unstacked during the reading process so that, in effect, the messenger "turns a corner" while it is read (Rich, 1974). This remains an exciting area for further research work.

Work by Erdmann and his colleagues (1973) has suggested that the T loop may become disengaged from the D loop inside the ribosome so that it can hydrogen-bond with the ribosomal 5S RNA. This interaction may be an important component in the translocation of tRNA from the aminoacyl site to the peptidyl site. Further, as mentioned above, the anticodon loop may differ in the A and P sites. Thus the tRNA molecule undergoes a conformational change within the ribosome. Determination of the nature of these events remains an important goal of research in this area. It is possible that this conformational change is triggered by codon-anticodon interactions, and this may be a consequence of the fact that the tRNA molecule as a whole exhibits long-range order (Rich and Rajbhandary, 1976). Altered interactions at one end of the molecule may give rise to a change in reactivity or in conformation at a more remote part of the molecule.

TRANSFER RNA AND THE ORIGIN OF PROTEIN SYNTHESIS

At present we have good models for understanding the early stages in the origin of life, including the accumulation of organic chemicals and the formation of nucleotides and polynucleotides. We do not have convincing models for the origin of polynucleotide-directed polypeptide formation; but it is reasonable to believe that the molecular mechanics of mRNA reading in the ribosome as well as peptide bond formation may provide a clue to its origin. We can assume that the conformation of the anticodon does not have a large change between the two ribosomal sites, but is similar to that shown in figure 5. We postulate a specific tRNA-tRNA interaction in the ribosome when they are in the aminoacyl and peptidyl sites, possibly using residue U33 in the ribosomal P site to bind to the tRNA in the A site. This interaction is such that it yields a spacing between the two anticodons that is directly responsible for the triplet code.

Thus, if two tRNAs and an mRNA form a transiently stabilized ternary complex, perhaps they would have the requisite geometry to bring aminoacylated ends of the tRNAs close together. An inefficient system of catalysis could then form a dipeptide, and the stage would be set for a repetition. Knowledge of the structural details of protein synthesis in contemporary systems may shed light on how this process evolved in the prebiotic era.

REFERENCES

Crick, F. H. C.: Codon-Anticodon Pairing: the Wobble Hypothesis. J. Molecular Biol., vol. 19, Aug. 1966, pp. 548–555.

Erdmann, V. A.; Sprinzl, M.; and Pongs, O.: The Involvement of 5S RNA in the Binding of tRNA to Ribosomes. Biochem. Biophys. Res. Commun., vol. 54, Oct. 1, 1973, pp. 942–948.

Fuller, W.; and Hodgson, A.: Conformation of the Anticodon Loop in tRNA. Nature, vol. 215, Aug. 19, 1967, pp. 817–821.

Holley, R. W.; Apgar, J.; Everett, G. A.; Madison, J. T.; Marquisse, M.; Merrill, S. H.; Penswick, J. R.; and Zamir, A.: Structure of a Ribonucleic Acid. Science, vol. 147, Mar. 1965, pp. 1462–1465.

Kim, S. H.; Suddath, F. L.; Quigley, G. J.; McPherson, A.; Sussman, J. L.; Wang, A. H. J.; Seeman, N. C.; and Rich, A.: Three-Dimensional Tertiary Structure of Yeast Phenylalanine Transfer RNA. Science, vol. 185, Aug. 2, 1974a, pp. 435–439.

Kim, S. H.; Sussman, J. L.; Suddath, F. L.; Quigley, G. J.; Wang, A. H. J.; Seeman, N. C.; McPherson, A.; and Rich, A.: The General Structure of Transfer RNA Molecules. Proc. Natl. Acad. Sci. USA, vol. 71, Dec. 1974b, pp. 4970–4974.

Kim, S. H.; Quigley, G. J.; Suddath, F. L.; McPherson, A.; Sneden, D.; Kim, J. J.; Seeman, N. C.; Weinzierl, J.; and Rich, A.: Three-Dimensional Structure of Yeast Phenylalanine Transfer RNA: Folding of the Polynucleotide Chain. Science, vol. 179, Jan. 19, 1973, pp. 285–288.

Kim, S. H.; Quigley, G.; Suddath, F. L.; and Rich, A.: High-resolution X-Ray Diffraction Patterns of Crystalline Transfer RNA that Show Helical Regions. Proc. Natl. Acad. Sci. USA, vol. 68, April 1971, pp. 841–845.

Kisselev, L. Lev; and Favorova, O. Olga: Aminoacyl-tRNA Synthetases: Some Recent Results and Achievements. Adv. Enzymol., vol. 40, 1974, pp. 141–238.

Lohrmann, R.; Bridson, P. K.; and Orgel, L. E.: Efficient Metal-Ion Catalyzed Template-Directed Oligonucleotide Synthesis. Science, vol. 208, June 27, 1980, pp. 1464–1466.

Quigley, G. J.; and Rich, A.: Structural Domains of Transfer RNA Molecules. Science, vol. 194, Nov. 19, 1976, pp. 796–806.

Rich, Alexander: How Transfer RNA May Move Inside the Ribosome. In: Ribosomes, M. Nomura, A. Tissieries, and P. Lengyel, eds. Cold Spring Harbor Laboratory, New York, 1974, pp. 871–884.

Rich, A.; and Rajbhandary, U. L.: Transfer RNA — Molecular Structure Sequence and Properties. Ann. Rev. Biochem., vol. 45, 1976, pp. 805–860.

Rich, A.; and Schimmel, P. R.: Structural Organization of Complexes of Transfer RNAs With Aminoacyl Transfer RNA Synthetases. Nucleic Acids Res., vol. 4, no. 5, 1977, pp. 1649–1665.

Robertus, J. D.; Ladner, J. E.; Finch, J. T.; Rhodes, D.; Brown, R. S.; Clark, B. F. C.; and Klug, A.: Structure of Yeast Phenylalanine tRNA at 3 Å Resolution. Nature, vol. 250, Aug. 16, 1974, pp. 546–551.

Seeman, N. C.; Rosenberg, J. M.; and Rich, A.: Sequence-Specific Recognition of Double Helical Nucleic Acids by Proteins. Proc. Natl. Acad. Sci. USA, vol. 73, March 1976, pp. 804–808.

Sprinzl, M.; Grueter, F.; Spelzhaus, A.; and Gauss, D. H.: Compilation of tRNA Sequences. Nucleic Acids Res., vol. 8, Jan. 11, 1980, pp. R1–23.

Wrede, P.; Woo, N. H.; and Rich, A.: Initiator tRNAs Have a Unique Anti-codon Loop Conformation. Proc. Natl. Acad. Sci. USA, vol. 76, July 1979, pp. 3289–3293.

Woo, N. H.; Roe, B.; and Rich, A.: Three-Dimensional Structure of Escherichia Coli Initiator tRNA$_f^{Met}$. Nature, vol. 286, July 24, 1980, pp. 346–351.

Emergence and Radiation of Multicellular Organisms

JAMES W. VALENTINE

The multicellular animals form a logical and elegant assemblage that excites admiration for their effective use of micro- and macroevolution. One can only hope that the generality of their features can someday be tested against another life system.

The origin and early diversification of multicellular organisms do not seem to mark planets in a fashion detectable from afar. Indeed, after generations of effort, the uncertainty associated with the dating of metazoan origins and early radiations on Earth is several hundred million years. Since we have by no means exhausted the possible observations, there is still hope that we shall further delimit the timing of these events from direct evidence. In the meantime we must rely on indirect evidence to support plausible models of the emergence and radiation of multicellular organisms on Earth.

One of the problems has been that metazoans with hard parts burst relatively suddenly into the fossil record, already well-differentiated into distinctive living phyla only distantly allied. Since evolution has been regarded as a gradual process, it has seemed doubtful to many that these phyla arose in a sudden burst of adaptive radiation. The Phanerozoic fossil record suggests a long Precambrian history for multicellular forms. Durham (1971) calculated that if Precambrian evolutionary rates were comparable to those displayed by Phanerozoic metazoans, then the morphological distance achieved by early Cambrian phyla implied that divergence from a founding metazoan lineage began a billion or more years earlier. Where are the traces or remains of those early animals?

Within the past two decades, evidence has accumulated to indicate that there was no such long period of gradual metazoan evolution culminating in

229

the ground plans of Phanerozoic phyla. The sudden advent of fossil phyla near the early Cambrian boundary probably coincides closely with their origin. Also within recent years, work in genetics and developmental biology has rendered plausible the sudden origination of such morphological novelties. Work in Phanerozoic paleontology has shown that even when the fossil record is most complete, most taxa display an abrupt rather than a gradual pattern of change. This has led to a downgrading of the role of microevolution in the origin and success of taxa.

In light of this work, an attempt is made here to assess the factors that led to the origin of major (phylum-level) multicellular grades and ground plans and that account for their diversity. Two particular questions are pursued: First, if time were turned back to the late Precambrian and multicellular grades were allowed to evolve again, how closely would this second fauna resemble the first, and how would it differ? Second, if this experiment were repeated under different environmental conditions, as on another planet, how closely would the results resemble and how differ from those on Earth?

FOSSIL RECORD OF METAZOAN BODY PLANS

Phyla — Living and Extinct

Figure 1 depicts the time of first appearance in the fossil record of each living animal phylum. The figure is fairly conservative, listing those appearances that seem at present to be most definitive. The intent is that most subsequent changes will involve only the discovery of earlier representatives of the phyla. Of course, it is possible that restudy of fossil material may discredit some of these occurrences. In most cases the earliest records are securely based on body fossils, but for a few phyla the remains of secreted structures (such as pogonophoran tubes) or other traces are tentatively accepted.

The earliest certain metazoan fossils occur in the Ediacaran interval, which began 100 million years or more before the start of the Cambrian (Glaessner, 1971). At least some elements of the Ediacaran fauna persisted into Cambrian time (Borovikov, 1976). The characteristic Ediacaran forms are soft-bodied. The only living phyla certainly represented are the Cnidaria and the Echiuroidea. There are also some vermiform fossils that might represent living phyla (perhaps the Annelida) but might equally well be the remains of extinct ones. Finally, there are enigmatic forms (such as *Tribrachidium*) which cannot be assigned to a living phylum.

Mineralized skeletons and other hard parts do occur during the Precambrian; most are minute denticles, plates, and other skeletal elements that

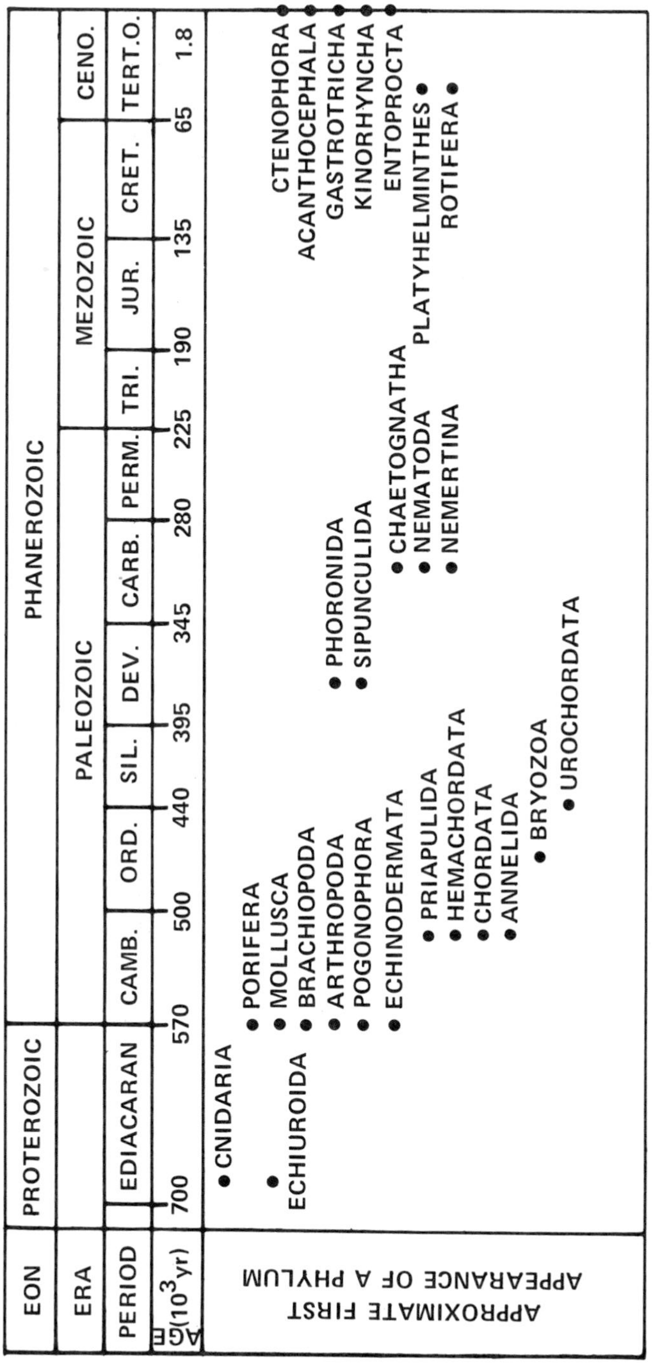

Figure 1. *Times of first definitive appearances of living phyla in the fossil record. Annelida are questionably reported from the Ediacaran.*

formed parts of larger organisms but cannot be associated with any living phylum (Matthews and Missarzhevsky, 1975). In the earliest Cambrian stage, the first extensive fauna with mineralized skeletons appears and includes skeletons identified as Mollusca. In succeeding Lower Cambrian stages, five more living phyla appear, so that all living phyla that have easily fossilizable hard parts appear during the Lower Cambrian except for the Bryozoa and the Chordata.

The next first appearances are of four living phyla in a remarkably preserved, soft-bodied fauna that also contains hard-part remains of animals typical of the time. This is the Middle Cambrian Burgess Shale fauna from British Columbia. It includes the first known chordate, a soft-bodied form somewhat resembling cephalochordates (Conway Moins and Whittington, 1979). Bryozoan skeletons first occur in the early Ordovician.

The remaining 13 living phyla are all soft-bodied; eight of them appear as fossils but are exceedingly rare. The other five are not known as fossils at all. Four of these are minute pseudocoelomates (commonly parasitic today), and one is the diploblastic comb jellies. None of these forms leaves characteristic traces, and they would be preserved only under the most unusual of circumstances. Their lack of a fossil record is thus not surprising. It is perhaps more surprising that all the living coelomate phyla, even soft-bodied ones, do have a fossil record of some sort.

The order of appearance of phyla in the fossil record seems to be based more on their fossilizability than on their order of origin. However, there is evidence that the appearance of phyla with mineralized skeletons near the Precambrian-Cambrian boundary is associated closely with their origins. The skeletal ground plans of these phyla are closely coadapted with their body plans, and in some cases the latter depend on the former (Cloud, 1948; Valentine, 1973). In these cases the advent of a mineralized skeleton indicates the origin of the phylum, and the record indicates that these events occurred near the Precambrian-Cambrian boundary.

The two phyla with durable skeletons that do not appear in Lower Cambrian time, the Chordata and Bryozoa, may have been represented then by soft-bodied lineages, for they have body plans that do not require durable skeletons. Indeed, the chordates appear in the Middle Cambrian, while durable chordate skeletons are not known until the Late Cambrian (Repetski, 1978). The Bryozoa first occur as relatively simple, calcified branching tubes, but they could have been represented by soft-bodied forms prior to mineralization of the tubes. Thus either of these phyla *could* have been present in the Lower Cambrian.

In short, it is possible to hypothesize that the entire array of living coelomate and pseudocoelomate phyla arose near the Precambrian-Cambrian boundary, and that their entry into the fossil record occurred as they acquired hard parts or as they were preserved by a fortuitous concomitance of circumstance. We will not argue this hypothesis at length here, but it does serve the facts and may be nearly, if not precisely, correct.

In addition to living phyla, the fossil record contains remains of extinct groups different from any living groups, and that probably represent extinct phyla. The anatomy of some unusual soft-bodied fossils from the Middle Cambrian Burgess Shale can be reconstructed in enough detail that we can be reasonably sure we are dealing with extinct phyla. Their body plans sometimes resemble those of living phyla but are different enough to suggest that they evolved separately. Ten forms are now known from the Burgess Shale that are likely to represent extinct phyla (Conway Moins and Whittington, 1979). Other soft-bodied fossils that probably represent extinct phyla are known from later rocks; a particularly distinctive form is found in the Late Carboniferous of Illinois (Richardson, 1966). The times of origin and extinction of these distinctive soft-bodied forms are unknown. It is possible that they all originated during the late Precambrian–Lower Cambrian interval and are part of a general radiation that gave rise to most or all living phyla.

Durable skeletons distinct in architecture from those of living phyla are also known in the fossil record. These fossils are sometimes placed into a living phylum that they resemble, but such assignments are often in dispute. Two of particular interest here are the Archaeocyatha and the Hyolithida. Both appear in the Lower Cambrian, and each has been considered an extinct phylum by some authorities. Since the soft-part anatomies of these forms are largely unknown, and since skeletal ground plans alone are not necessarily definitive of relationship at the phylum level, their status remains unclear.

If a gigantic radiation of phyla did indeed occur near the end of the Precambrian, how can we account for it? What triggered it, and even more importantly, what processes operated to produce so much fundamental morphological divergence so quickly? Reconstruction of the fauna of the Late Precambrian *before* the suggested radiation is an obvious first step toward an answer.

Animals Before Phyla

The heading above is somewhat facetious — because of the convention by which animals are classified, all must belong to one phylum or another. Nevertheless most living phyla represent clades that diversified well after the metazoans had originated. The morphological distinctiveness believed appropriate to phyla is based on experience with these clades, although some, such as the Arthropoda (Tiegs and Manton, 1958), may be polyphyletic. While Cnidarians and Platyhelminthes may be ancestral to other phyla, the fossil record suggests that most phyla have distinctive morphologies because they were adapted to different modes of life rather than because they represent a chain of ancestors and descendants arranged in some *scala naturae*. Put another way, the ground plans of the phyla probably did not evolve gradually as steps or divisions on a scale of complexity, but rather evolved relatively suddenly in response to broad adaptive opportunities that may all have been present contemporaneously. To be sure, some phyla descended from others, but the pattern of novel morphology results from the quality of adaptive opportunity.

Something of the character of animals ancestral to living phyla can be inferred from the architecture of the latter. The body plans of living phyla fall into several distinctive groups. These are particularly well discussed by Hyman (1940) and Clark (1964) and are summarized in table 1. Note that the times of appearance of these groups as fossils bear little relationship to their presumed phylogenies. Some groups represent different grades of construction: the diploblastic Cnidaria, the triploblastic but acoelomate Platyhelminthes, and the pseudocoelomates, for example. Others are on the same grade of construction but have distinctive body plans; groups of coelomates, for example, can be identified on the basis of their coelomic architectures (table 1).

Presumably the Late Precambrian fauna included diploblastic and triploblastic acoelomate phyla (Cnidaria and probably Platyhelminthes at least), primitive pseudocoelomates, and a few coelomate groups ancestral to each group of living phyla that differs by coelomic plan. The Cnidaria and Platyhelminthes, probably largely pelagic and epifaunal, would not ordinarily leave much of a record of their presence (we are indeed fortunate to have the Ediacaran body fossils). Pseudocoelomates and coelomates, however, possess body cavities that were probably evolved as hydrostatic skeletons, an important primitive function of which was to aid in burrowing (Clark, 1964). Burrows that disturb bedding fossilize rather easily, so traces of early burrowing phyla might be expected. Both horizontal and vertical burrows are reported from the Late Precambrian. They are not all accurately dated but certainly occur during the Ediacaran interval and possibly earlier. These burrows are relatively rare in the Precambrian, but they increase significantly in kind and number in the Cambrian (Crimes, 1974). It is reasonable that the Edicaran burrows represent pseudocoelomate or (perhaps more likely because of their larger size) coelomate stocks that are ancestral to modern phyla (Valentine, 1973).

The exact timing of the rise of multicellularity as a new grade of organization cannot be deduced from the fossil record. There is little dispute as to the advantages of this condition: increased size, longevity, homeostasis, and ultimately complexity of cellular differentiation can stem from a multicellular condition, permitting development of life modes unavailable to unicells. Multicellularity arose numerous times from unicellular lineages (Stebbins in

TABLE 1.— MAJOR GRADES AND BODY PLANS OF LIVING METAZOAN
PHYLA

Grade or plan	Phylum
Tissue	Porifera (descended separately from other phyla)
Diploblastic (two germ layers)	Cnidaria Ctenophora
Triploblastic acoelomate (three germ layers, no body cavity)	Platyhelminthes Nemertina
Triploblastic pseudocoelomate (three germ layers, body cavity between gut tissue and body wall)	Rotifera Acanthocephala (many minute, often Gastrotricha parasites) Kinorhyncha Nematoda Entoprocta (colonial) Priapulida (may be coelomate)
Triploblastic coelomates (three germ layers, body cavity lined with mesoderm)	
Amerous (unsegmented and regionated coelom)	Sipunculida (probably not closely Echiuroidea related)
Metamerous (segmented coelom, organs serially repeated)	Annelida Arthropoda
Pseudometamerous (coelom unsegmented but some organs repeated serially)	Mollusca
Oligomerous (coelom regionated into two or three divisions)	Phoronida Brachiopoda (tentaculate body plans) Bryozoa Chaetognatha Pogonophora Echinodermata Hemichordata Urochordata (body plans with gill Chordata slits or homologues)

Dobzhansky et al., 1977). This is clear evidence of its adaptive advantage, although the precise adaptive pathways followed by the earliest metazoans are much disputed.

On present evidence, a plausible descriptive model of the origin of metazoan grades and ground plans has the following features (fig. 2). Multi-

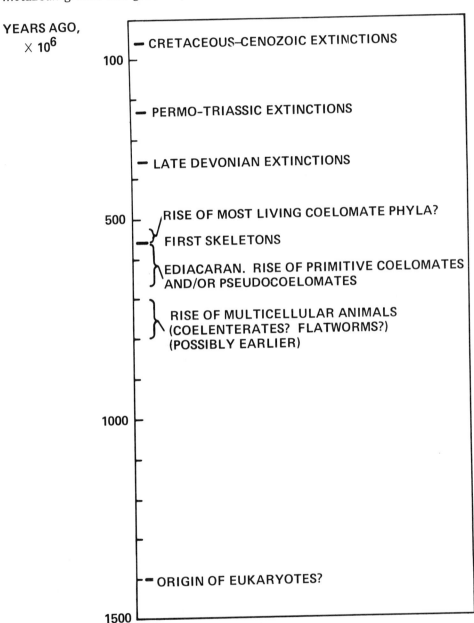

YEARS AGO,
× 10⁶

Figure 2. *Some key events in the rise of animal phyla. Many phyla that are not known to be living may have originated between about 600 and 570 million years ago and perhaps have become extinct during one or another of the mass extinction events.*

cellular organisms arose from protozoan ancestors, and it is easy to imagine a wide variety of pelagic and benthic forms evolved at a simple grade of construction (Valentine, 1977). This early multicellular radiation must have occurred some time after the appearance of eucaryotes (perhaps near 1400 million years ago) but before 700 million years ago; very likely it was nearer the younger age since meiosis may have had to be perfected in eucaryotic unicells. From products of this early radiation, diploblastic and acoelomate triploblastic grades arose, possibly independently but probably by evolution of the latter from the former. These events may have occurred very near 700 million years ago. Then body cavities evolved in triploblastic lineages at least twice and probably a few times, radiating to form a variety of pseudocoelomate and coelomate clades. The coelomate radiation produced the metamerous, oligomerous, amerous, and pseudometamerous vermiform "superphyla" stocks from which the phyla of the Lower Cambrian radiated in turn (or perhaps in some cases the Cambrian phyla represent Late Precambrian holdovers). The coelomate superphyla may have developed during the hundred-million-year interval preceding the Cambrian. The bulk of living phyla originated in radiations from the coelomate superphyla and from pseudocoelomate stocks probably near 570 million years ago at the beginning of the Cambrian. The entire sequence of the origin of metazoa and the development of phyla could have been compressed inside the 150 million years preceding the Cambrian (although it may have taken longer). At any rate the radiation associated with the origin of most of the clades that we now call phyla seems to have taken only a few tens of millions of years at most.

MACROEVOLUTION

Microevolution and Macroevolution

Microevolution usually refers to the changes consequent on the selection of one genotype over another *within* the gene pool of a population or of an entire species. Macroevolution usually refers to the origin of taxa and of evolutionary trends above the level of species. This term will be used here in a somewhat more specialized sense to refer to the changes consequent on selection *between* gene pools. Genotypes that increase in frequency by microselection are said to be more *fit;* populations or species that increase through macroselection are said to have *adaptive advantage.* Both fitness and adaptive advantage are relative measures, the former between genotypes, the latter between taxa. Adaptation or *adaptive value* is the ability of an individual or taxon to cope with a given environment; it is measured on an absolute rather than a relative scale (Ayala, 1969). Of course, all three of these features — fitness, adaptive advantage, and adaptive value — vary in relation to the environment.

Microevolution can certainly be responsible for the origin of some new species, but it acts too slowly to account for the origin of many species and of most higher taxa (Stanley, 1975, 1976, 1978). Since fitness must by definition be enhanced by new genes or gene combinations for microevolution to occur, large heterozygous genetic changes that upset the biochemistry of development are usually lethal. Eldredge and Gould (1972) and Stanley (1975) have shown that many fossil species (morphospecies) evolved only slightly if at all during their existence; the morphological changes that differentiate them from their ancestors must have been associated with their origins. The morphological distance achieved during the speciation process must often be considerable, equivalent to the morphological differences between many genera. Indeed, fossil genera appear much too rapidly to have originated through microevolutionary processes alone (Stanley, 1978). They must be products of macroevolution and perhaps often of a single speciation event. This argument has even more force when applied to higher taxa such as orders, classes, and phyla; the time required for microevolutionary processes to produce such distinctive morphologies as distinguish these higher taxa (which sometimes possess different body plans altogether) is far greater than the time available according to the fossil record.

Macroevolution does not depend on fitness and indeed must often circumvent it. If this can be done, then novelties considerably different morphologically from their ancestors can arise rapidly; their survival depends only on their adaptive value. An example is the rapid quantum speciation possible in organisms (chiefly plants) that are self-fertilizing. A large mutation can produce a descendant that is infertile with members of its parental species, including its parents. If it can fertilize itself, however, it may propogate and thrive; whether it does or not depends on its adaptive value. Indeed, if it should happen to have a distinct adaptive advantage over its parent, or any other species for that matter, due to a higher reproductive rate, greater hardiness, or some such feature, it may eventually replace an older species. Microevolution has had very little to do with the origin or the success of this "instant species."

In lineages that are not self-fertilizing, the perpetuation of a marked novelty is more difficult, for it must backcross with "normal" members of its species and pass through at least one heterozygous generation to produce homozygotes. Since the heterozygotes would usually have much lower fitness than normal individuals, they must evade the winnowing effects of microselection to produce offspring. That such evasions do happen has long been postulated. For example, Simpson (1944, 1953) and Mayr (1942, 1963) have both appealed to genetic "drift" in small populations as a means of achieving novel gene combinations in macroevolution. Fixation of chromosomal mutations in small populations of mammals, isolated by breeding patterns and subject to drift because of their small effective sizes, has been reported by Bush (1975) and Bush et al. (1977). In these mammalian examples, there appears to be little morphological difference between the normal and mutant phenotypes, but they do indicate how novel genotypes can be fixed despite probable heterozygote disadvantage.

Developmental and Phylogenetic Morphogenesis

The creation of form through developmental processes has long served as an analogy to the rise of form through evolutionary processes. Even though ontogeny does not recapitulate phylogeny in the classic sense, there is clearly much to be learned of evolutionary processes from the study of individual development. In both cases, a single cell can be elaborated into a complex organism. A complex individual may have 200 different types of cells, such as nerve and muscle cells, each functioning in tissue and organ systems as different as brains and biceps. Each cell (excluding gametes and with unimportant exceptions) contains an identical genotype. Differentiation occurs because different sets of genes function in different cells: the genomes in each cell contain all the information necessary to construct an entire organism, but instead they produce only a particular cell phenotype. Switching of gene activities on and off to obtain products from the appropriate set of genes in the proper amounts and in the correct sequence to make the proper number of cells of each type arrayed in a desirable form requires an elegant regulatory system.

Two main gene functions involve the production of polypeptide products (via *structural* genes) and control of the pattern of gene activity (via *regulatory* genes). Results of regulatory gene activity are seen in differentiation among genetically identical cells (Britten and Davidson, 1969). Some genes have both regulatory and structural functions. All eucaryotic cells require a rather similar set of structural genes to produce substances required for cell growth and general cell metabolism. The chief differences between the genomes of morphologically distinctive metazoa seem to reside in the genetic regulatory apparatus. Although some differences in structural functions between members of different phyla are obvious, their distinctive body plans must arise from different patterns of activities of their structural genes rather than from differences in the structural genes themselves. These patterns of gene expression are determined by the regulatory gene system (Britten and Davidson, 1971; Wilson, 1975; King and Wilson, 1975; Valentine and Campbell, 1975).

Molecular details of genetic regulation in eucaryotes have not been extensively defined, but there are enough observational data to give some idea of how the regulatory systems operate. In individual organisms, growth and development are mediated by hormones. Some hormones are polypeptides and thus are coded by structural genes, while others are removed from direct gene action along what Stebbins (1968) has called *informational relays*. Hormones may switch on or off large sets of genes. In some cases, two or more hormones control an array of processes according to their relative concentrations. Matsuda (1978) has reviewed the activity of some hormones in arthropod development. Morphogenesis is triggered by the hormone ED (ecdysone), the effects of which vary according to the availability of a second hormone JH (juvenile hormone). ED induces larval molt in the presence of large amounts of JH, pupal molt with less JH, and imaginal molt in the absence of JH. The inhibiting effect of JH is clear, and it has other well-studied effects. For example, overproduction of JH late in postembryonic development causes precocious egg maturation, resulting in developmental neoteny. Mutations that reduce or stimulate the production of this hormone at unusual levels or times can result in abnormal metamorphosis.

Unusual environmental stimuli can alter the level of hormone secretion and produce abnormalities similar to those resulting from mutations. The morphological consequences of these developmental abnormalities can be gross, ranging from changes in limb size and number and wing reduction or loss to the maturation of larvae with body plans distinct from those of adults. Many examples of regulatory gene abnormalities are known from representatives of other phyla, including humans.

The potential for large and sudden morphological change thus clearly exists within metazoan genomes. The problem is to translate this potential into a rapid evolutionary event that produces descendants quite different from their ancestors but ones that are viable and reasonably well adapted. Genetic drift is a possible mechanism, as we have seen. Genomes suffering extensive changes in developmental pathways to produce "monster" morphologies, however, are likely to produce lethal heterozygotes when crossed with normal genomes; at least this is the experience of laboratory studies (Dobzhansky, 1971). Matsuda (1978) has suggested how this problem may be surmounted. If a population is subjected to a novel environment that induces unusual hormone production, many similar abnormal phenotypes will result. This is a nongenetic origin for a novel morphology. These novel individuals would presumably have no special trouble in reproducing, assuming they happen to be well adapted. The morphology could thus be propagated. Later genetic changes could then fix the new morphology in phylogeny. These genetic changes might be favored by microevolution because they would stabilize a phenotype with a large adaptive value.

SCENARIOS OF QUANTUM EVENTS IN EVOLUTION

There are thus several scenarios to explain the occurrence of sudden large morphological jumps — quantum events — in evolution. One is micro-evolutionary, when mutations with large morphological effects prove fit enough in heterozygotes that homozygotes, which are fitter still, become common; eventually the mutation is fixed. The case when this seems most likely to occur is neoteny, for the developmental pathways of a mutant and of its normal associates may remain nearly identical up to the point of adventitious reproduction in the mutant. Heterozygotes might therefore develop viably; if mutant homozygotes were fitter they would eventually replace the ancestral type.

The other scenarios are all primarily macroevolutionary in the sense that it is the adaptive value of the novelty that permits it to endure, even though it is less fit than the normal members of its parent species within the parental population. One of these scenarios involves genetic drift in small, reproductively isolated populations. Geographic isolation would be the most common type in invertebrates, and perhaps social isolation (inbred kin groups or harems) would be common among some higher vertebrates. Even though heterozygotes for a mutant would be less fit than the parental geno-types, drift would sometimes permit the appearance of homozygotes by chance. Their subsequent success would depend on their level of adaptation.

A third scenario involves organisms that are self-fertilizing and do not require drift to create homozygotes. For metazoans this scenario must have limited application.

A fourth scenario involves the creation of a new morph by environmen-tal induction of developmental changes, perhaps by the switching of regula-tory genes through hormonal activity. Individuals belonging to the new morph should not have unusual genotypes, and if the new morph happens to be well adapted there may be no reduced fertility. Subsequent genetic changes, presumably microevolutionary, are required to fix the novelty.

Whatever the scenario, microevolution and macroevolution play differ-ent roles (Stanley, 1975). The more significant novelties, in which the great-est morphological changes are involved, must be a product of a macroevolu-tionary mode. This should occur most commonly when the population is suddenly faced with a distinctively new environment. Large changes would then be indicated because small ones would not usually suffice to maintain a viable adaptive value. The sorts of genetic changes known to underlie highly distinctive morphologies are those of the regulatory gene system; many structural loci may have their expressions changed by a mutation at a single regulatory locus. The morphological results of regulatory changes will not be scaled to the amount required by environmental change; they could easily be greater than necessary. Regardless of the size of the morphological change, the survival of the novelty is first of all a function of its adaptive value when founded. Next it would be a function of the ability of microevolution to maintain the adaptive value at a suitably high level to prevent extinction. This may be critical and is the main importance of microevolution. The duration of lineages should often be a measure of the success of microevolutionary processes.

ADAPTIVE KALEIDOSCOPE

History of Life Models

As an aid in thinking about the evolution of phyla, it is useful to model the process in terms of a changing adaptive landscape and some simple rules for diversification. Adaptive landscapes have been used for this purpose for many years (see McCoy, 1979, for some early examples); Simpson (1944, 1953) used a model of adaptive zones to excellent effect in considering macroevolution. None of the previous visualizations is quite right for our purposes, however. We shall use a topographically featureless plain demarcated into small irregularly shaped patches. Each patch represents a fairly homogeneous environment. The boundaries between patches represent environmental discontinuities so that conditions change across patch boundaries but are relatively constant between them. There are several classes of boundary strengths; we shall arbitrarily specify that each succeeding increase in boundary strength doubles the environmental change across the boundary. In effect, the plain represents an environmental mosaic, and each patch is an individual tessera.

Through time the environment changes; tesserae enlarge or shrink, or coalesce as boundaries fade. Other boundaries change their positions and intensities, and new boundaries may appear across some tesserae. The environmental mosaic becomes an asymmetric kaleidoscope. Heterogeneous environments have numerous small tesserae separated by a high average boundary class; more homogeneous environments have large tesserae separated by a low average boundary class.

On this sort of kaleidoscopic game board, we shall model the diversification of the animal kingdom. At some location on the board, we position a small sphere, representing the first metazoan lineage. From this locus, other small spheres (potential descendant lineages) are episodically broadcast into the air to fall back onto the board. The direction in which they travel is selected at random. The distance they travel depends on two factors: their original impetus and the strength of the tessera boundaries they happen to cross. The original impetus may be expressed in boundary units: an impetus of 10 will take a sphere across two class five boundaries, 5 class two boundaries, or some such mixture. The sphere then settles on the last tessera, the airspace of which it can penetrate. The original impetus is selected at random (with replacement) from a pool in which the frequency distribution is previously determined (see below). The spheres are fragile, and there are only certain tesserae scattered randomly across the board on which any given sphere may settle without bursting. Furthermore, only one sphere can be accommodated on any tessera; late newcomers burst. There is one last rule: each sphere that successfully settles on a tessera becomes a locus for broadcasting new spheres under conditions identical to those at the original locus.

Before starting the game we must choose a pool of impetuses (fig. 3). We shall play three games, each with a different pool. One involves small impetuses (fig. 3(a)); they must be greater than 1, however, or they cannot penetrate any of the tesserae. A second game is played with a pool of very large impetuses (fig. 3(b)). Finally, there is a game from a pool in which most impetuses are small but a very few are large (fig. 3(c)). The spheres are obviously analogous to species. The impetuses are analogous to morphological distances. Clusters of spheres with little morphological distance represent genera or families; spheres that are very different morphologically represent different phyla.

We begin game 1 using pool 1, small impetuses only (fig. 4). By definition, the spheres that successfully colonize tesserae near their source are morphologically similar to their parent sphere. Tesserae distant from the

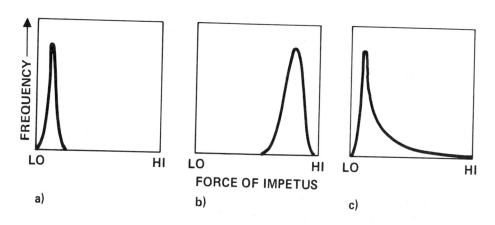

Figure 3. *Pools of impetuses for three phylogenetic games on a kaleidoscopic board. Impetuses represent morphogenetic events; small morphological changes result from low impetuses, large changes from high impetuses. Other pools of impetuses are, of course, possible; for example, impetuses may be normally distributed with an intermediate mode.*

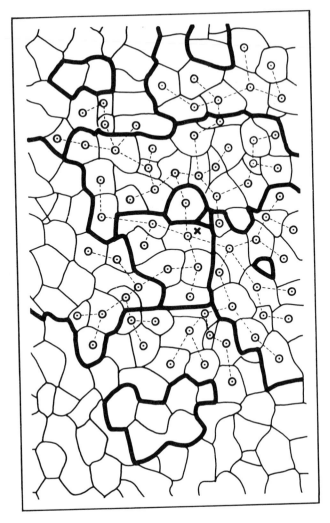

Figure 4. *Tesserae colonized by spheres (species) propelled by low impetuses (small morphological changes) only; game incomplete. The spreading pattern tends to be roughly radial from the first occupied tessera (marked with an X). Different branches will be somewhat distinctive as they inherit the peculiarities of their first divergent form. Since all the morphological changes are small, however, the groups would have a common general appearance.*

original locus can be reached only by the colonization of intervening tesserae since all impetuses are small; the occupied region of the board therefore spreads gradually about the source. Tesserae inhospitable to spheres from the parent locus may accept later ones originating from other loci. When the tesserae near the original parent fill up, the parent can no longer produce viable offspring spheres.

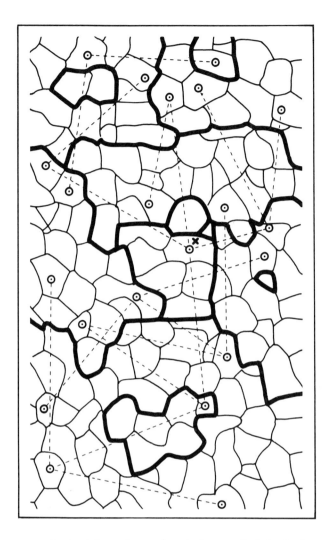

Figure 5. *Tesserae colonized by spheres (species) propelled by high impetuses (large morphological changes) only; game incomplete. Occupants of adjacent tesserae must be morphologically quite distinctive since at least two large morphogenetic events must separate them.*

Eventually, the spread of spheres will reach all edges of the board. The game would now be over except that, as the tesserae change kaleidoscopically, some spheres inevitably burst and some new tesserae appear. These unoccupied sites provide fresh opportunity for colonization. A balance is eventually reached between the rate of sphere production and the rate of extinction, establishing an equilibrium population size of spheres or one that oscillates with random accelerations and decelerations in rate of environmental change as the kaleidoscope operates. This recalls the equilibrium species

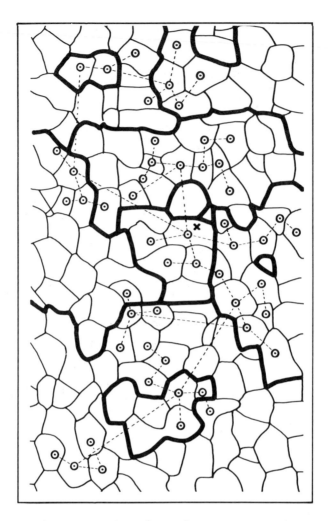

Figure 6. *Tesserae colonized by spheres (species) drawn from a skewed frequency distribution of impetuses, most being near the low end of the scale; game incomplete. Centers of spreading occur around novel morphologies that result from high impetuses, leading to distinctive groups of species. The groups often contain distinctive subgroups.*

diversity postulated in island biogeographic theory (MacArthur and Wilson, 1967). Morphologically, the entire population of the board is rather similar, since all the spheres are variations on the ancestral theme, with nearly all morphologies connected across small gaps only. A few larger gaps may be created by extinctions. Such an assemblage of morphologies might all be placed within a single phylum, with clusters of similar morphologies (classes, orders, and so forth) defined chiefly by extinction patterns. This is clearly not a close simulation of evolution on Earth.

For game 2 we use only large impetuses (fig. 5). Vast adaptive distances can be crossed rapidly by hopping across the board, many tesserae at a hop, so that the game board would soon contain a scattering of spheres. Gradually, the density of spheres would rise and the frequency of successful colonizations decline. As open tesserae became rare, they would be harder to fill than in the preceding game, for the target circumference of new spheres would be larger with large impetuses than with small. An equilibrium sphere number would eventually be achieved. Morphologically, a given sphere would be very different from its parents and even more so from its neighbors, especially from those originating in sources far from the source of the given sphere. In the extreme case, such an assemblage might have to be broken up into as many phyla as there are spheres. Obviously, this does not approximate evolution on Earth either.

Finally, we use the pool of impetuses that is strongly skewed so that most are small, a few are of intermediate strength, and a rare few are large (fig. 6). As play begins, the original source morphology begins to spread as in game 1, but a few different morphologies occasionally appear on outlying tesserae; in some rare cases, descendants from distant tesserae may appear back near the original source. Each very distinctive morphology becomes surrounded by a spreading zone occupied by spheres with minor modifications, making it increasingly difficult for the rare spheres from distant sources to find a suitable open tessera. Eventually, the board fills to an equilibrium occupation level. As vacant tesserae appear, their chances of being colonized from a distant source are vanishingly small; they might rarely be occupied from a source at an intermediate distance, but will most likely be filled from a closely neighboring source. To increase the colonization chances from a far or intermediate source, a large area of the board comprising many tesserae must be cleared of spheres. Then, although spreading from marginal tesserae will gradually fill the cleared region from the edges, the central part might well receive immigrant spheres from some distance. The larger the cleared patch, the more likely this would be.

Morphologically, the spheres could be divided hierarchically, first into clusters of spheres descended from the founding spheres with large impetuses; then into subclusters descended from separate intermediate-impetus spheres, daughters or granddaughters of the founding sphere; and then perhaps into tertiary clusters within these and quaternary clusters within those and so on. Clearly, among the games we have played, this most closely represents the hierarchical pattern of morphology on Earth.

Game 3 could be modified in a number of obvious ways to bring it into closer conformity with life history. For example, the density of tesserae that could be successfully colonized might be decreased with distance from each source tessera, or some spheres could be permitted to occupy more than one tessera at certain times or places, and so forth. However, the game rules already rival those of Monopoly, and such complications do little to clarify the basic questions of the diversification pattern of higher taxa.

Application to Life History on Earth

The adaptive kaleidoscope involves features that may not closely simulate life history. For example, the morphogenetic events analogous to the large impetuses of the model may not often or ever result from single mutations. They remain the most mysterious parts of metazoan history. Also the model implies that morphological and adaptive distances go hand in hand; this is clearly not the case. Nevertheless, the model does contain some features similar to those indicated by the fossil record, and thus it suggests interesting hypotheses.

The most striking aspect of game 3 is the large role played by chance in determining the morphological pattern of the phyla (and of lower taxa as well). A given tessera, or a given region of tesserae that has low-class boundaries internally but is demarcated by higher class boundaries, may be colonized by a descendant of any number of ancestral spheres; that is, it may support any number of body plans. The potential body architectures that can function in any given habitat on Earth are clearly very great. In the game, the first arrival that does not burst takes possession of a tessera. If the source tessera is somewhat remote, then the immigrant sphere must be rather novel. Empty nearby tesserae are most likely to be colonized by descendants of the novelty since minor novelties are so common. Therefore, the chance success of a major novelty will commonly result in the occupation by spreading of a large area of the gameboard by organisms with a similar body plan. Raup and Gould (1974) have demonstrated that patterns of random spreading of morphologies in situations much like this game result in quite plausible family trees. If the area had been colonized from a very different source, the major novelty of the founding lineage would have been very different, and the subsequent inhabitants of the area, deployed by spreading from this major novelty, would be quite different.

Thus, in the game, the key to body-type pattern is the chance of colonization, which is based on two factors: the arrival of an immigrant from a distance and the viability of the immigrant on the tessera. In nature, the latter is chiefly a matter of chance preadaptation, but microevolutionary processes, operating after colonization, may contribute significantly to the continuing success of the immigrant. The origin of the immigrant morphology is not related to the conditions at what becomes the occupied tessera. In addition to the morphological features that prove useful in the tessera, the immigrant may possess many other features that have no special relation to the particular conditions that distinguish the tessera or the neighborhood of the tessera from other places. These features may nevertheless contribute to the harmonious functioning of the organism at the time of colonization and may be retained or perhaps modified as the immigrant is engineered to conditions on the tessera by microevolution.

From the foregoing, it appears that prediction from first principles of the morphology of the occupant of any given tessera would not be possible, even if one had a knowledge of the pool of morphologies from which it must arise (assuming a reasonably large diversity of morphology within the pool). Consider a neighborhood of tesserae that represents infaunal burrowing in marine bottom sediments. Among living phyla, a wide variety of body plans have burrowing representatives. Most use a hydraulic system (usually associated with a hydrostatic skeleton) to burrow. Some Cnidaria (sea anemones) burrow by using their gut as a hydraulic skeleton. Some phyla burrow by puncturing the substrate with probelike organs, pulling themselves after (some sipunculids use probosci operated by coeloms; some bivalve mollusks use a foot operated by the haemocoel). Other phyla use peristaltic waves (some nematodes use a pseudo-coel; some annelids use a coelom). In some phyla, such as arthropods, burrowers use appendages, and still other methods exist.

Some of the morphologies associated with burrowing are elegantly engineered as burrowing machinery; annelids are a fine example (Clark, 1964). Others seem to be jury-rigged, with improvised structures and functions tacked onto basic morphologies adapted to some different habit altogether. The burrowing inarticulate brachiopods certainly convey this impression; although equipped with elongate posterior pedicles containing a coelomic cavity, they burrow anteriorly by a scissorslike motion of their shells, aided by setae for transporting sediment particles (Thayer and Steel-Petrovic, 1975). This is hardly an elegant approach to burrowing, yet the fossil record indicates that these forms developed their burrowing behavior early in the Phanerozoic and have maintained it for several hundred million years. A theoretical morphologist might predict that an annelidan sort of burrowing could evolve, but he would hardly predict the inarticulate system. There are several phyla that do not and presumably have not burrowed in any significant way (Chaetognatha and Bryozoa, for example). However, it is not safe to conclude that their body plans preclude a burrowing habit.

The convergence of distinctive morphologies on similar functions is not unique to burrowing functions; indeed it is a widespread phenomenon among living phyla. A single subtidal rock may support sessile suspension feeders belonging to several phyla and feeding on roughly similar resources. For example, Mollusca (mussels), Brachiopoda (articulates), Bryozoa, Arthropoda (barnacles), and Echinoderma (crinoids) have representatives with such a life mode. Of these, bryozoans and articulate brachiopods are elegantly adapted to this habit, as are crinoids; mussels were nicely preadapted, while barnacles have passed along a tortuous adaptive pathway. One would hardly predict a barnacle as an outcome of arthropod radiation, even though they are perfectly plausible. The potential modes of life available to primitive arthropods, given modifications extensive enough to develop barnacles, are so broad as to preclude the prediction of any special one that might be realized.

Clearly, many important functions can be realized by animals with any of a wide array of body plans. The model of game 1, filling the board gradually in small morphological steps so that the entire planetary fauna has in the end the same basic body plan, seems plausible. It requires an evolutionary process that does not produce large viable morphological jumps. The modification of a body plan to permit life in most tesserae would seem to present no special problem, although, of course, there may be internal constraints in some cases.

The model of game 2, filling the adaptive landscape with forms that diverge by large morphological jumps only, seems plausible also. There is certainly no unique fit of body plan to tessera, and if there are general limits on modification to body plans, they have not yet been defined.

The number of major ground plans realized on Earth as viable animals is not known. A reasonable guess might be between 40 and a few hundred on the level of the phylum. An important regulator of this number is obviously the frequency distribution of the distinctiveness of morphological novelties. The number of phyla we have had reflects this distribution as it existed in Late Precambrian and Lower Cambrian times. Another possible regulator of phylum number is the rapidity and extent of environmental change, measured against the effectiveness of microevolution in maintaining adaptation. If environmental change is so rapid as to cause large-scale extinctions, then large areas of the adaptive landscape may become available for reoccupation. A phylum with low diversity may become extinct. The probability of appearance of new phyla in the vacated tesserae then depends on the size of the vacant area and the frequency distribution of novel morphological distance. On Earth we have lost a number of phyla to extinction. New classes or, more commonly, new orders have appeared, presumably filling tesserae vacated by extinction, but it seems that new phyla have rarely, if ever, been evolved after the Cambrian. This does not necessarily indicate that large regions of the adaptive landscape were never emptied by extinction. It may mean, in part, that lineages from which major new body plans could be given off became swamped by an increasing number of lineages that could invade open regions of the adaptive landscape with less significant modification, recognized not as phyla but as lower taxa (Valentine, 1973). Even the invasion of the terrestrial environment failed to produce a novel phylum. The probability of a new phylum arising probably becomes lower as the old phyla diversify.

SEARCH FOR THE GENERAL CASE OF MULTICELLULAR EVOLUTION

Reruns

From the adaptive kaleidoscope game, one can gain insight into general features of the diversification of multicellular animal-like organisms. One approach is to imagine multiple runs of each game. Figure 7 depicts typical life trees resulting from each game. Reruns would resemble each other more

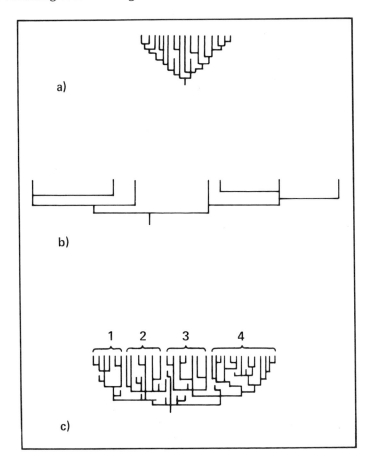

Figure 7. *Highly diagrammatic representations of phylogenetic trees resulting from kaleidoscopic games depicted in figures 4, 5, and 6. (A) Low-impetus pool (fig. 4) results in a compact spreading pattern. (B) High-impetus pool (fig. 5) results in an open spreading pattern (only a few orientations are shown because of the limitations of space and dimensionality). (C) Skewed impetus pool (fig. 6) results in a hierarchical pattern. The number of impetus events is underrepresented in the low (early) portion of the figure to show a few high-impetus events, otherwise difficult due to limitations of space and dimensionality.*

closely in game 1 (fig. 7(a)) than in the others. First, the morphology of the organisms, arising gradually from a similar unicellular ancestor and diverging gradually as descendant lineages spread outward from the ancestral tessera, would not diverge as much from the simple multicellular forms. Second, although details would vary, the usual pattern of morphological variation would be similar: irregularly concentric with respect to the ancestor. More extreme cases would arise when descendants happened to be given off chiefly in one direction, producing a chain of morphologies extending from the founding ancestor and snaking across the gameboard. The end member of the chain would be more different from the founding member than in other spreading patterns; if extinction eliminated segments of the chain, one might recognize two or more phyla instead of just one.

In game 2 (fig. 7(b)), with only major morphological jumps, the pattern of morphological variation would also tend to be similar from run to run, with great heterogeneity among morphologies in any region but with the average degree of divergence increasing away from the founding tessera. Extreme spreading patterns — for example, radial until the board margin is reached, then counterradial back to the inner tesserae, then radial again, and so on — would create different patterns of divergence, but these would be rare. A richness of morphology would appear in all games. Occasional close similarities of body plans in two different games would not be surprising, but the arrays of morphologies in each game might be quite distinctive. It is likely that in separate games the body plans of the occupants of any particular tessera would be derived from different ancestral body plans, even though they would display similar functions. Similarities in results from reruns of game 2 would be greater than expected if strong constraints were placed on the colonization pattern of a given lineage by its morphology. Some such constraints must exist, but whether they would force a great convergence among successive runs is uncertain. This is an important problem.

Game 3 (fig. 7(c)) would also display a similarity of patterns from run to run: a hierarchy of morphologies would appear, varying with the frequency of large vs small branching events (and capable of some extreme configurations). The actual body plans representing the major morphological branches would usually differ greatly from game to game. The major branches that arise earliest would be more likely than later ones to resemble each other in repeated games. The rate of major branching in time should resemble the classic S-shaped logistic curve, but it should be compressed into a short time interval early in the game (see Sepkoyski, 1978). In this respect, early game 3 runs resemble game 2 runs. When the tesserae finally fill up, so that those opened by extinction are chiefly filled from neighboring tesserae, game 3 runs resemble game 1 runs. The occasional large patches of tesserae cleared by mass extinction permit game 3 to develop its characteristic family tree at lower taxonomic levels.

The richness of variation possible within such trees is indicated by the clade simulations of Raup et al. (1973), in which splitting, continuation, and termination of lineages are determined as random choices but at fixed rates. The patterns of growth and diminution of diversity within these simulated clades resemble patterns documented by fossils in actual clades. Here we are interested in the pattern of distinctness among the branches. The characteristic tree of game 3 (fig. 7(c)) is most like the pattern documented by fossils, and it would display a richness in detailed variation in reruns.

Life Games on Other Worlds

In the kaleidoscope games, the quality of macroevolution is the most important factor in structuring the morphological variation. It is probably a safe assumption that macroevolution is a universal feature of life. Macroevolutionary factors should affect multicellular diversification patterns on other worlds, much as they do on Earth in one or another kaleidoscope game, although the results would, of course, vary with environmental conditions. Microevolution, on the other hand, might be an entirely different matter elsewhere. It would certainly be different in detail. We could not expect the same enzyme systems and probably not the same genetic materials or cell behaviors as we find on Earth. Meiosis as we know it would not be likely to develop in another life system. Earthly microevolution is closely tied to the planet's particular genetic system.

The pattern of morphological variation of animal-like organisms on other worlds can run the gamut from the most homogeneous sort resulting from game 1 to the most heterogeneous and chaotic sort resulting from game 2. The character of the morphologies is not strictly predictable. Certain qualities may be likely. Cellular differentiation would seem to be a prerequisite of any true complexity on the multicellular level. Complexity is a property that usually develops when evolving organisms are faced with a series of environmental challenges. It is often easier to modify by adding something to an established structure, thus producing a complex elaboration, than by redesigning the structure from a simpler developmental stage. Also, given a morphology, there are more potential variants of it that are complex than that are simple.

Wherever we find it, complexity seems always to be organized hierarchically (Simon, 1962). Therefore we might expect complex morphologies to be based on a hierarchical organization, with cells or other such modules at the base, and higher levels analogous to tissues, organs, and organ systems. Adaptations for maintaining a spatial place in nature — locomotory, sessile, or floating adaptations — have shaped the body plans of most of Earth's phyla, and presumably it would be much the same elsewhere. Nearly all such body plans require a contractile system; movement is essential.

Although it is amusing and not too difficult to continue in this vein, listing the attributes required of presentable multicellular organisms, it turns out that one is simply tabulating the basic features of multicellular animals here on Earth. They form a logical and elegant assemblage that excites admiration for evolution and gives aesthetic pleasure in its results. One can only hope that the generality of their features can someday be tested against another life system.

ACKNOWLEDGMENT

I thank Dr. S. Conway Moins, Open University, for a discussion of the manuscript and for data on the fossil record of certain phyla, and Dr. C. A. Campbell for reviewing the manuscript.

REFERENCES

Ayala, F. J.: An Evolutionary Dilemma; Fitness of Genotypes Versus Fitness of Populations. Canadian J. Genet. Cytol., vol. 11, 1969, pp. 439–456.

Borovikov, L. I.: First Find of Fossil Dickinsonias in Lower Cambrian Sediments in the USSR. Dokl. Akad. Nauk. SSSR, vol. 231, 1976, pp. 1182–1184.

Britten, Roy J.; and Davidson, Eric H.: Gene Regulation for Higher Cells: A Theory. Science, vol. 165, July 1969, pp. 349–357.

Britten, Roy J.; and Davidson, Eric H.: Repetitive and Non-Repetitive DNA Sequences and a Speculation on the Origins of Evolutionary Novelty. Quarterly Rev. Biol., vol. 46, 1971, pp. 111–133.

Bush, G. L.: Modes of Animal Speciation. Ann. Rev. Ecology Systematics, vol. 6, 1975, pp. 339–364.

Bush, G. L.; Case, S. M.; Wilson, A. C.; and Patton, J. L.: Rapid Speciation and Chromosomal Evolution in Mammals. Proc. Natl. Acad. Sci. USA, vol. 74, Sept. 1977, pp. 3942–3946.

Clark, Robert B.: Dynamics in Metazoan Evolution; The Origin of the Coelom and Segments. Clarendon Press, Oxford, 1964.

Cloud, P. E., Jr.: Some Problems and Patterns Exemplified by Fossil Invertebrates. Evolution, vol. 2, no. 4, 1948, pp. 351–374.

Conway Moins, Simon; and Whittington, H. B.: The Animals of the Burgess Shale. Sci. American, vol. 241, no. 1, July 1979, pp. 122–133.

Crimes, T. P.: Colonisation of the Early Ocean Floor. Nature, vol. 248, March 22, 1974, pp. 328–330.

Dobzhansky, Theodosius: Genetics of the Evolutionary Process. Columbia Univ. Press, N.Y., 1971.

Dobzhansky, Theodosius; Ayala, F. J.; Stebbins, G. L.; and Valentine, J. W.: Evolution. W. H. Freeman, San Francisco, 1977.

Durham, J. W.: The Fossil Record and the Origin of the Deuterostomia. Proceedings North American Paleontol. Convention, Chicago, IL, Allen Press, Lawrence, Kansas, 1971, pp. 1104–1131.

Eldredge, N.; and Gould, S. J.: Punctuated Equilibria: An Alternative to Phyletic Gradualism. In: Models in Paleobiology, Thomas J. M. Schopf, ed., W. H. Freeman, San Francisco, 1972, pp. 82–115.

Glaessner, Martin F.: Geographic Distribution and Time Range of the Ediacara Precambrian Fauna. Bull. Geol. Soc. Am., vol. 82, Feb. 1971, pp. 509–514.

Hyman, L. H.: The Invertebrates: Protozoa Through Ctenophora. McGraw-Hill, New York, 1940.

King, Mary-Claire; and Wilson, A. C.: Evolution at Two Levels in Humans and Chimpanzees. Science, vol. 188, April 1975, pp. 107–116.

MacArthur, R. H.; and Wilson, E. O.: The Theory of Island Biogeography. Princeton Univ. Press, 1967.

Matsuda, R.: Abnormal Metamorphosis and Arthropod Evolution. In: Arthropod Phylogeny, A. D. Gupta, ed., Van Nostrand, N.Y., 1978, pp. 137–256.

Matthews, S. C.; and Missarzhevsky, V. V.: Small Shelly Fossils of Late Precambrian and Early Cambrian Age: Review of Recent Work. J. Geol. Soc. London, vol. 131, May 1975, pp. 289–304.

Mayr, Ernst: Systematics and the Origin of Species. Columbia Univ. Press., N.Y., 1942.

Mayr, Ernst: Animal Species and Evolution. Harvard Univ. Press, 1963.

McCoy, J. W.: The Origin of the "Adaptive Landscape" Concept. Am. Naturalist, vol. 113, 1979, pp. 610–613.

Raup, D. M.; and Gould, S. J.: Stochastic Simulation and Evolution of Morphology — Towards a Nomothetic Paleontology. Syst. Zool., vol. 23, 1974, pp. 305–322.

Raup, David M.; Gould, Stephen J.; Schopf, Thomas J. M.; and Simberloff, Daniel S.: Stochastic Models of Phylogeny and the Evolution of Diversity. J. Geol., vol. 81, Sept. 1973, pp. 525–542.

Repetski, John E.: A Fish from the Upper Cambrian of North America. Science, vol. 200, May 5, 1978, pp. 529–531.

Richardson, Eugene S., Jr.: Wormlike Fossil from the Pennsylvanian of Illinois. Science, vol. 151, Jan. 7, 1966, pp. 75–76.

Sepkoyski, J. J., Jr.: A Kinetic Model of Phanerozoic Taxonomic Diversity. I. Analysis of Marine Orders. Paleobiology, vol. 4, 1978, pp. 223–251.

Simon, H. A.: The Architecture of Complexity. Proc. Am. Phil. Soc., vol. 106, 1962, pp. 467–482.

Simpson, George, G.: Tempo and Mode in Evolution. Columbia Univ. Press, 1944.

Simpson, George G.: The Major Features of Evolution. Columbia Univ. Press, 1953.

Stanley, S. M.: A Theory of Evolution Above the Species Level. Proc. Natl. Acad. Sci. USA, vol. 72, Feb. 1975, pp. 646–650.

Stanley, Steven M.: Stability of Species in Geologic Time. Science, vol. 192, April 1976, pp. 267–269.

Stanley, Steven M.: Chronospecies' Longevities, the Origin of Genera, and the Punctuational Model of Evolution. Paleobiology, vol. 4, 1978, pp. 26–40.

Stebbins, G. L.: Integration of Development and Evolutionary Progress. In: Population Biology and Evolution, R. C. Lewontin, ed., Syracuse Univ. Press, 1968, pp. 17–36.

Thayer, Charles W.; and Steel-Petrovic, H. Miriam: Burrowing of the Lingulid Brachiopod *Glottidia pyramidata:* Its Ecologic and Paleoecologic Significance. Lethaia, vol. 8, 1975, pp. 209–221.

Tiegs, O. W.; and Manton, S. M.: The Evolution of the Arthropoda. Biol. Rev., vol. 33, 1958, pp. 255–337.

Valentine, James W.: Evolutionary Paleoecology of the Marine Biosphere. Prentice-Hall, Englewood Cliffs, N.J., 1973.

Valentine, James W.: General Patterns of Metazoan Evolution. In: Patterns of Evolution as Illustrated by the Fossil Record, A. Hallam, ed., Elsevier, Amsterdam, 1977.

Valentine, James W.; and Campbell, Cathryn A.: Genetic Regulation and the Fossil Record. Am. Sci., vol. 63, Nov.-Dec. 1975, pp. 673–680.

Wilson, A. C.: Evolutionary Importance of Gene Regulation. Stadler Symp., vol. 7, 1975, pp. 117–133.

Speculations on the Evolution of Intelligence in Multicellular Organisms

DALE A. RUSSELL

The abundance of inorganic nutrients is postulated to have a major effect on rates of encephalization; the origin of life is therefore a much more probable event than is the origin of higher intelligence. Further human encephalization may well occur.

The edifice we call science is built of bricks contributed by many, many workers, laid according to a design too large to be comprehended by any one reviewer. — *Virginia Trimble*

During the past two decades of my life I have been fortunate to study the skeletal remains of the great reptiles, popularly known as dinosaurs, which once abounded on the surface of our planet. Until recently, these fascinating creatures seemed to me to be utterly removed from the professional concerns of my colleagues who probe the stuff of the Universe using tools of electromagnetic radiation. One cloudy August day, a western Canadian rancher led me to the fragmented bones of a small dinosaur that died in Alberta 75 million years ago. The creature evidently walked on its hind legs and possessed three-fingered hands in which the outer finger closed against the other two like a thumb. Its eyes were large and directed toward the front of the skull, suggesting a stereoscopic field of vision. Most significantly, the cerebral hemispheres of the animal were enlarged, exceeding those of any living reptile in relative size and equaling those of some living mammals. This small dinosaur was a manifestation of the widespread tendency of animate organisms to become more intelligent through geologic time.

An understanding of the factors that can affect the evolution of intelligence is of obvious importance in evaluating the Cosmos as a home for intelligent organisms. Metazoan creatures now living on the surface of our planet,
as well as the fossil record of life preserved within its crust, must reflect the
effects of these factors. Unfortunately, the existing body of relevant information is both imprecise and incomplete. Because of this and because of my
own limitations, the remarks that follow necessarily contain a large speculative component. It is my belief, however, that they indicate a new area of
interdisciplinary research that is ripe for exploitation.

We shall begin with a working biological definition of intelligence. The
highest levels of intelligence attained by organisms through geologic time follow a pattern which suggests that environmental conditions affect the rate of
increase of intelligence. Ten factors will be listed which may be correlated
with intelligence, as deduced from the geographic and temporal distributions
of intelligent organisms. These factors may in turn serve as a basis for evaluating the incidence of human or greater levels of intelligence within the
Cosmos.

The brain is the organ of intelligence. It becomes more differentiated
and increases in relative size as one ascends the *scala naturae.* The mass of
neural tissue in the brain is generally related to the amount of information it
can process (Jerison, 1973, 1975a). Much of this information is related to
body function. Succinctly put, larger organisms have larger organs, which
contain more cells, and which must, in turn, be regulated by a larger number
of nerve cells (Jerison, 1973). Brain size is related to body size according to

$$E = kP^{2/3}$$

where E is brain weight (in cgs units) and P is body weight (Jerison, 1973).
It is reasonable to assume that the relative intelligence of an organism is
proportional to the amount of brain tissue in excess of that necessary for
body function. This is measured by the proportionality constant k in the
foregoing equation. If, after removal of the body-related factor, the brain of
one species of organism contains a residue of nervous tissue twice that of
another species, it may be considered to have twice as much neural tissue
available for sensory integration and behavioral responses. The first organism
would thus be twice as encephalized and, for the purposes of the present
discussion, would be considered twice as intelligent. This postulate is supported by the work of Riddell and Corl (1977), who have documented in
mammals a direct relationship between the number of neurons present in the
brain in excess of the body component and learning ability. There is evidence of a correlation between brain size and intelligence even within our
own species (Van Valen, 1974). These concepts are reviewed in the publications of Jerison (1973, 1975a, 1976) and Passingham (1975a).

MAXIMUM ENCEPHALIZATION RATES

The intelligence sought in the search for extraterrestrial intelligence (SETI) is of a human or superior level (Morrison et al., 1977). On Earth, more than 700 million years elapsed between the appearance of multicellular metazoans and *Homo sapiens*. In order to discover to what extent encephalization rates may have changed during this interval, it may be useful to examine the distribution of the highest levels of encephalization through time. Previous authors have taken the average level of encephalization of living mammals as a standard for comparison. Here the level of encephalization of lower invertebrates, such as annelid worms, is taken as unity.

The comparability of encephalization indices derived from the entire range of animate organisms could be questioned. In cephalopods such as squid and octopi, relative brain size and learning ability seem to be related in the same manner as they are in backboned animals. However, it is virtually impossible to find published brain vs body weight data for cephalopods, and cerebral indices have never, to the best of my knowledge, been compared to learning ability in this group. Parenthetically, it would also be useful to know if an as yet undetected relationship exists between the size of the body component of the brain and metabolic rate in organisms. The most highly encephalized creatures on Earth have consistently been backboned animals since their appearance about 500 million years ago, thereby diminishing somewhat the uncertainties of cross-phyletic comparisons. The highest relative levels of encephalization known for various times are listed in table 1. The data are plotted on logarithmic coordinates in figure 1.

Several unanticipated inferences can be drawn from these data:

1. The increase in encephalization rates leading toward man is in accord with a trend established over 200 million years ago.

2. The increase in maximum encephalization rates was altered about 230 million years ago (in temporal proximity to the Permo-Triassic boundary crisis).

3. According to the older rates of increase, a human level of encephalization would have been attained 60 million years ago.

4. According to the younger rates, a span of 20 billion years would have been necessary to attain a human level of encephalization from that characteristic of primitive metazoans.

Were it not for the initial, relatively higher increase in encephalization rates, which followed the appearance of multicellular metazoans, humanoid intelligence could not have evolved on Earth. The earlier points were based on data drawn from marine organisms, suggesting that between 700 and 230 million years ago the oceans provided a more favorable environment for encephalization than did subsequent continental environments. These later

TABLE 1.– RELATIVE LEVEL OF ENCEPHALIZATION THROUGH
GEOLOGIC TIME OF MOST ENCEPHALIZED ORGANISM KNOWN

Encephalization level (EL)	Approximate increase from reference EL	Approxi-mate rela-tive EL	Approximate age of oldest record, millions of years
1. Insects, annelids, primitive mollusks (Crile and Quiring, 1940, p. 221).	---	1	700 (Runnegar, 1977).
2. Level of lower vertebrates (earliest jawed fishes: Jerison, 1973, p. 136).	4 X level 1	4	400 (Thomson, 1971; Van Eysinga, 1975).
3. Mammal-like reptiles (*Diademodon,* Jerison 1973, p. 153).	3 X level 2	12	225 (Kitching, 1977; Retallack, 1977; Van Eysinga, 1975).
4. Primitive mammals (*Triconodon:* Jerison, 1973, pp. 213–214; 1975a, p. 403).	4 X level 2	16	150 (Jerison, 1973; Van Eysinga, 1975).
5. Advanced dinosaurs (*Stenonychosaurus:* Russell, 1969).	5 X level 2	20	75 (Russell, 1969; Gill and Cobban, 1973).
6. Archaic primate (*Plesia-dapis:* Radinsky, 1975a).	0.47 X level 9	19	55 (Radinsky, 1975a; Berggren et al., 1978).
7. Ancient ungulate (*Heptodon:* Radinsky, 1978, p. 822).	0.6 X level 9	24	50 (Radinsky, 1978).
8. Ancient primate (*Necrolemur:* Radinsky, 1975a, p. 84).	0.79 X level 9	32	40 (Radinsky, 1978).
9. Living mammals (*Rooneya:* Radinsky, 1975a, p. 85).	10 X level 2	40	35 (Radinsky, 1975b; Berggren et al., 1978).
10. Primitive anthropoid apes (*Aegyptopithecus:* cf. Radinsky, 1975a, p. 85).	1.24 X level 9	50	27 (Radinsky, 1975b).
11. Anthropoid apes (*Dryopithecus:* Radinsky, 1974, p. 21).	1.91 X level 9	76	18 (Radinsky, 1974).

Table 1.— CONCLUDED

Encephalization level (EL)	Approximate increase from reference EL	Approximate relative EL	Approximate age of oldest record, millions of years
12. Primitive hominid (*Australopithecus:* Jerison, 1975b, p. 45).	3.39 X level 9	136	5 (Tobias, 1973).
13. Primitive *Homo* (*H. habilis:* Jerison, 1975b, fig. 1).	1.3 X level 12	177	2.3 (Johansen and White, 1979).
14. *Homo erectus* (Jerison, 1975b, fig. 1).	1.8 X level 12	245	1.5 (Johansen and White, 1979).
15. *Homo sapiens* (Jerison, 1975b, fig. 1).	2.6 X level 12	354	0.25 (Johansen and White, 1979).

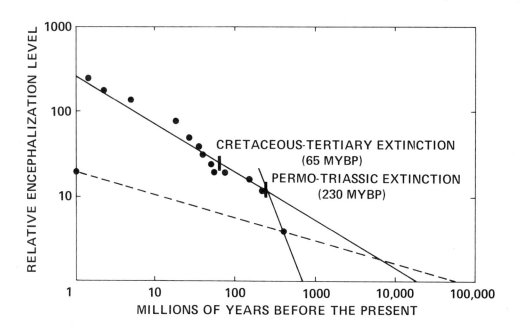

Figure 1. *Maximum encephalization levels through geologic time (see table 1). The dotted line is drawn through the highest encephalization levels attained by modern sharks (cf. Crile and Quiring, 1940; Bauchot et al., 1976) and of primitive jawed fishes at the time of their appearance.*

terrestrial environments were in turn more hospitable sites for encephalization than were contemporaneous oceans (see dotted line in fig. 1).

The foregoing trends are difficult to explain. They parallel in an interesting manner organic diversity trends plotted for the same general span of
time (Valentine, 1972, fig. 10-3). For the purpose of stimulating discussion,
it is suggested that the establishment of terrestrial ecosystems during the
latter part of the earlier period may have impaired the flow of terrestrial
nutrients to the oceans, producing the observed pattern of encephalization
trends as well as a net lowering of rates of maximum increase of encephalization on the planet.

BIOLOGY OF ENCEPHALIZATION

The ecology and biogeography of encephalization have not yet been
studied systematically. Available information is still much too incomplete
and too poorly defined to support statistical analyses. Accordingly, the following factors, which may be related to level of encephalization (intelligence), are more the result of my own fallible intuition than of an existing
consensus among scholars. They are listed in order of presumed effectiveness
during the course of development of life on Earth:

Temperature

Mean annual temperatures on the surface of Earth generally increase
from the poles toward the equator. This is accompanied by an increase in the
abundance of highly encephalized fishes (Bauchot et al., 1977) and reptiles
(Hopson, 1977) as well as primates, among mammals. Learning in poikilo-
thermic lizards becomes more rapid as body temperatures approach the
range 35° to 40°C (Brattstrom, 1978, p. 174). However, brain malfunction
and death occur in both cold- and warm-blooded vertebrates above 40°C
(Wheeler, 1978), suggesting a general thermal optimum in the range
25°–35°C.

Time

Encephalization is an attribute that does not evolve rapidly. In the
Hominidae, the most rapidly encephalizing phylum known, the average level
of encephalization has doubled during the last 2 million years (Jerison,
1975b; Johanson and White, 1979). In mammals, the average level has only
doubled during the last 45 million years (Radinsky, 1978). Encephalization,

although apparent in many other groups of organisms, generally proceeds at much slower rates (Jerison, 1973).

Biogenic Environmental Diversity

It is axiomatic that complicated responses tend to be required of organisms inhabiting complex environments. The rates at which organic processes occur should also affect the rapidity and precision of behavioral responses. Greater plant diversity increases competitive interactions, enhancing selection for rapid growth strategies (Regal, 1977) and thereby plant productivity. Thus biogenic environmental diversity should produce high rates of encephalization (Regal, 1978). It has been suggested that the need to locate packets of energy-rich foods that are erratically distributed in time and space has produced higher levels of encephalization in fruit-eating than in insect-eating bats (Regal, 1978; Eisenberg and Wilson, 1978). Factors that depress diversity, such as environmental instability, aridity, unavailability of essential elements, and ionizing radiation (Whittaker and Woodwell, 1971; Valentine, 1972; Parker, 1976; Connell, 1978; May, 1978), may also depress rates of encephalization. As a possible example of the latter phenomenon, marine organisms have generally not achieved levels of encephalization comparable to those of their more highly encephalized terrestrial counterparts. The most highly encephalized marine organisms are birds and mammals that reinvaded the seas from the land, indicating that a niche for intelligent creatures was present there but underexploited. Neither are marine environments generally as productive as terrestrial environments (Rodin et al., 1975), perhaps partly as a result of the capture of nutrients derived from subaerial erosion by terrestrial ecosystems (cf. Tappan, 1968, pp. 195–197). Could the lower levels of encephalization of marine organisms be in part a consequence of long-term (>100 million years) suboptimal nutrient availability?

Area

South America and Australia were isolated from other large land masses during much of the past 60 million years. Indigenous land-dwelling mammals showed little increase in encephalization (Jerison, 1973, pp. 320–339). It has been postulated (Andrews in Jerison, 1975a, p. 405) that this was due to very low levels of biotic interchange with vast, intermittently connected land areas of great environmental diversity. The primitive nature of the terrestrial biota of New Zealand and New Caledonia (Raven and Axelrod, 1972), isolated on small land masses for 80 million years, may be cited in support of this suggestion. A relatively well-studied, direct relationship exists between area and biotic diversity (May, 1978).

Mass Extinctions

A brief but acute environmental crisis precipitated the sudden extinction of the dinosaurs about 65 million years ago, together with more than half the species of organisms then in existence (Russell, 1979). Consequently, the mean level of encephalization in terrestrial vertebrates may have effectively remained stable for 10 million years (table 1) until the basic diversification of replacing mammalian groups was completed. The effects of previous mass extinctions are not clear, although temporary reductions in biogenic environmental diversity would presumably have had a negative effect on encephalization rates.

Adaptation

Organisms that defend themselves by means of static defense mechanisms, such as armor, spines, or poison, tend to be poorly encephalized. The coincidence has been noted in bony fishes (Bauchot et al., 1977) and in dinosaurs (Hopson, 1977). Sessile creatures such as corals and oysters possess relatively simple nervous systems (Bullock and Horridge, 1965), and most vascular plants are devoid of organs analogous to nerves in function.

Trophic Level

Active, foraging creatures tend to occupy relatively higher positions in the food chain and to be relatively more encephalized. This has been noted in invertebrates (Bullock and Horridge, 1965), sharks (Okada et al., 1969; Bauchot et al., 1976), lizards (Regal, 1978), and dinosaurs (Hopson, 1977). Avian and mammalian herbivores are approximately as encephalized as carnivores within their respective classes (cf. data in Crile and Quiring, 1940; Radinsky, 1978), possibly because of their active, evasive, or aggressive strategies of defense.

Metabolism

In man, the brain is an expensive organ to maintain, requiring 20% of the energy consumed by the body when resting (Sokoloff, 1976). This fraction approaches the overall metabolic needs of active reptiles of comparable body size (fig. 4 in Farlow, 1976). A large brain is therefore incompatible with a reptilian metabolism. The highest levels of encephalization occur in birds and mammals, where an enlarged brain is supported by high metabolic rates. These organisms have colonized the cooling, diminishingly productive polar regions in the same manner as they invaded the sea, with no apparent effect on their levels of encephalization (data in Portmann, 1947a; Crile and Quiring, 1940). Some highly encephalized predatory sharks also have relatively high metabolic rates (Stevens and Neill, 1978).

Dimensional Complexity

The physical milieu forms an element of the environmental complexity of an organism's habitat, as well as influencing the speed at which an animate creature moves. This is exemplified by a supplementary neurophysiological capacity in flying insects and birds (Jerison, 1973, pp. 156–161). Pelagic, free-swimming sharks (Okada et al., 1969) and bony fishes (data in Tuge et al., 1968; Bauchot et al., 1977) are more highly encephalized than their counterparts inhabiting essentially two-dimensional environments on the sea floor. However, pelagic cephalopods are not more highly encephalized than benthonic cephalopods (data in Wirz, 1959), possibly because their habitat is more complicated than the less dimensionally restricted environment preferred by their nektonic relatives.

Parental Care

In relatively highly encephalized organisms, a large proportion of the brain contains information that is derived from experience rather than inherited. The young of highly encephalized birds often require a correspondingly greater investment of parental care (Portmann, 1947b). Our own species may be cited as an example of this phenomenon.

CONCLUSIONS

What planetary conditions would most effectively promote the development of humanoid intelligence from primitive multicellular metazoans? Earth has probably not been an ideal nursery for intelligent organisms. However, the general environmental framework of the planet should resemble that of Earth during the past half-billion years. Inorganic nutrients should be in sufficient abundance to sustain high encephalization rates in the oceans after the establishment of terrestrial ecosystems. The thermal regime should be such that the largest possible area of the planetary surface enjoys mean annual temperatures of about 30°C. Rotational rates must be neither so rapid that steep latitudinal thermal gradients are produced (Hunt, 1979), nor so slow that diurnal thermal fluctuations become extreme. Epicontinental seas should be shallow and extensive to facilitate the growth of biogenically and dimensionally complex environments. Seasonality should be minimal, and the rotational poles nearly vertical to the planetary orbit. Planetary size should be the maximum consistent with long-term environmental stability. Given these conditions, it would not be surprising if octopoid creatures (cf. Bullock and Horridge, 1965, p. 1434), with humanoid levels of encephalization and the capacity to manipulate objects, were derived from primitive multicellular metazoans in about 500 million years. These creatures could be expected to have relatively high metabolic rates, low reproductive rates, and long life spans, much of which would be invested in caring for their young.

The oceans, deserts, and polar regions of Earth may not have produced organisms of a high level of encephalization. Rather, they have been invaded by highly encephalized creatures that evolved in biologically more productive terrestrial environments. Thus the process of encephalization, and accompanying physiologic homeostasis, may tend to free organisms from constraints that limit the distributions of their less encephalized predecessors.

In this general context, it might be useful to consider briefly the future of our own species. A great range of possible social behaviors and a rapidly growing technological capacity have generated an undercurrent of pessimism concerning the probability of long-term human survival. Whether these behavioral contradictions are typical only of man or are endemic to any organism that attains a humanoid level of intelligence is a moot point. It is conceivable, however, that behavioral selective pressures will operate in a manner such that imminent human extermination will be avoided. A trend toward improvement of brain function, manifested in both brain volume and organization, would continue, although at times organization may change more rapidly than size, and conversely. Assuming that the curve in figure 1 usefully describes encephalization increase over such a short interval of time, it can be crudely projected to a point 900,000 years in the future (fig. 2) by adding this figure to each of the age records tabulated in table 1.

At such a time, our brain would exceed its present volume by a factor of about 3, and life spans would nearly double (Sacher, 1978). The upper part of the vertebral column would be rather more powerfully constructed to support the additional weight of the head. The young would be born after a shorter term and in a more immature state neurologically, so that the larger head could more easily pass through the somewhat wider pelvic canal (cf. Sacher and Staffeldt, 1974; Passingham, 1975b). The enlarged brain could consume nearly 1/3 the total adult daily energy requirements (cf. Tuttle and Schottelius, 1965, pp. 340, 342, 510), implying that present metabolic rates would then be taxed to maintain bodily functions. Nevertheless, a brain growth of this magnitude does not seem physiologically impossible.

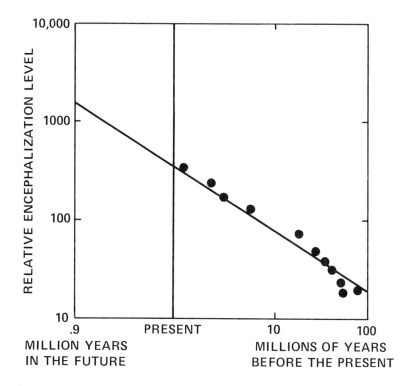

Figure 2. *Maximum encephalization level projected 900,000 years into the future. The projection intersects the ordinate at a relative encephalization level of about 1,100, representing an increase of three in brain weight over the level typical of* Homo sapiens, *but no increase in body weight. For further explanation see text.*

When human brain size merely doubled during the last 2 million years, man acquired language, libraries, and a technology that has enabled him to begin to explore the planets. The projected increase in brain size could not unreasonably be postulated to produce an analogous expansion in human mental abilities. Compared to the nearly 4-billion-year record of life on Earth, the next 900,000 years is not a large span of time. Until recently, the terrestrial biosphere was assumed to be "closed" in the sense that it would permanently be confined to the surface of Earth. There it would eventually become simpler and vanish as the Sun entered its red giant phase. If man is already able to establish ecologically self-sustaining colonies in space, as O'Neill (1976) has suggested, then the biosphere of Earth will become "open" in the sense that its ultimate fate need not be linked to that of the Sun. It is difficult to imagine why creatures of human or superior levels of intelligence would not rapidly colonize stellar regions close to the planetary systems within which they originated. The SETI program need not be limited to stars that are as stable or as old as the Sun.

The physiological and astrophysical implications of the generalizations tentatively proposed here far exceed my area of competence (cf. Carr and Rees, 1979). In a mysterious manner, metazoan encephalization rates even on Earth have during certain periods been such that, if sustained, humanoid levels of intelligence could only appear on a time scale comparable to the age of the Universe. It would seem that the origin of life is intrinsically a much more probable event than the origin of higher intelligence.

ACKNOWLEDGMENTS

The memory of assistance generously given by others is an enduring source of pleasure. I am, accordingly, fortunate to express my gratitude to my colleagues at the National Museum of Natural Sciences (Canada) for their help in this project. Drs. Donald E. McAllister and Henri Ouellet provided valuable material on relative brain size in fishes and birds, respectively. The manuscript has been much improved through the constructive comments of Dr. Pierre Béland. Within the larger academic community of Ottawa, Dr. Anthony Devine, M.D., supplied interesting information relating to the metabolic requirements of the human brain, as did Dr. Nancy M. McAllister with regard to those of lower vertebrates. I have also benefited from the perspectives of Drs. E. O. Dodson of the University of Ottawa and R. E. Morlan of the National Museum of Man. Drs. H. J. Jerison, of the University of California at Los Angeles, and L. B. Radinsky, of the University of Chicago, reviewed the manuscript in light of their extensive paleoneurological experience. I am very grateful to them. Finally, and particularly, I am grateful to my wife, Dr. Janice Alberti Russell, for many hours shared discussing the human ramifications of the biology of intelligence.

REFERENCES

Bauchot, R.; Bauchot, M. L.; Platel, R.; and Ridet, J.-M.: Brains of Hawaiian Tropical Fishes; Brain Size and Evolution. Copeia, 1977, pp. 42–46.

Bauchot, R.; Platel, R.; and Ridet, J.-M.: Brain Weight, Body Weight Relationships in Selachii. Copeia, 1976, pp. 305–310.

Berggren, W. A.; McKenna, M. C.; Hardenbol, J.; and Obradovich, J. D.: Revised Paleogene Polarity Time Scale. J. Geol., vol. 86, Jan. 1978, pp. 67–81.

Brattstrom, B. H.: Learning Studies in Lizards. In: Behavior and Neurology of Lizards, N. Greenberg and P. D. MacLean, eds., U.S. National Institute of Mental Health (DHEW Publication no. (ADM)77-491.), 1978, pp. 173–181.

Bullock, T. H.; and Horridge, G. A.: Structure and Function in the Nervous Systems of Invertebrates. W. H. Freeman, San Francisco, 1965.

Carr, B. J.; and Rees, M. J.: The Anthropic Principle and the Structure of the Physical World. Nature, vol. 278, Apr. 12, 1979, pp. 605–612.

Crile, G.; and Quiring, D. P.: A Record of the Body Weight and Certain Organ and Gland Weights of 3690 Animals. Ohio Journal of Science, vol. 40, 1940, pp. 219–259.

Connell, Joseph H.: Diversity in Tropical Rain Forests and Coral Reefs. Science, vol. 199, March 24, 1978, pp. 1302–1310.

Eisenberg, J. F.; and Wilson, D. E.: Relative Brain Size and Feeding Strategies in the Chiroptera. Evolution, vol. 32, 1978, pp. 740–751.

Farlow, J. O.: Consideration of Trophic Dynamics of a Late Cretaceous Large-Dinosaur Community (Oldman Formation). Ecology, vol. 57, 1976, pp. 841–857.

Gill, J. R.; and Cobban, W. A.: Stratigraphy and Geologic History of the Montana Group and Equivalent Rocks, Montana, Wyoming, and North and South Dakota. United States Geological Survey Professional Paper 776, 1973.

Hopson, J. A.: Relative Brain Size and Behavior in Archosaurian Reptiles. Ann. Rev. Ecology Systematics, vol. 8, 1977, pp. 429–448.

Hunt, B. G.: The Effect of Past Variations of the Earth's Rotation Rate on Climate. Nature, vol. 281, Sept. 20, 1979, pp. 188–191.

Jerison, H. J.: Evolution of the Brain and Intelligence. Academic Press, N.Y., 1973.

Jerison, H. J.: Evolution of the Brain and Intelligence. Current Anthropol., vol. 16, 1975a, pp. 403–426.

Jerison, H. J.: Fossil Evidence of the Evolution of the Human Brain. Ann. Rev. Anthropol., vol. 4, 1975b, pp. 27–58.

Jerison, H. J.: Paleoneurology and the Evolution of Mind. Sci. Amer., vol. 234, Jan. 1976, pp. 90–101.

Johanson, D. C.; and White, T. D.: A Systematic Assessment of Early African Hominids. Science, vol. 203, Jan. 26, 1979, pp. 321–330.

Kitching, J. W.: The Distribution of the Karoo Vertebrate Fauna. Bernard Price Institute of Paleontological Research Memoir 1, 1977.

May, R. M.: The Evolution of Ecological Systems. Sci. Amer., vol. 239, Sept. 1978, pp. 160–175.

Morrison, P.; Billingham, J.; and Wolfe, J.: The Search for Extraterrestrial Intelligence. NASA SP-419, 1977.

Okada, Y.; Aoki, M.; Sato, Y.; and Masai, H.: The Brain Patterns of Sharks in Relation to Habit. J. für Hirnforschung, vol. 11, 1969, pp. 347–365.

O'Neill, G.: The High Frontier: Human Colonies in Space. Bantam Books, Des Plaines, IL, 1976.

Parker, R. H.: Classification of Communities Based on Geomorphology and Energy Levels in the Ecosystem. In: Structure and Classification of Paleocommunities, R. W. Scott and R. R. West, eds., Dowden, Hutchinson and Ross, Stroudsburg, PA, 1976, pp. 67–86.

Passingham, R. E.: The Brain and Intelligence. Brain, Behavior and Evolution, vol. 11, 1975a, pp. 1–15.

Passingham, R. E.: Changes in Size and Organization of the Brain in Man and His Ancestors. Brain, Behavior and Evolution, vol. 11, 1975b, pp. 73–90.

Portmann, A.: Etudes sur la Cérébralisation chez les oiseaux. II. Les Indices Intra-cérébraux. Alauda, vol. 15, 1947a, pp. 1–15.

Portmann, A.: Etudes sur la Cérébralisation chez les oiseaux. III. Cérébralisation et mode ontogénétique. Alauda, vol. 15, 1947b, pp. 161–171.

Radinsky, L.: The Fossil Evidence of Anthropoid Brain Evolution. Am. J. Phys. Anthropol., vol. 41, 1974, pp. 15–27.

Radinsky, L.: Early Primate Brains: Facts and Fiction. Am. J. Phys. Anthropol., vol. 42, no. 2, 1975a, p. 324.

Radinsky, L.: Primate Brain Evolution. American Scientist, vol. 63, 1975b, pp. 656–663.

Radinsky, L.: Evolution of Brain Size in Carnivores and Ungulates. American Naturalist, vol. 112, 1978, pp. 815–831.

Raven, P. H.; and Axelrod, D. I.: Plate Tectonics and Australasian Paleobiogeography. Science, vol. 176, June 30, 1972, pp. 1379–1386.

Regal, P. J.: Ecology and Evolution of Flowering Plant Dominance. Interplay of Seed and Pollen Dispersal Systems May Explain Angiosperm vs Gymnosperm Dominance. Science, vol. 196, 1977, pp. 622–629.

Regal, P. J.: Behavioral Differences Between Reptiles and Mammals: An Analysis of Activity and Mental Capabilities. In: Behavior and Neurology of Lizards; An Interdisciplinary Colloquium, Neil Greenberg and Paul D. MacLean, eds., U.S. Department of Health, Education and Welfare, Rockville, MD, 1978, pp. 183–202.

Retallack, G. J.: Reconstructing Triassic Vegetation of Eastern Australasia: A New Approach for the Biostratigraphy of Gondwanaland. Alcheringa, vol. 1, 1977, pp. 247–278.

Riddell, W. I.; and Corl, K. G.: Comparative Investigation of the Relationship Between Cerebral Indices and Learning Abilities. Brain, Behavior and Evolution, vol. 14, 1977, pp. 385–398.

Rodin, L. E.; Bazilevich, N. I.; and Rozov, N. M.: Productivity of the World's Main Ecosystems. In: Productivity of the World Ecosystems; D. E. Reichle, J. F. Franklin, and N. N. Rozov, eds., U.S. National Academy of Sciences, 1975, pp. 13–26.

Runnegar, B.: Evidence for Precambrian Eukaryota. Alcheringa, vol. 1, 1977, pp. 311–314.

Russell, D. A.: A New Specimen of *Stenonychosaurus* from the Oldman Formation (Cretaceous) of Alberta. Canadian J. Earth Sci., vol. 6, 1969, pp. 595–612.

Russell, D. A.: The Enigma of the Extinction of the Dinosaurs. Ann. Rev. Earth Planetary Sci., vol. 7, 1979, pp. 163–182.

Sacher, G. A.: Longevity and Aging in Vertebrate Evolution. Bioscience, vol. 28, Aug. 1978, pp. 497–501.

Sacher, G. A.; and Staffeldt, E. F.: Relation of Gestation Time to Brain Weight for Placental Mammals: Implications for the Theory of Vertebrate Growth. American Naturalist, vol. 108, 1974, pp. 593–615.

Sokoloff, L.: Circulation and Energy Metabolism of the Brain. In: Basic Neurochemistry, 2nd edition, G. J. Siegel, R. W. Alberts, R. Katzman, and B. W. Arganoff, eds., Little, Brown, and Co., Boston, 1976, pp. 388–413.

Stevens, E.; and Neill, W. H.: Body Temperature Relations of Tunas, Especially Skipjack. In: Fish Physiology, W. S. Hoar and D. J. Randall, eds., Academic Press, N.Y., 1978, pp. 315–359.

Tappan, H.: Primary Production, Isotopes, Extinctions and the Atmosphere. Palaeogeography, Palaeoclimatology, Palaeoecology, vol. 4, 1968, pp. 187–210.

Thomson, K. S.: The Adaptation and Evolution of Early Fishes. Quarterly Rev. Biol., vol. 46, June 1971, pp. 139–166.

Tobias, P. V.: Implications of the New Age Estimates of the Early South African Hominids. Nature, vol. 246, Nov. 9, 1973, pp. 79–83.

Tuge, H.; Uchihashi, K.; and Shimamura, H.: An Atlas of the Brains of the Fishes of Japan. Tsukiji Shokan Publishing Co., Tokyo, 1968.

Tuttle, W. W.; and Schottelius, B. A.: Textbook of Physiology, 15th edition, C. V. Mosby Co., St. Louis, Mo., 1965.

Valentine, J. W.: Conceptual Models of Ecosystem Evolution. In: Models in Paleobiology, T. J. M. Schopf, ed., W. H. Freeman, San Francisco, 1972, pp. 192–215.

Van Eysinga, F. W. B.: Geological Time Table. Elsevier, Amsterdam, 1975.

Van Valen, L.: Brain Size and Intelligence in Man. Am. J. Phys. Anthropol., vol. 40, 1974, pp. 417–423.

Wheeler, P. E.: Elaborate CNS Cooling Structures in Large Dinosaurs. Nature, vol. 275, Oct. 5, 1978, pp. 441–443.

Whittaker, R. H.; and Woodwell, G. M.: Evolution of Natural Communities. Oregon State University, Biological Colloquium, vol. 31, 1971, pp. 137–159.

Wirz, K.: Etude biométrique du système nerveux des céphalopodes. Bulletin Biologique France et Belgique, vol. 93, 1959, pp. 78–117.

Evolution of Technological Species

BERNARD CAMPBELL

Technology is adaptive, cumulative, and generally progressive. At its simplest, it is older than reason; at its most advanced, it is the product of cooperative undertakings by large numbers of highly intelligent organisms.

Though the word is now used loosely, technology can most conveniently be defined as the systematic study of tools: a technological species is any species of animal that uses tools. Tools must be clearly distinguished from artefacts. *Artefacts* are common in the animal kingdom — birdnests, beehives, and beaver lodges are among the finest examples — and they are made by some creatures quite low in the evolutionary scale (such as the caddis fly larva, that makes cases and nets). *Tools* are distinct from artefacts because they can be used to make other objects or to facilitate activities such as resource extraction. They are not by themselves of immediate and direct use.

If we are to assess the probability of technology developing on other planets, it would appear most useful to discuss the origins of terrestrial technology, its simplest forms, and the factors that made possible its extraordinary development.

It is now well known that the use and modification of tools is not confined to man. Examples include the Galápagos finch *Cactospiza,* which uses a cactus (prickly pear) spike to poke out insects embedded in the branches or trunks of trees. Where there are no cacti, *Cactospiza* breaks off a short stiff twig from a tree. A second example is the British Greater Spotted Woodpecker *(Dendrocopos major),* which pecks out a V-shaped cleft in a tree trunk and uses it to anchor pine cones, oak-apples, etc., while it extracts seeds or insects. These are two examples of tool modification found among birds, but there may be many others.

Tool use (though not modification) has been identified among insects. Most striking is the use by the solitary burrowing wasp *Ammophila* of a small pebble to firm up the sealed entrance to its burrow. When the wasp has filled the burrow with eggs and a food supply of caterpillars, it seals the burrow and covers it with sand grains hammered firmly into place. In the end, all trace of the nest is obliterated.

These examples from the insects and birds illustrate the two stages of what I shall call prototechnology (table 1).

TABLE 1.– FOUR STAGES OF EARLY
TECHNOLOGY

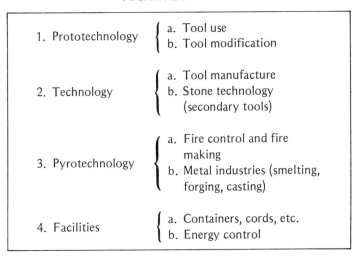

1. Prototechnology	a. Tool use
	b. Tool modification
2. Technology	a. Tool manufacture
	b. Stone technology (secondary tools)
3. Pyrotechnology	a. Fire control and fire making
	b. Metal industries (smelting, forging, casting)
4. Facilities	a. Containers, cords, etc.
	b. Energy control

Category 1a includes the use of found objects as tools. To our introductory examples we can add the famous example of the sea otter *(Enhydra),* which floats on its back with a stone on its chest, opening shellfish such as abalone by smashing their shells with the stone. These examples of tool use imply that certain qualities belong to the species that uses them: the existence of manipulative organs (a beak and forelimbs in our examples), a good visual sense, and a behavioral flexibility that permits discovery and invention — although at a very simple level. The activity is in most cases a solution to a problem of obtaining environmental resources, but in the case of the wasp it presumably improves reproductive success. In the instances of the finch, the woodpecker, and the otter, important new food resources are made available as a result of tool use.

It is important to stress this point: the motive must be present as well as the potential. Tools allow these animals to tap food resources not otherwise available to them, and tools give them great advantage in that they then have a monopoly or near monopoly of that particular resource. This freedom from competition is a very desirable and rare biological state. The phenomenon of tool use is not confined to mammals and therefore does not require a very highly evolved central nervous system.

The second stage of prototechnology (category 1b) involves the natural object specially modified for tool use. The classic example here is the chimpanzee, which modifies grass stems or twigs for termite fishing. The twigs are collected, any leaves are removed, and the length is adjusted to a standard. Less well known is the chimpanzee's preparation of leaves by chewing to make them act like a sponge to absorb water in order to drink from an awkward place such as a hole in a tree trunk. This category clearly represents a distinct advance in the history of technology and implies not only discovery and imagination but foresight. Thus the preparation of a twig by breaking one of an appropriate length by *Cactospiza* is a remarkable fact. One might have expected that a highly evolved central nervous system was essential for this development. It is certainly very rare among animals. Among chimpanzees it is transmitted by learning: this is highly significant since it allows rapid modification of technique and thus rapid development of technique. Whether it has become a programmed behavior among finches, as distinct from a learned one, which is how it must have begun, is not known.

Stage 1a does not easily lead to 1b and indeed does not logically imply such development. Stage 1b, however, has immense potential, and while in the chimpanzee it has not led to further advances, there seems every reason to believe that it will give its practitioner the potential to develop true technology (stage 2).

The earliest evidence we have of humans *making* tools is, of course, lithic. This, however, should not lead us to suppose that stone tools were the first human tools. Modern tribal peoples only have a small percentage of material culture that will be preserved in an archaeological site. The vast majority of the technology is biodegradable: besides stone, simple tools are made of bone, antler, horn, teeth, wood, and leather.

The difference between categories 1b and 2a may seem very slight at first, but in practice there is a world of difference between modifying a twig and removing its leaves (for the twig existed in the first place and could be clearly seen) and making a pebble into a knife-like object for chopping or cutting. This surely requires far greater imagination. However, stone choppers and stone flakes do occur naturally and can be produced by dropping one stone on another by chance. Sharp stone flakes were in our environment a billion years before hominids found a use for them. When they did — where the motive was present and the brain sufficiently developed — then the step from the use of existing objects to tool manufacture might have been rapid. Whatever other biodegradable technology such as digging sticks, rakes, and levers they may have had, the use of stone cutters and choppers would have opened up a vast new food resource — not free of competition but otherwise unavailable to a hominid lacking a carnivorous dentition or even powerful canine teeth. The food resource took the form of large mammals.

The manufacture of stone tools depended essentially on motive and manipulative skill, neither of which at first need have been highly developed. But it led to a most important development: because stones were so hard, they could be used to cut other softer (yet quite hard) materials such as wood and bone which, in turn, could be used as tools.

Thus simple stone tools such as flakes and choppers could become, almost immediately, secondary tools (stage 2b). In this way, human technology began its extraordinary development.

Let us stop here for the moment and review the essential factors involved in this development:

1. A lead-in from the prototechnology of stone; that is, the use of naturally occurring flakes and choppers.

2. Observation of natural examples of stone tool-making by percussion, that is, the effects of falling or dropped stones.

3. The existence of a motive related to cutting tool use, namely, the addition of large mammals as a food resource.

4. The availability of raw materials in certain areas.

5. The hardness of raw materials, allowing development of tool-making, as well as good cutting edges.

6. The generation by the central nervous system of a good visual sense, imagination and foresight, behavioral flexibility, learning potential, and motor skill.

From approximately 2 million years ago, we can see the development of lithic technology in the archaeological record. As this development accelerated — and it accelerated incredibly slowly from our viewpoint — we find evidence of the unfolding of human skill, imagination, and ingenuity. The stone technology was no doubt accompanied by a wonderful proliferation of biodegradable tools. With few exceptions these have not been preserved from the earliest days, and for most of human prehistory technological data are confined to stone tools. However, we should record the following: first prepared bone point, Olduvai (Tanzania), about 1.75 million years ago; first prepared wood spear point, Clacton (England), about 300,000 years ago; first bone tool kit, Choukoutien (China), 400,000 to 500,000 years ago. Such evidence is meager but suggestive of the extent of such tool kits.

Although stone tools are not known much before 2 million years ago, tool manufacture must be very much older, and tool modification and tool use older still. We have no good evidence of this earlier phase of human prototechnology, although Louis Leakey claimed that he had evidence of bone bashing with a rounded and battered cobble, on an ancient land surface at the site of Fort Ternan, dated about 14 million years ago. The progression 1a → 1b → 2a → 2b seems an eminently reasonable assumption.

The next landmark is probably the most important one since it introduced the possibility of sophisticated metal technology. I refer to the capture, control, and eventual making of fire (stage 3a). Pyrotechnology probably had a very slow beginning: the generation of wild fire through lightning, the Sun's rays, oil seeps, etc., is rare enough, but its capture must have been even less common. Following capture, the fire needed to be controlled and fed. Why should the task ever have been undertaken? Two motives seem compelling: curiosity and the need for warmth. Curiosity might well have led men to capture fire and play with it, but only the need for warmth could lead man to capture and feed fires for weeks, months, and even years.

The earliest evidence of fire in the archaeological record is important. The two sites with the clearest hearths are those of Vértesszöllös in Hungary and Choukoutien near Peking. In both areas, humans were hunting and gathering in a temperate environment with cold winters. The Hungarian site is open, the Chinese site, a cave used for habitation. The dates are probably both in the period 400,000 to 500,000 years ago. At Choukoutien there is also evidence of burnt bone, which implies roast meat. At Vértesszöllös, the hearths are not very large: the site is a butchery site. At Choukoutien, the cave deposits are immensely deep and the hearths contain many meters of ash superimposed: evidently the fires were permanently maintained throughout the year. (The validity of earlier evidence of fire from L'Escale Cave in the south of France about 700,000 years ago has been questioned.)

Fire is not recorded from Africa until very much later: about 60,000 years ago, at Kalambo Falls. While there is no shortage of earlier archaeological living sites in Africa, fire is always absent. There was evidently no reason to capture and control it. It seems fairly certain that the development of pyrotechnology was a correlate of a temperate climate. It led to new techniques of toolmaking and hunting, and it also gave protection from predators (especially the cave bear that competed for living space); it gave light and a means of cooking. But almost certainly the most important factor was warmth.

In the manufacture of tools, fire has many uses, from the breaking of large stones by heating and sudden cooling to the hardening of spear points (stage 3b). By far the most important technological development was the smelting of metal ores (copper and tin), first recorded in eastern Europe about 4,500 years ago, some 500,000 years after the first indication of fire use. Here again a motive must have existed to produce copper in large quantities, and such need could hardly have developed until naturally occurring metallic gold, silver, and copper had been in use for some time. Copper smelting at its simplest merely implies putting some malachite in a hot fire, but recognition of the value of the product was needed to bring about regular production. This is a general rule: a species will not make a tool unless its value has been clearly established. The idea of tool research and development is very recent in human history. Metals were not smelted in Africa south of the Sahara, Australia, or America until the technique was introduced from Europe. Europeans were most probably the inventors of smelting and alloying, and they have since led the rest of the world in metal technology.

The geographer Philip Wagner defined a class of objects he termed *facilities* as objects that restrict or prevent motion or energy exchanges (stage 4). Examples include containers, dams, boats, fences, cords (stage 4a), and anything that insulates and retains heat, such as clothing, shelters, or tents (stage 4b). This is an extremely important class of artefact distinct from simple tools; but since many tools operate by altering energy flow in some way, the distinction is not absolute.

What is important is that the simplest facilities so defined were unquestionably of the greatest importance in human evolution, particularly in temperate zones. They become an essential component in any evolving technology, and the principles involved in this class of artefact become quite important in human technology generally.

A further subcategory of tools is *machines,* which are complex tools with moving parts. Either the spear thrower (about 14,000 years ago) or the bow and arrow (about 10,000 years ago or more) might qualify as the earliest examples.

Finally, toys may be defined as tools used for play, that is, for experimental purposes. Play, which is seen in many mammals, especially carnivores and primates, has had an extremely important part in human evolution. Play depends on leisure, or freedom from a constant concern with the search for resources. In play, young primates test their own physique, the social responses of their peers and elders, and their environment. When tools are used, the young will play with them and even attempt to make them. Play tools or toys are made for experimental and exploratory activities. Play is the activity in which innovative behaviors are established and new artefacts are incorporated into a culture.

SUMMARY

In this brief review of terrestrial technology, we have traced a few simple but important steps that have led to the manufacture of radiotelescopes. In human evolution, each of the stages in groups 1 and 2 has followed the last in succession; but as we have seen, the development of this series is by no means inevitable. In insects and birds it is clearly blocked by inadequate manipulative organs and insufficient intelligence. The development of facilities and fire was critical in the establishment of a metal technology. Adequate resources derived from agriculture were essential for the development of advanced technology. At all stages, the ratio of cost to reward had to be small and the technological development had to follow an appropriate cultural preadaptation. Play would unquestionably have been important in technological innovation. The history of toys has yet to be written, but it may be a key to an understanding of the progressive development of human technology.

DISCUSSION

Having reviewed the nature of the development of simple terrestrial technology, we can return to examine its place in terrestrial prehistory. What factors made this development possible? The evolution of social groups as distinct from aggregations of individuals is by no means uncommon in terrestrial evolution. Primate-like social groups are evidently not essential preconditions for the evolution of tool use, and learned behavior patterns can be passed from parent to offspring in their absence. Such social groups do, however, facilitate certain aspects of technological development.

Social groups can be expected to occur whenever some or all of the following conditions are fulfilled:

1. Multiple channels exist for communication between individuals.

2. Food is either widespread and common or clumped.

3. Social cooperation is at a premium for some reason, such as protection from large predators.

4. The rapid spread of learned behavior is at a premium.

For technological progress, social groups have the following advantages:

1. More exploratory behavior and play.

2. Better learning conditions, with rapid spread of learned behavior patterns through a population.

3. Potential to exploit large "locked" food resources.

Primate-like social groups are clearly essential for the later and more advanced phases of technological development in which the cooperation of a number of individuals is essential to make and use tools. The important factors are the following:

1. Social groups should be well integrated.

2. The reward for tool development must be available in terms of valued resources.

3. The material resources needed to make the tool must be available (suitable stone, wood, etc.).

4. A social situation must exist which encourages invention and skill.

5. Individuals must have a good visual sense and manipulative skill.

It seems that the hominid adaptation of toolmaking bipeds was a product of social primates with the following essential characteristics:

1. An advanced central nervous system, on both sensory and motor sides.

2. Some prototechnology.

3. A strongly motivated exploratory drive.

4. A rich and complex environment, with an extensive forest/savanna ecotone.

5. A potential for habitat and niche expansion, given the key (cutting tools) to a huge resource of savanna mammal meat.

The use of stone tools, recorded since 2 million years ago and associated with the genus *Homo*, at first evolved slowly, while humans were confined to tropical and subtropical regions. With the evolution of language and the development of the control of fire, humans entered temperate and eventually arctic regions, and these remarkable adaptations ushered in the appearance of *Homo sapiens* about 300,000 years ago. The development of technology from that time has allowed a vast expansion in the range and biomass of the species and the generation of an immense surplus of wealth (resources) in some regions. This surplus, based on agriculture and labor, has made possible the construction of radiotelescopes which themselves do not increase or facilitate resource extraction: they must therefore be defined as research tools or toys.

In an extraterrestrial context, we may predict that prototechnology is by no means uncommon where animals have evolved, but that full metal technology is likely to depend on the existence of a suitable social species. Should organisms with radiotelescopes exist, it seems reasonable to predict that they would carry most of the following characteristics:

1. Animal-like mobility.

2. Bilateral symmetry.

3. Heterotrophic nutrition.

4. Nervous systems.

5. Organs of manipulation.

6. Social groups.

7. Ability to learn and communicate efficiently.

8. Play and exploration.

9. Evolved visual sense.

10. Large and structured central nervous system.

11. Flexible behavioral response capability (intelligence).

12. Adaptations to rich, varied, complex environments.

13. An environment with seasonal variations in temperature.

Moreover, they would have suitable raw materials present: metal ores, energy sources, and surplus food resources.

Technology is adaptive, cumulative, and generally progressive. At its simplest, it is older than reason. At its most advanced, it is the product of cooperative undertakings by large numbers of highly intelligent organisms.

Astrophysical Crises in the Evolution of Life in the Galaxy

WALLACE H. TUCKER

Aspiring young life forms seeking a safe habitat in the Milky Way should find a good stable star of 0.8 to 1.2 solar masses, settle down, avoid crowds, and above all, stay away from hot, young, heavenly bodies.

The fate of intelligent life in the Galaxy may be in many ways analogous to the fate of an individual on Earth. He must survive the embryonic stage and the early years; then, if war, famine, pestilence, or just plain bad luck do not get him, the natural evolutionary process of aging and death will. Similarly, if complex life is to develop and thrive on a given planet, there must be no miscarriage because of too much or too little starshine. It must survive supernovae, giant flares, interstellar dust clouds, and large-scale galactic convulsions, only to succumb eventually when the parent star burns out.

My purpose is to discuss these cosmic vicissitudes of life and to show that, in spite of them, life should persist in abundance in the Galaxy for essentially the same reason that communication with that life will be so difficult: the vast interstellar distances that at the same time protect and imprison us.

EVOLUTION OF PARENT STAR

Broadly stated, the critical cosmic conditions for the rise of intelligent life on a planet are that the planet must receive the proper amount of energy from the parent star for the proper length of time. Calculations of model atmospheres (Hart, 1978, 1979, and references) show that the proper energy

287

flux can, with some degree of confidence, be assumed to be the flux received at Earth from the Sun, $F_\odot$, plus or minus 10% give or take a few percent.

The parameter that determines the evolution of a star is its mass. For stars having masses in the range 0.4–7 solar masses $(M_\odot)$, the luminosity during the hydrogen burning, or main-sequence phase, varies with approximately the fifth power of the mass: $L \propto M^5$. The energy supply E of a star is proportional to the star's mass, so the main-sequence lifetime t_{MS} is of the order $t_{MS} \approx E/L \propto M/L$ or $t_{MS} \propto M^{-4}$. Since the Sun with mass $M_\odot$ has a main-sequence lifetime of about 10 billion years, we find

$$t_{MS} \approx 10^{10} (M/M_\odot)^{-4} \text{ yr} \tag{1}$$

for stars in the mass range 0.4–7 $M_\odot$.

At the end of this time, the core of the star, having exhausted its supply of hydrogen, collapses until it is stabilized by a new set of fusion reactions that convert helium into carbon. These reactions produce energy much more rapidly than do the hydrogen fusion reactions. To accommodate this increased flow of energy, the outer shell of the star must expand. The radius and luminosity of the star increase by a factor of a 100 or more. The star has become a red giant. The main-sequence phase has ended and, with it, in all likelihood, any life in orbit around the star.

In terms of the evolution of life we can say that, if intelligent life has not developed around a star by the end of the star's main-sequence lifetime, it never will. If we assume that the time for evolution of intelligence is at least 1.5 billion years, then by equation (1) it could only happen around stars with masses less than about 1.6 $M_\odot$.

Stars much less massive than the Sun have very long main-sequence lifetimes, but they also have low luminosities. If life is to evolve on a planet around such a star, it must be very close to the star, say as close as Mercury is to the Sun, in order to receive an adequate amount of starlight. But if the planet is too close, the star will raise huge tides on the planet. These tides will tend to bring the planet's rotation into synchronism with its orbit so that the same side will face the star at all times. The oceans on the near side will boil, and on the far side they will freeze, making it impossible for intelligent life to evolve. This argument eliminates stars with masses less than about 0.7 $M_\odot$ (Dole, 1970; Project Cyclops, 1973).

More restrictive conditions have been derived from consideration of the evolution of planetary atmospheres. Computer simulations taking into account all the major processes affecting the bulk composition of the atmosphere and the mean surface temperature of the planet show that the parent star must evolve in a manner very similar to that of the Sun to avoid either a runaway greenhouse effect or runaway glaciation. Low-mass stars evolve too slowly. If the planet were far enough away from such a star to avoid overheating early in its evolution, it would be too far away to avoid glaciation later since the luminosity of the star will not have increased sufficiently in the intervening eons. Conversely, massive stars evolve too quickly and produce too much ultraviolet light.

In summary, the only stars that have continuously habitable zones around them are ones in the range 0.8 to 1.2 $M_\odot$. Bearing in mind the uncertainties in atmospheric models and the sensitivity of the results to the planet size, this range should perhaps be expanded somewhat, but it does reinforce our conviction that only stars similar to the Sun are likely to support the rise of intelligent life.

EFFECT OF EVOLUTION OF NEARBY STARS

The evolution of a nearby star to the red giant phase will not significantly affect the overall energy budget of a planet in orbit around a main-sequence star. If Sirius, at a distance of 8 light years, were to become a red giant with a luminosity of a thousand Suns, then the flux incident at the top of our atmosphere would be changed by only about 30 millionths of 1%. To alter the total energy input significantly, a star would have to attain the luminosity of a thousand Suns and be located roughly at the orbit of Pluto. The only places where we find stars this close together are in binary systems or in the nuclear core of the Galaxy, both of which regions are unlikely to support life for other reasons. Of course, gross energy flux is not the whole story. The climatic stability and the continued good health of living things depend on the spectral distribution as well as the magnitude of the incident radiation. In particular, a large (thousandfold) increase in high-energy radiation could be fatal either indirectly through the changes it would cause in the upper atmosphere and hence the climate or directly by radiation damage to living tissue.

EFFECT OF A NEARBY SUPERNOVA EXPLOSION

The most probable cause of a large increase in high-energy radiation is a supernova explosion, although giant stellar flares or explosions in the galactic nucleus are also possibilities. The cause of a supernova outburst is still the subject of intense investigation and considerable controversy, but it is generally agreed that the onset of the explosion is ultimately related to instabilities in the structure of the star that arise when the supply of nuclear fuel in the core is exhausted. These instabilities occur only in stars that arrive at the end of the red giant phase with masses greater than about 1.5 $M_\odot$. The density and temperature in the central core in such stars exceed the critical values beyond which stability is impossible. The star collapses, and a supernova explosion ensues.

For almost 2 weeks, the supernova radiates more energy than a billion Suns and ejects matter at velocities approaching the velocity of light. The expanding shell of debris creates a nebula of hot gas and high-energy particles that, for hundreds or even thousands of years, radiates vigorously in both the x-ray and radio regions of the spectrum. In addition, the cataclysm may leave behind a small, extremely dense core known as a neutron star. These stars have been detected as pulsating sources of radiation — pulsars.

About one star in a hundred will eventually become a supernova of type II. (Type I supernovas seem to be associated with mass-exchange processes in close binary systems.) These are the hot, bright young stars that go through their evolution quickly because of their large mass (see eq. (1)). They are located primarily within the spiral arms of the Galaxy, and it is there that the frequency of supernova explosions is highest. Clark et al. (1977) have shown that a supernova is likely to occur within 30 light years of the Sun approximately once every 100 million years during the Sun's passage through a spiral arm.

In discussing the extent to which a nearby supernova explosion could change the ionizing radiation environment, I shall start with changes we can be sure of and work my way gradually out on a limb as I discuss more speculative possibilities.

The normal cosmic contribution to the ionizing radiation environment consists of cosmic rays, which come from the Galaxy and perhaps beyond, and ultraviolet, x, and gamma radiation, which comes from the Sun. Table 1 summarizes these contributions.

TABLE 1.– NORMAL COSMIC RADIATION BACKGROUND

Type of radiation	Approximate flux at top of atmosphere, erg/cm^2/sec*
Cosmic rays	3×10^{-3}
Visible light (3000–6000 Å)	10^6
Ultraviolet (2000–3000 Å)	10^5
X-ultraviolet (100–1000 Å)	1
X rays (100–0.1 Å)	0.1

*A flux of 1 erg/cm^2/sec implies a radiation dose rate of about 10 roentgen/yr.

The characteristic signature of a supernova remnant is the emission of polarized radio waves. It is well established that this radio emission is produced by very high-energy electrons trapped in the magnetic field of the remnant. From the observations we can obtain reliable estimates of the cosmic ray content of supernova remnants. For remnants between a few hundred and a few thousand years old, the cosmic ray content is in the range 10^{49} to 10^{50} ergs (Woltjer, 1972), assuming that 10 times as much energy goes into cosmic ray protons as into the electrons.

This estimate does not take into account the cosmic rays that have escaped the remnant. Considerations of galactic cosmic ray and gamma ray fluxes indicate that on the order of 10^{50} to 10^{51} ergs of cosmic rays are produced by each supernova during its explosion and subsequent evolution (Lingenfelter, 1969; Stecker, 1975; Higdon and Lingenfelter, 1975; Chevalier et al., 1976).

For a supernova occurring at a distance of 30 light years, the cosmic ray cloud would take 10^3 to 10^4 yr to sweep over Earth. During this period the cosmic ray flux on Earth would increase by a factor of 1000.

A more speculative possibility is that large quantities (3×10^{50} ergs) of cosmic rays are expelled during the early phases of supernova explosions, producing a relativistic blast wave that would sweep over Earth in about 100 years. In this event the cosmic ray background would be increased by a factor of 10,000. This possibility is more speculative because as yet we have no confirmation of the theoretical models. However, in quasars and other violently active extragalactic sources, explosive events in which 10^{51} ergs of high-energy particles are released on a time scale of months are commonly observed. It is possible that these are just the early phases of supernova explosions, which have been suggested to occur at the rate of 1 to 10 per year in the nuclei of active galaxies (see Arons et al., 1975).

The biologic effects of a nearby supernova can be grouped into two broad categories: (1) climatic effects and (2) effects due to changes in the ionizing radiation environment on the surface of Earth.

The effects of an enhanced flux of high-energy radiation on the ozone layer have been estimated by a number of authors (Ruderman, 1974; Reid et al., 1978; Whitten et al., 1976). In particular, Reid et al. (1978) show that the ozone layer would be 90% depleted, and the column density of NO_2 would be increased by a factor of 100, leading to *decreased* absorption of ultraviolet light and *increased* absorption of visible light. These changes in the radiation budget of the atmosphere could lead to a drop in the average surface temperature of about 3 K, a decrease in global precipitation, and a worldwide drought, which could be catastrophic.

As a result of the depletion of the ozone layer, the biologically effective ultraviolet radiation dose would be increased by a factor of 10 or more. The ultimate biologic consequences due to cell deaths and increased mutation rate are unknown, but a large-scale extinction of life on the exposed planet is certainly feasible.

The normal cosmic ray flux produces a radiation dose rate of about 0.03 roentgen/yr (Herbst, 1964), so a thousandfold increase in the cosmic ray flux would imply a dose rate of 30 roentgens/yr. The lethal dose for most laboratory animals is in the range 200–700 roentgens (Terry and Tucker, 1968, and references); so life would not be killed outright, but the accumulated dose could well be fatal. A 10,000-fold increase in dose rate would imply a dose of 300 roentgens/yr and 6000 roentgens in 20 years, which would almost certainly be fatal. As discussed above, such an increase could occur if a relativistic blast wave were produced by a supernova, or if the supernova were to occur at a distance of only 10 to 15 light years.

These numbers show that a nearby supernova could well cause a mass extinction of life on Earth. Perhaps if it were to occur now, we would have the technology to go underground and survive somehow. But if such an event had occurred a few thousand years ago, the human race probably would not have survived. Indeed it has been suggested by a number of authors dating back to Schindewolf (1954) that a supernova explosion was responsible for the mass extinctions that occurred 60 million years ago in connection with the disappearance of the dinosaurs (Krasovski and Shklovsky, 1957; Terry and Tucker, 1968; Ruderman, 1974; Russell and Tucker, 1971; Reid et al., 1978).

More to the point, what are the implications for the search for extraterrestrial life? First of all, we could not expect civilizations to thrive in an environment where a nearby supernova occurred every 10 million years or so. Large radiation-sensitive species simply would not have time to establish themselves. Of course, it is possible that radiation-resistant populations such as insects might develop intelligence under these conditions. If we ignore this possibility, then we must eliminate those regions of the Galaxy where the density of massive stars and hence of supernovae is high: the spiral arms and the region within a few thousand light years of the galactic center. Stars similar to the Sun, which pass through the spiral arms only once every 100 million years, taking 10 million years to make the passage, would be expected to be just marginally safe.

EXPLOSIONS IN THE GALACTIC NUCLEUS

The galactic nucleus may be an inhospitable place for more reasons than one. Compared to some other galaxies and to quasars, the power output of our galactic nucleus is unimpressive. However, the nucleus shows unmistakable evidence of a violent past history, the most direct evidence being clouds of gas streaming out of the center at high velocity. The largest of these features, which is detectable some 10,000 light years from the galactic nucleus, implies an explosion 500 million years ago which released the energy equivalent of a hundred million supernovae (see Oort, 1977). If other active galaxies are a guide, this did not occur all at once, but in a thousand smaller explosions over the period of a few million years. Thus any x- or gamma-ray flash would have a luminosity of 10^{45} ergs/sec at most, and would produce a noticeable but not catastrophic enhancement at Earth. Most of the cosmic ray energy would escape the Galaxy in a direction perpendicular to the plane, and so would have little effect either. In conclusion, explosions in the galactic nucleus could wreak havoc within a few thousand light years of the nucleus, but are unlikely to cause problems further out.

LARGE SOLAR FLARES

Reid et al. (1978) have shown that a very large flare on our Sun (100–1000 times the largest ever observed) would affect the ozone layer in much the same way as a supernova occurring at a distance of about 30 light years. Some stars do give off flares of this magnitude, but they have large turbulent convective zones that are presumably responsible for the greatly enhanced surface activity. Presently accepted solar models do not predict such large convection zones; but these models also predict that neutrinos should be emitted by the nuclear reactions in the core and that these neutrinos should be detectable at Earth (see Ulrich, 1975). They are not, which has led to speculation that the Sun undergoes periods of large-scale convective mixing, which could in turn allow the possibility of large solar flares. If the Sun does it, then so should other stars similar to the Sun, and these superflares should be detectable over distances of a 1000 light years. Eventually, data from x-ray observatories should allow a test of this hypothesis. Until then, I feel we cannot rule it out.

INTERSTELLAR DUST CLOUDS

An encounter between the Solar System and a dense cloud of interstellar material during its passage through a spiral arm of the Galaxy might produce a climatic change by shielding Earth from the solar wind, thus changing the quantity of energetic particles reaching Earth from the Sun (McCrea, 1975; Begelman and Rees, 1976). It is doubtful, however, that the effects would be as catastrophic as those discussed previously.

CONCLUSION

Life on any given planet can be adversely and even fatally affected by a variety of cosmic catastrophies. My advice to any aspiring young life forms seeking a safe place in the Galaxy would be similar to the advice a father might give his son as he prepared to venture into the world: find a good stable star, settle down, avoid crowds, and above all, stay away from hot, young, heavenly bodies.

REFERENCES

Arons, J.; Kulsrud, R. M.; and Ostriker, J. P.: A Multiple Pulsar Model for Quasi-Stellar Objects and Active Galactic Nuclei. Astrophys. J., vol. 198, 1975, pp. 687–705, 707.

Begelman, M. C.; and Rees, M. I.: Can Cosmic Clouds Cause Climatic Catastrophes? Nature, vol. 261, May 27, 1976, pp. 298–299.

Chevalier, R. A.; Scott, J. S.; and Robertson, J. W.: Cosmic-Ray Acceleration and the Radio Evolution of Cassiopeia A. Astrophys. J., vol. 207, 1976, pp. 450–459.

Clark, D. H.; McCrea, W. H.; and Stephenson, F. R.: Frequency of Nearby Supernovae and Climatic and Biological Catastrophes. Nature, vol. 265, Jan. 1977, pp. 318–319.

Dole, S. H.: Habitable Planets for Man. 2nd ed., American Elsevier, New York, 1970.

Hart, M. H.: The Evolution of the Atmosphere of the Earth. Icarus, vol. 33, Jan. 1978, pp. 23–29.

Hart, M. H.: Habitable Zones About Main Sequence Stars. Icarus, vol. 37, Jan. 1979, pp. 351–357.

Herbst, W.: Investigations of Environmental Radiation and Its Variability. In: Natural Radiation Environment, J. Adams and W. Lowder, eds. Univ. of Chicago Press, 1964, pp. 781–796.

Higdon, J. C.; and Lingenfelter, R. E.: The Origin of Cosmic Rays and the Vela Gamma-Ray Excess. Astrophys. J., vol. 198, May 1975, pp. L17–L20.

Krasovski, V. I.; and Shklovsky, I.: Supernova Explosions and Their Possible Effect on the Evolution of Life on the Earth. Dokl. Akad. Nauk SSR, vol. 116, 1957, pp. 197–199.

Lingenfelter, R.: Pulsars and Local Cosmic Ray Prehistory. Nature, vol. 224, Dec. 20, 1969, pp. 1182–1186.

McCrea, W. H.: Ice Ages and the Galaxy. Nature, vol. 255, June 19, 1975, pp. 607–609.

Oort, J. H.: The Galactic Center — Structure and Radiation Characteristics. Ann. Rev. Astron. Astrophys., vol. 15, 1977, pp. 295–362.

Project Cyclops: A Design Study of a System for Detecting Extraterrestrial Intelligent Life. NASA CR-114445, 1973.

Reid, G. C.; McAfee, J. R.; and Crutzen, P. J.: Effects of Intense Stratospheric Ionisation Events. Nature, vol. 275, Oct. 1978, pp. 489–492.

Ruderman, M. A.: Possible Consequences of Nearby Supernova Explosions for Atmospheric Ozone and Terrestrial Life. Science, vol. 184, June 1974, pp. 1079–1081.

Russell, D.; and Tucker, W.: Supernovae and the Extinction of the Dinosaurs. Nature, vol. 229, 1971, pp. 553–554.

Schindewolf, O. H.: Uber die Moglichen Ursachen der grossen erdeschichtlichen Faunenschnitte. Neues Jahrb. Geol. Paleontaol. Monatsh., vol. 10, 1954, pp. 457–465.

Stecker, F. W.: Origin of Cosmic Rays. Phys. Rev. Letters, vol. 35, July 1975, pp. 188–191.

Terry, K. D.; and Tucker, W. H.: Biologic Effects of Supernovae. Science, vol. 159, Jan. 26, 1968, pp. 421–423.

Ulrich, R.: Solar Nutrinos and Variations in the Solar Luminosity. Science, vol. 190, Nov. 14, 1975, pp. 619–624.

Whitten, R. C.; Cuzzi, J.; Borucki, W. J.; and Wolfe, J. H.: Effect of Nearby Supernova Explosions on Atmospheric Ozone. Nature, vol. 263, Sept. 30, 1976, pp. 398–400.

Woltjer, L.: Supernova Remnants. Ann. Rev. Astron. Astrophys., vol. 10, 1972, pp. 129–158.

Biochemical Keys to the Emergence of Complex Life

KENNETH M. TOWE

The development of entirely new amino acids was the key opening the way to complex multicellularity. It permitted evolution of the structural proteins needed to provide mechanical support for increased size and greater morphogenetic experimentation.

If there is intelligent life elsewhere in the Universe with whom we might communicate, it is unlikely that such life is unicellular or primitively multicellular. It is more likely to be of the complex multicellular type — the so-called higher forms of life. The purpose of this paper is to emphasize the fundamental importance of two "rare" protein amino acids to the development of complex multicellular life on Earth.

There are only 20 amino acids found in proteins. A statement similar to this occurs in almost every biochemistry text. As a generalization to most proteins this statement is certainly true, but from the standpoint of the structural proteins and of the evolutionary history of higher organisms it can be rather misleading. Despite the repetitive, degenerate nature of the genetic code, there are two important protein amino acids that are uncoded: hydroxyproline and hydroxylysine. These two amino acids play critical roles in the formation of structural glycoproteins in both plants and animals. In an evolutionary context, the absence of these hydroxyamino acids from the genetic code implies that they are relative "newcomers" on the biological scene, and it is fair to conclude that their functional importance must therefore outweigh the complexities required for their formation.

One or both of these hydroxyamino acids occur in all metazoans (Bairati, 1972; Adams, 1978), higher plants, most higher algae (Lamport, 1977), and certain types of fungi (LéJohn, 1971). They are absent (or very rare) in "higher" fungi, procaryotes, protozoans, and red algae. In other words, these amino acids are widely distributed among the higher forms of life but are generally absent from lower organisms. The higher fungi may be the exception that proves the rule. The Ascomycetes and Basidomycetes, like the red algae from which they were probably derived (Demoulin, 1974), lack both hydroxyamino acids.

The biosynthesis of both hydroxyproline and hydroxylysine follows a pathway whereby the parent amino acids proline and lysine are hydroxylated only after incorporation into polypeptide linkage (see Miller and Matukas, 1974, for a review). The free amino acids cannot act as substrates. The steps in the formation of these two uncoded, posttranslational hydroxyamino acids are both specific and complicated. Hydroxylation is mediated by the enzymes proline and lysine hydroxylase, respectively. Both enzymes meet the same four requirements: (1) molecular oxygen, (2) ferrous iron, (3) ascorbic acid, and (4) α-ketoglutarate. Moreover, the same basic pathway is found to exist in both plants and animals (Chrispeels, 1976), indicating a common evolutionary origin in some primitive, and probably photosynthetic, ancestor (Aaronson, 1970; Lamport, 1977).

The peptide-bound hydroxyproline and hydroxylysine in metazoans appear in the structural protein *collagen*. In the higher plants and green algae, hydroxyproline occurs ultimately in a family of glycoproteins termed *extensins* (Lamport and Miller, 1971). The ultimate location of peptide-bound hydroxyproline or hydroxylysine in the other organisms in which it occurs is not known for certain.

The requirement for α-ketoglutarate by the hydroxylase enzymes is absolute and specific (Hutton et al., 1967). Since α-ketoglutarate is a tricarboxylic acid (Krebs) cycle intermediate, it may not be coincidental that those organisms in which the TCA cycle may be limited (Klein and Cronquist, 1967, p. 189) — photosynthetic bacteria, blue-green algae, and red algae — are also among those organisms generally lacking in hydroxyproline or hydroxylysine. α-Ketoglutarate is also a requirement in the synthesis of lysine via the aminoadipic acid pathway. Interestingly, the few organisms that synthesize lysine by this pathway — red algae, euglenoids, and some higher fungi — also lack these two hydroxyamino acids.

The enzyme requirement for molecular oxygen is also absolute. It has been demonstrated that the oxygen of the hydroxyl group is derived from molecular oxygen rather than from water in collagen hydroxyproline (Fujimoto and Tamiya, 1962; Prockop et al., 1962), in collagen hydroxylysine (Hausmann, 1967), and in extensin hydroxyproline (Lamport, 1963). Indeed, all collagens appear to require molecular oxygen, even those in the so-called anaerobic, parasitic worms (Fujimoto, 1967; Smith, 1969).

Although the basically similar hydroxylation reactions are important in both extensin and collagen, additional strength is provided through subsequent steps. It is the linkage between carbohydrate and protein that is crucial. In plants, the extensin hydroxyproline is covalently bonded to a sugar (arabinose) through an O-glycosidic linkage (Lamport, 1967). In metazoans, the situation is different. Collagen hydroxyproline, in association with glycine, forms and stabilizes the unique triple-stranded structure of the protein macromolecule (Berg and Prockop, 1973; Jimenez et al., 1973). It is the hydroxylysine in collagen that is linked to the carbohydrate (Butler and Cunningham, 1966). This material is further strengthened into true collagen fibrils by intermolecular crosslinks that involve lysine and hydroxylysine mediated by another oxygen-dependent enzyme, lysyloxidase (Traub and Piez, 1971). Hydroxylysine is essential to collagen crosslinks (Bailey et al., 1974). Among the numerous vertebrate and invertebrate collagens studied, only the nematode *Ascaris* is an exception (McBride and Harrington, 1967). The crosslinks are important to the high strength and chemical resistance necessary for the functioning collagen fibrils.

Is there significance in all of this for the origin and development of complex multicellular life? In plants, the precise function of extensin as a structural protein is not known, although its extracellular position and its association with periods of active growth in higher plants suggest an important role in the regulation of growth and development (Lamport, 1977). Similarly, collagen is also involved in developmental tissue interactions through extracellular matrices (see Lash and Burger, 1976). But in addition to this role, its primary role as the "tape and glue" of the metazoan world is undeniable. It has no substitute in this function as a connective-tissue protein. Collagen is the principal organic matrix component of vertebrate bone. It holds the calcitic plates of echinoderm skeletons together. Worm cuticles and body walls are collagenous, as is the mesoglea of jellyfish. It is a major component of sponges, and the muscles of bivalves are held to their shell with its help. These few examples serve to illustrate its importance.

The evolution of a functional substitute for collagen using only some combination of the 20 amino acids in the genetic code has not yet been achieved. And it is unlikely that it was ever achieved at any time in the past when one considers the lengths to which metazoans must go to manufacture this critical substance. First of all, a mixed-function oxygenase is required to synthesize hydroxyproline and hydroxylysine. Oxygenases are known to be comparatively inefficient and to compete with conventional respiratory pathways for free oxygen (Kaufman, 1962; Hayaishi, 1962). Second, the all-important intermolecular crosslinks that provide the mechanical strength also need oxygen (lysyloxidase) and in addition require more lysine — an amino acid that animals are unable to synthesize and must obtain through food sources. Add to this the evolution of the posttranslational pathway with four enzyme cofactors and you have a protein that is complicated and expensive to manufacture. Thus, if any organism in the past had found a simpler way to make this structural protein, or even a suitable substitute, it would have had a clear adaptive advantage. The importance of hydroxyproline and hydroxylysine is therefore indelibly expressed through their wide multicellular distribution in nature. Evolutionary increase in size and complexity through multicellularity was accomplished through oxidative metabolism, the eucaryotic condition, and structural proteins using new amino acids that placed a premium on the availability of free oxygen. Only the higher fungi have managed a small measure of success in increasing size and complexity without the new amino acids. But they lack mechanical strength, and this is reflected in their poor fossil record.

If the first 2.5 billion years of Earth history were dominated by an environment in which free oxygen was generally unavailable to the biosphere because of primitive reduced inorganic acceptors or "sinks," then one of the principal strategies of life would have been to adapt to this environment. Neither collagen nor extensin would have been possible in the absence of free oxygen, and the fossil record shows that during this period few significant increases in body size were made. After the exhaustion of most primordial inorganic oxygen sinks about 2 billion years ago (Cloud, 1976; Schopf, 1978) and the development of the eucaryotic cell and oxygenic photosynthesis, the major environments of the world began to turn from predominantly reducing to predominantly oxidizing conditions. Multicellularity would have been a priority development and collagen one of its certain prerequisites. The geologic evidence shows that the transition to an oxygenic world did not take place rapidly; if it had, there would be major deposits of reduced carbon in the sediments of the period, and none is found. The buildup of free oxygen is more likely to have been rather slow. With some limited free oxygen present in the environment, limited collagen (and extensin) could have evolved in some organisms where the wide-open ecological niches available and the adaptive advantages being conveyed outweighed the complexities and expense involved in its manufacture. I have already discussed how this may have affected the fossil record (Towe, 1970), but it is worth repeating that in this initially low-oxygen environment, the competition for the low levels of free oxygen by oxidative metabolisms would have restricted most collagen use to small, thin, diffusion-limited organisms unlikely to be found preserved as fossils. As Boaden (1977) has emphasized, the world at this time may have been like the modern thiobios — a sulfide-rich habitat dominated by mostly microscopic, interstitial meiofaunal elements.

The biochemistry of collagens from modern near-anaerobic nematodes is instructive in making further comparisons with the Late Precambrian fossil record. Both the cuticle and body wall of *Ascaris lumbricoides* contain collagen. The body-wall collagen has large amounts of hydroxyproline and hydroxylysine. Adapted to low oxygen tensions, their formation through the hydroxylating enzymes is actually inhibited by too much oxygen (>1%) in the environment (Fujimoto and Prockop, 1969). The cuticle collagen lacks hydroxylysine, contains little hydroxyproline, and appears to be strengthened by disulfide crosslinks (McBride and Harrington, 1967), another adaptation to low-oxygen tensions.

Given this information, one can speculate that the early use of collagen in Late Precambrian low-oxygen environments may have eventually produced "worms" with similarly adapted collagen metabolism, which permitted some of them to attain much larger sizes than the remaining interstitial faunal elements. Perhaps the enigmatic coiled fossils from the billion-year-old Greyson Shale (Walter et al., 1976) or from the Little Dal Group (Hofmann and Aitken, 1979) are rare body fossils of such worms. Or perhaps some of the Late Precambrian burrows were produced by similar worms who had become adapted to the low-oxygen environment and were, like their modern thiobiotic descendents, burrowing to avoid the increasing oxygen tensions that must inevitably have taken place. Burrowing to avoid oxygen at the sediment-water interface in this very early stage of metazoan history seems more likely than burrowing to avoid predators, the types of which are unknown and the fossil evidence for which is otherwise nonexistent.

Ultimately, the further increase in availability of free oxygen as the result of increasing oxygenic photosynthesis would have brought an end to such adaptations, and therefore any experiments toward developing collagens completely free of an oxygen requirement were terminated. At the same time, the high competitive priority of respiratory events for oxygen would have been moderated, allowing many more morphological experiments with collagen to take place in other previously limited metazoan phyla. Even the sclerotization of the arthropod cuticle, which is also inhibited by lack of atmospheric oxygen (Richards, 1951), would have been improved. All this would have caused a dramatic worldwide increase in the size and hence ready preservability of numerous organisms. The Late Precambrian–Early Cambrian fossil record can then be interpreted as an explosion of fossils rather than as a sudden eruption of metazoan phylogenesis with highly evolved, diverse, and morphogenetically advanced forms appearing suddenly side by side around the world, few of which have any plausible immediate ancestors as fossils.

What does the hydroxyproline-hydroxylysine connection with multicellular evolution have to tell us about phylogeny? Although the data are very incomplete and hence subject to change, there are some interesting observations that presently invite speculation and comment. These speculations fall into the category of the "wild and wooly," but they may serve to provoke thought and further data collection.

It is conventional to consider that the origin of the Metazoa lies somewhere among the protozoans (animal-like protists) and probably among those that are nonpseudopodial (Hanson, 1976). There is a problem with this viewpoint. It concerns the origin of collagen, which, although it occurs in all metazoans, including the Porifera (Garrone, 1978), has not been found in *any* protist. As pointed out by Delmer and Lamport (1977), "the origin of collagen predicates the origin of the Metazoa." And collagen requires hydroxyproline and hydroxylysine biosynthesis.

Hydroxyproline has been found in only one protozoan to date: the shell of an amoeboid foraminifera (Hedley and Wakefield, 1969). Hydroxylysine has never been reported in any protozoan. While it is true that many more protozoans need to be examined carefully (especially the choanoflagellates), it is still fair to ask those who will derive the Metazoa from animal-like protists to explain how collagen could have been evolved from organisms not equipped with the complex pathways required to synthesize either of the basic building blocks. To argue for the independent discovery in some protozoan of the complicated biosynthetic pathways involved is to suggest a highly unlikely convergence. Lamport (1977) recognized this problem and resurrected the alga *Volvox* as a branchpoint. It contains abundant hydroxyproline and has the appearance of a blastula. But the close derivation of the Metazoa from any alga requires, in addition to the origination of collagen, the loss of lysine synthesis and the replacement of phototrophy by ingestion. The ability to synthesize chitin and ferritin might also be a problem since the former occurs in few algae and the latter in none.

Implausible as it may sound, it is instructive to examine the potential of the fungi (fungus-like protists) in this regard. Considered as a broad group, fungi have the ability to synthesize hydroxyproline, chitin, cellulose, and even ferritin. They have already replaced phototrophy with heterotrophic absorption. And fungi were surely not derived from protozoa, simply because to do so would require that the protozoan ancestor, already having lost the ability to synthesize lysine, rederived it later — a peculiar step indeed. Whittaker (1977) has already noted that "metazoans with digestive tracts have probably evolved from absorptive flagellates and, in this evolution, internalized the process of food absorption and added it to the process of ingestion." Could some lower fungal type have given rise to a protometazoan by some neotenous or paedomorphic transformation during a flagellate stage? And if it were one of the fungi that had utilized the aminoadipic acid pathway for lysine synthesis, the loss of this capability might have conserved α-ketoglutarate for proline hydroxylation. There would have been nothing to lose but the ability to synthesize lysine and everything to gain in evolutionary potential.

It took over 2 billion years for life on Earth to evolve the capacity for complex multicellular development. This length of time alone would seem to be good evidence that, with or without free oxygen, no combination of the 20 coded amino acids could be found that could produce a structural protein comparable to the extensin-collagen family. And while free oxygen was pivotal for efficient respiration and the eucaryotic condition, these advances were also insufficient to allow for the increase in size and complexity we find in metazoans and higher plants. The development of entirely new amino acids was the real key to opening the way for complex multicellularity. This permitted the evolution of the structural proteins necessary to provide the mechanical support for increased size and morphogenetic experimentation. Only then could the long-vacant niches begin to be filled by larger, rapidly diversifying organisms better suited to fossilization. If there had remained only 20 amino acids in all proteins, Earth would probably still be dominated, as it was for much of the Precambrian, by procaryotes and smaller eucaryotes, and fossils would be rare everywhere.

ACKNOWLEDGMENTS

I thank Dr. James W. Valentine for agreeing to read a draft of the manuscript and for making helpful suggestions for its improvement.

REFERENCES

Aaronson, S.: Molecular Evidence for Evolution in the Algae: A Possible Affinity Between Plant Cell Walls and Animal Skeletons. Ann. N.Y. Acad. Sci., vol. 175, art. 2, 1970, pp. 531–540.

Adams, Elijah: Invertebrate Collagens. Science, vol. 202, 1978, pp. 591–598.

Bailey, A. J.; Robins, S. P.; and Balian, G.: Biological Significance of the Intermolecular Crosslinks of Collagen. Nature, vol. 251, 1974, pp. 105–109.

Bairati, A.: Collagen: An Analysis of Phylogenetic Aspects. Bull. Zool., vol. 39, 1972, p. 205.

Berg, R. A.; and Prockop, D. J.: The Thermal Transition of a Non-Hydroxylated Form of Collagen. Evidence for a Role for Hydroxyproline in Stabilizing the Triple-Helix of Collagen. Biochem. Biophys. Res. Comm., vol. 52, 1973, pp. 115–129.

Boaden, P. J. S.: Thiobiotic Facts and Fancies. Mikrofauna Meersboden, vol. 61, 1977, p. 45.

Butler, William T.; and Cunningham, Leon W.: Evidence for the Linkage of a Disaccharide to Hydroxylysine in Tropocollagen. J. Biol. Chem., vol. 241, 1966, pp. 3882–3888.

Chrispeels, Maarten J.: Biosynthesis, Intracellular Transport, and Secretion of Extracellular Macromolecules. Ann. Rev. Plant Physiol., vol. 27, 1976, pp. 19–38.

Cloud, Preston: Beginnings of Biospheric Evolution and Their Biogeochemical Consequences. Paleobiol., vol. 2, 1976, p. 351.

Delmer, D. P.; and Lamport, D. T. A.: Cell Wall Biochemistry Related to Specificity in Host-Plant Pathogen Interactions. In: Cell Wall Biochemistry, B. Solheim and J. Raa, eds. Proceedings of a symposium held at the University of Trowis, Trowis, Norway, Aug. 1976. Columbia Univ. Press, N.Y., 1977, pp. 85–104.

Demoulin, V.: Origin of Ascomycetes and Basidiomycetes — Case for a Red Algal Ancestry. Botanical Rev., vol. 40, 1974, pp. 315–345.

Fujimoto, Daisaburo: Biosynthesis of Collagen Hydroxyproline in *Ascaris.* Biochem. Biophys. Acta, vol. 140, 1967, pp. 148–154.

Fujimoto, D.; and Tamiya, N.: Incorporation of ^{18}O from Air into Hydroxyproline by Chick Embryo. Biochem. J., vol. 84, 1962, pp. 333–335.

Fujimoto, Daisaburo; and Prockop, Darwin J.: Protocollagen Proline Hydroxylase from *Ascaris lumbricoides.* J. Biol. Chem., vol. 244, 1969, p. 205.

Garrone, Robert: Phylogenesis of Connective Tissue. Morphological Aspects and Biosynthesis of Sponge Intercellular Matrix. Frontiers of Matrix Biology, vol. 5, 1978, p. 1.

Hanson, E. D.: Major Evolutionary Trends in Animal Protists. J. Protozool., vol. 23, 1976, pp. 4–12.

Hausmann, E: Cofactor Requirements for the Enzymatic Hydroxylation of Lysine in a Polypeptide Precursor of Collagen. Biochem. Biophys. Acta, vol. 133, 1967, pp. 591–593.

Hayaishi, Osamu: History and Scope. In: Oxygenases, O. Hayaishi, ed., Academic Press, N.Y., 1962, pp. 1–29.

Hedley, Ronald H.; and Wakefield, James: Fine Structure of *Gromia oviformis* (Rhizopodea: Protozoa). Bull. Brit. Mus. Nat. Hist., vol. 18, 1969, p. 67.

Hofmann, H. J.; and Aitken, J. D.: Precambrian Biota from the Little Dal Group, Mackenzie Mountains, Northwestern Canada. Canadian J. Earth Sci., vol. 16, 1979, p. 150.

Hutton, John J., Jr.; Tappel, A. L.; and Udenfriend, Sidney: Cofactor and Substrate Requirements of Collagen Proline Hydroxylase. Arch. Biochem. Biophys., vol. 118, 1967, p. 231.

Jimenez, Sergio; Harsch, Margaret; and Rosenbloom, Joel: Hydroxyproline Stabilizes the Triple Helix of Chick Tendon Collagen. Biochem. Biophys. Res. Comm., vol. 52, 1973, pp. 106–114.

Kaufman, Seymour: Aromatic Hydroxylations. In: Oxygenases, O. Hayaishi, ed., Academic Press, N.Y., 1962, pp. 129–180.

Klein, Richard M.; and Cronquist, Arthur: A Consideration of the Evolutionary and Taxonomic Significance of Some Biochemical, Micromorphological, and Physiological Characters in the Thallophytes. Quart. Rev. Biol., vol. 42, no. 2, June 1967, pp. 105–296.

Lamport, Derek T. A.: Oxygen Fixation into Hydroxyproline of Plant Cell Wall Protein. J. Biol. Chem., vol. 238, no. 4, 1963, pp. 1438–1440.

Lamport, Derek T. A.: Hydroxyproline-O-glycosidic Linkage of the Plant Cell Wall Glycoprotein Extensin. Nature, vol. 216, 1967, pp. 1322–1324.

Lamport, Derek T. A.: Structure, Biosynthesis and Significance of Cell Wall Glycoproteins. Recent Adv. Phytochem., vol. 11, 1977, pp. 79–115.

Lamport, Derek T. A.; and Miller, David H.: Hydroxyproline Arabinosides in the Plant Kingdom. Plant Physiol., vol. 48, 1971, pp. 454–456.

Lash, J. W.; and Burger, M. M., eds.: Cell and Tissue Interactions. Soc. Gen. Physiol. Ser., vol. 32, 1976.

LéJohn, H. B.: Enzyme Regulation, Lysine Pathways and Cell Wall Structures as Indicators of Major Lines of Evolution in Fungi. Nature, vol. 231, 1971, pp. 164–168.

McBride, O. Wesley; and Harrington, William F.: *Ascaris* Cuticle Collagen: on the Disulfide Cross-Linkages and the Molecular Properties of the Subunits. Biochemistry, vol. 6, 1967, pp. 1484–1498.

Miller, Edward J.; and Matukas, Victor J.: Biosynthesis of Collagen. Fed. Proc., vol. 33, May 1974, pp. 1197–1204.

Prockop, D.; Kaplan, A.; and Udenfriend, S.: Oxygen-18 Studies on the Conversion of Proline to Hydroxyproline. Biochem. Biophys. Res. Comm., vol. 9, 1962, pp. 162–166.

Richards, A. Glenn: The Integument of Arthropods. The Chemical Components and Their Properties, the Anatomy and Development, and the Permeability. Univ. of Minnesota Press, 1951.

Schopf, J. William: The Evolution of the Earliest Cells. Sci. American, vol. 239, 1978, pp. 110–138.

Smith, Malcolm H.: Do Intestinal Parasites Require Oxygen? Nature, vol. 223, 1969, pp. 1129–1132.

Towe, Kenneth M.: Oxygen-Collagen Priority and the Early Metazoan Fossil Record. Proc. Natl. Acad. Sci. USA, vol. 65, 1970, pp. 781–788.

Traub, Wolfie; and Piez, Karl A.: The Chemistry and Structure of Collagen. In: Advances in Protein Chemistry, vol. 25, 1971, pp. 243–352.

Walter, M. R.; Oehler, John H.; and Oehler, Dorothy Z.: Megascopic Algae 1300 Million Years Old from the Belt Supergroup, Montana: A Reinterpretation of Walcott's *Helminthoidichnites.* J. Paleontol., vol. 50, 1976, pp. 872–881.

Whittaker, R. H.: In: Protozoa of Medical and Veterinary Interest, J. P. Kreier, ed., Academic Press, N.Y., 1977.

Gravity, Lignification, and Land Plant Evolution

S. M. SIEGEL, B. Z. SIEGEL, and JUNG CHEN

Only that algal group commonly believed ancestral to land plants produces lignin from monomeric precursors. The lignin content of common plants can be reduced by gravity compensation and increased by mechanical load. Evolutionary opportunism?

At some point in early Paleozoic time — the Upper Ordovician perhaps — vascular plants began their occupation of the wetlands interfacing both terrestrial and marine environments. Whether this movement was a bona fide invasion driven by increasing population pressures in the marine littoral, an accident of gradual subsidence, or the aftermath of a period of exceptional tides will probably never be known, but the impact on the terrestrial landscape and on the course of subsequent evolution is unmistakable.

Despite their juxtaposition, these early populations, save for thalloid forms appressed closely to rocks, can hardly be considered to occupy a safe transition zone. The early Psilophytean derivatives, with their emergent vertical, branched axes, left behind an existence in the comfortable protective sea and immediately faced problems of support, water supply, gas exchange, and radiation exposure. To be sure, solutions calling for minimal adaptation exist within the time dimensions of tidal periodicity, but these must have limited the deployment of terrestrial life forms to the strand and coastal marsh — the continental fringe — with no guarantee of permanence.

We are here concerned with those forms able to cope with a more or less modern atmosphere, strong sunlight, little water, and unrelieved gravity, and with the ways in which they may have differed constitutionally from their Chlorophycean ancestors. If we neglect the host of minor and inconsistent variations seen among modern green algae (and the more organized

307

Charophytes) and their most primitive vascular descendants, one of the most consistent differences resides in the abundance of phenolics, especially lignins, nonnitrogenous polymers which may comprise 10, 30, 50% or more of the total dry matter in trees (Brauns, 1952; Wise, 1952) and are significant constituents, with important exceptions, even of small herbaceous vascular plants. In contrast, the phenolics of the green algae consist principally of tyrosine and related aromatic amino acids, which make up a small to minute percentage of dry matter, almost completely in proteins and peptides.

To what extent, then can lignins account biophysically and biochemically for the adaptive processes involved in the emergence of the vascular land plant? And how is this process of lignification programmed and regulated?

Can the antiquity of the lignins be verified? Spectrochemical and chemical analysis of Devonian and Pennsylvanian fossils verifies the presence of phenyl methyl ethers, characteristic of ligning; and in the case of *Calamites* and *Lepidodendron*, extractable lignin-like fractions have been obtained in small amounts (Manskaya, 1959; Siegel et al., 1958; Siegel, 1968). More recent specimens (Pliocene) contain large amounts of lignins with reactive aldehyde groups as weil as methyl ethers (Siegel, 1968).

In other respects, answers to questions relating to the phylogeny and adaptive significance of the lignins must depend on experiments and observations using contemporary plant material. The remainder of this paper will summarize a series of such observations.

Observation: Conversion of orthosubstituted para-hydroxycinnamic acid to lignin requires the enzyme peroxidase (Higuchi, 1959; Siegel, 1968; Siegel et al., 1960). Although forms of this enzyme occur and even abound in many algae (table 1), only those present in the Chlorophyta can utilize the lignin precursor as a substrate (Siegel and Siegel, 1970).

Significance: Only that algal group commonly accepted to be ancestral to land plants can produce lignin from monomeric precursors.

Observation: When peroxidase-loaded fibers are incubated with lignin precursors, polymer is deposited on polysaccharides but not on proteins. And less-crystalline, lower-molecular-weight polysaccharides are more "efficient" by one or two orders of magnitude (table 2).

Significance: Fiber models show that the polymerization process is selective of the plant cell surface, but that secondary-wall polysaccharides (pectin and equivalents) are favored. This corresponds to established histochemistry (Siegel, 1956, 1957, 1959; Wardrop and Bland, 1959).

Observation: All terrestrial vascular plants, angiosperms (and gymnosperms), ferns, horsetails, etc., contain lignin (table 3), whereas aquatic and marine vascular forms and ordinary nonvascular plants do not (Siegel et al., 1972). Nevertheless, all plants tested that had Chlorophycean affinities can

TABLE 1.– CAPACITY OF VARIOUS ALGAE AND ALGA-LIKE FORMS
TO PEROXIDIZE MONOMERIC LIGNIN PRECURSORS

Group and examples	Relative ability to peroxidize:		
	Simple phenol	p-hydroxy, unsubstituted	Cinnamic acid, O-substituted
Tracheophytes (pea, cucumber, ferns, etc.)	1.0	1.0	1.0
Cyanophytes *(Nostoc, Oscillatoria)*	0–1.0	0–1.0	0
Rhodophytes *(Porphyra, Gelidium, Rhodymenia)*	0.5–1.5	0.8–1.6	0
Phaeophytes *(Sargassum, Postelsia, Laminaria)*	0.6–1.5	0.5–2.0	0
Chlorophytes *(Nitella, Valonia, Ulva, Chara)*	0.6–1.3	0.8–1.9	1.0–1.5

TABLE 2.– SURFACE COMPOSITION
REQUIRED FOR LIGNIN POLYMER
DEPOSITION: FIBER MODEL
SYSTEMS

Surface material	Yield of lignin, mg/kg
Cotton	10–45
Purified cellulose	12–25
Acetylated cellulose	<0.1
Alkali methyl cellulose	>1000
Pectin	>1000
Agar	Active
Chitin (molluscan)	<0.1
Deacetylated chitin	73
Starch	>1000
Hair	<0.1
Keratin	<0.1
Fibrin	<0.1
Collagen	<0.1

synthesize lignin from p-hydroxycinnamic acid-type monomers such as
caffeic acid or coniferaldehyde. Excluded are the red and brown algae, blue-
green forms, and the fungi (with rare exceptions only among wood-rotting
types).

TABLE 3.– GROUP AND HABITAT RELATIONS IN LIGNIFICATION

Group and examples	Lignification of axis	
	Endogenous source	Exogenous source
Angiosperms		
Terrestrial (bamboo, maize, bean, cucumber)	+	+
Aquatic *(Elodea, Vallisneria,* Lotus)	–	+
Marine *(Halophila, Zostera)*	–	+
Ferns and other Vascular Plants		
Terrestrial *(Polypodium, Equisetum, Selaginella)*	+	+
Aquatic *(Ceratopteris, Marsilea)*	–	+
Nonvascular Plants		
Terrestrial		
Mosses (normal size)	–	+
Liverworts	–	+
Fungi	–	rare
Aquatic-marine		
Green algae (chlorophytes)	–	+
All other algae	–	–

Significance: The inability to realize a potential for lignification means that (a) it never developed or (b) it is not now operational. The aquatic and marine vascular forms are generally regarded as "cetaceans" of the plant kingdom with a fully developed land ancestry (Siegel, 1979) followed by loss or "shutdown" of the required capabilities for continued existence out of water (option (b)).

Among the nonvascular forms in aquatic-marine habitats, the nongreen algae are assigned to option (a), but the status of the green algae is more conjectural. Originally, the Chlorophyta were assumed to be in option (a), but this may be the modern condition. Lignified fibers, lacking resiliency, confer brittleness on pliant rubbery algal tissues, and lignified algae anchored along rocky coastlines may well have disintegrated under wave impact. Conceivably, the capability for lignification only appeared among evolving wetlands or intertidal populations in latent (repressed?) form and was subsequently activated.

Is there then any evidence for the presence of the requisite information for lignification in latent and/or facultative form?

Observation:　In mosses, the sporophyte (2N) generation contains semi-lignified structures whose function is tied to its hydrophobic character. The common 1–10 cm gametophyte (1N) axis contains less than 0.3% lignin, if any (fig. 1). Gametophytes of 40–60 cm may contain 5–6% lignin, and the axes of the giant New Zealand mosses, up to 100 cm in height, contain 10–12% lignin (Siegel et al., 1972).

Significance:　Even the smallest moss contains genetic information for lignification, but only those extended sufficiently above the surface plane encounter mechanical stresses sufficient to induce its redeployment into the upright axis.

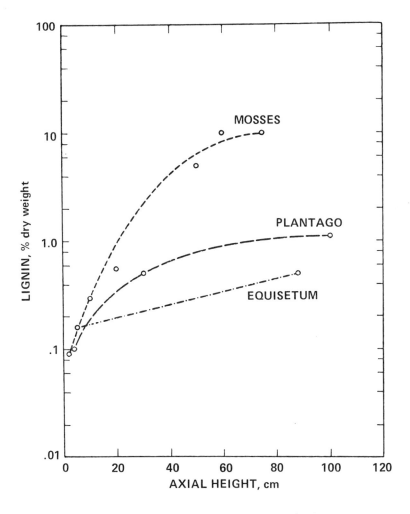

Figure 1. *Lignin content as a function of axial height in related moss species, plantain* (Plantago) *ecotypes and horsetail* (Equisetum) *ecotypes.*

Observation: In addition to mosses, dwarf, ordinary, and giant ecotypes of common plantain and horsetail vary in total lignification with axial height (fig. 1). Even within the axis of a semiwoody plant, the lignin content from tip to base is closely related to mechanical loading (pressure of the axis on itself; fig. 2).

Significance: Either the original (nonvascular) ability to synthesize lignin was not size-related but became so in the mosses as their axes were extended upward, or the mosses and their relatives are degenerate forms of vascular origin. In vascular plants, lignin is largely facultative and mechanical-load-dependent. It may thus be a function of the gravity-mass interaction.

Observation: By the use of clinostat, flotation, and centrifuge techniques (Chen et al., 1980; Siegel, 1968; Siegel et al., 1972, 1977; Waber et al., 1975), the lignin content of common plants can be reduced under conditions of gravity compensation and increased with mechanical load (table 4). Of particular note is the response of *Elodea,* an angiosperm returned to the aquatic habitat. We have shown that *Elodea,* without growth, can be induced to form lignin from endogenous sources while on the centrifuge for 6 days (Chen et al., 1980).

Significance: The inference drawn (earlier) from size-lignin relations that lignification is a function of gravitational load is here confirmed using *g* as the variable. However, the original conclusion (Siegel, 1953) that *Elodea*

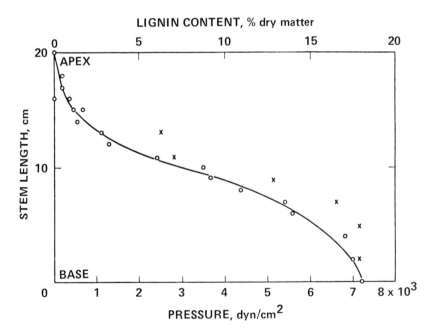

Figure 2. *Lignin content as a function of compression force generated on the axis of a semiwoody plant* (Euphorbia) *by its own mass.*

TABLE 4.– REGULATION OF LIGNIN CONTENT IN SIX SPECIES BY
SIMULATED GRAVITY

Species	Condition	Relative lignin content for a g-value (relative) of:				
		0	1	25	50	75
Anagallis	Clinostat/centrifuge	0.32	1.00	2.07	3.11	2.60
Cucumber	Clinostat, flotation/centrifuge	.46	1.00	1.69	2.30	3.12
Elodea	Centrifuge	1.00	1.00	---	---	3.74
Bean Cotyledon	Centrifuge	---	1.00	---	---	1.81
Mangrove	Centrifuge	---	1.00	6.78	10.22	12.82
Marigold	Clinostat, flotation	.64	1.00	---	---	---

Note: The g-values are relative measures of gravity: $g = 1$ corresponds to the present terrestrial gravitational load.

and related aquatic or marine forms had lost the capacity for monomer, but not polymer, synthesis, is refuted. The alternative is gravity-induced derepression on induction of a long-dormant program for monomer synthesis.

We assume that the gravitational "switch" operates on phenylalanine/tyrosine α-deaminases, which convert the aromatic amino acids into cinnamic and p-coumaric acids, respectively (Brown, 1964; Conn, 1964; Neish, 1964). These in turn, after relatively minor modifications, become monomers.

Observations: Withholding water from air-grown cucumber seedlings enhances lignification (table 5). Similar effects can be obtained with water-grown plants in a saline medium. An O_2 deficiency has the opposite effect. Both cucumber and bean seedlings grown in atmospheres of 5% O_2 have about half the normal lignin content (Siegel, 1953).

Significance: Lignins are hydrophobic, resinous substances and have long been recognized as waterproofing agents for vascular and epidermal cells. The lignins thus overlap with the polyester epidermal covering known as "cutin," not only in the waterproofing role, but often in the outer walls where the spectrochemical properties of the lignins, strong extinction of near ultraviolet radiation, can be expressed (Eglinton and Hamilton, 1967). Their function in smaller moss sporophytes, as noted above, is also related to this hydrophobic character. Reductions either in bulk water supply or in water activity apparently comprise or generate a signal which in turn promotes lignin synthesis.

The role of O_2 in this system is simple and direct. It is needed for synthesis of specific monomers from phenylalanine or tyrosine, but its principal

TABLE 5.— OXYGEN AND WATER RELATIONS
IN CUCUMBER LIGNIFICATION

Experimental condition	Lignin content (relative)	
	Root	Shoot
Water		
Planted in soil (21 days)		
Daily water	100	100
Daily, first 5 days only	187	151
Flotation		
On water	57	62
On 1% NaCl	88	95
Oxygen, %		
21 (air)	100	100
10	90	66
5	76	45
1	38	>5

role is in monomer dehydrogenation to the lignin polymer (Siegel et al., 1972). The system is first order in O_2 pressure; thus any variations in oxygen concentrations will be reflected directly in rates of peroxidation.

The differences between modern Chlorophyta and vascular land plants cannot be explained in full on the basis of lignification alone. Nevertheless, the data point to the emergence of the primitive land populations into an oxygen-rich terrestrial world where the need for mechanical support, water conservation, and, to some degree, radiation protection could be met by a single aerobic biochemical process connected to essential aromatic amino acids likely to be found in every cell. If there is such a thing as evolutionary opportunism, this must be a good example.

ACKNOWLEDGMENTS

This work was supported in part by National Aeronautics and Space Administration grants NASw-767 and NGR 12-001-053, and Contracts NAS2-6624 and NAS2-8687.

REFERENCES

Brauns, Friedrich E.: The Chemistry of Lignin. Academic Press, N.Y., 1952.

Brown, S. A.: Lignin and Tannin Biosynthesis. In: Biochemistry of Phenolic Compounds, J. Harborne, ed., Academic Press, New York, 1964, pp. 361-398.

Chen, N.; Siegel, S. M.; and Siegel, B. L.: Gravity and Land Plant Evolution — Experimental Induction of Lignification by Simulated Hypergravity and Water Stress. In: Life Sciences and Space Research, vol. 18, 1980, pp. 193-198.

Conn, E. E.: Enzymology of Phenolic Biosynthesis. In: Biochemistry of Phenolic Compounds, J. Harborne, ed., Academic Press, New York, 1964, pp. 399-456.

Eglinton, Geoffrey; and Hamilton, R.: Leaf Epicuticular Waxes. Science, vol. 156, June 9, 1967, pp. 1322-1335.

Higuchi, Takayoshi: The Biosynthesis of Lignin. Proc. 4th Intern. Vienna Congr. Biochem., vol. 2, 1959, pp. 161-188.

Manskaya, S. M.: Phylogenesis of the Lignins. Proc. 4th Intern. Congress Biochem., vol. 2, 1959, pp. 215-226.

Neish, A. C.: Major Pathways of Biosynthesis of Phenols. In: Biochemistry of Phenolic Compounds, Academic Press, New York, 1964, pp. 295-359.

Siegel, B. Z.: and Siegel, S. M.: Anomalous Substrate Specificities Among the Algal Peroxidases. Amer. J. Bot., vol. 57, 1970, pp. 285-287.

Siegel, S.; Speitel, T.; Shiraki, D.; and Fukumoto, J.: Effects of Experimental Hypogravity on Peroxidase and Cell Wall Constituents in the Dwarf Marigold. In: COSPAR, Life Sciences and Space Research, vol. XVI, R. Holmquist, ed., Pergamon Press, 1977, p. 105.

Siegel, S.: In: Subcellular Particles, Teru Hayashi, ed. A symposium held during the meeting of the Soc. of General Physiologists, Ronald Press, New York, 1959.

Siegel, S. M.: Biosynthesis of Lignins. Physiol. Plant., vol. 6, 1953, pp. 134–139.

Siegel, S. M.: The Biosynthesis of Lignin: Evidence for the Participation of Celluloses as Sites for Oxidative Polymerization of Eugenol. J. Amer. Chem. Soc., vol. 78, 1956, p. 1753.

Siegel, S. M.: Non-enzymic Macromolecules as Matrices in Biological Synthesis: The Role of Polysaccharides in Peroxidase-Catalyzed Lignin Polymer Formation from Eugenol. J. Amer. Chem. Soc., vol. 79, 1957, p. 1628.

Siegel, S. M.: Biochemistry of the Plant Cell Wall, Ch. 1. In: Comprehensive Biochemistry, Marcel Florkin and Elmer H. Stotz, eds., Elsevier, N.Y., vol. 26A, 1968, pp. 1–48.

Siegel, S. M.: Gravity as a Biochemical Determinant. In: COSPAR, Life Sciences and Space Research, vol. XVII, R. Holmquist, ed., Pergamon Press, 1979, pp. 147–160.

Siegel, S. M.; Carrol, Paula; Umeno, Iram; and Corn, Carolyn: Evolution of Lignin; Experiments and Observations. Recent Adv. Phytochem., vol. 4, 1972, pp. 223–238.

Siegel, S. M.; Frost, P.; and Porta, F.: Effects of Indoleacetic Acid and Other Oxidation Regulators on In Vitro Peroxidation and Experimental Conversion of Eugenol to Lignin. Plant Physiol., vol. 35, 1960, pp. 163–167.

Siegel, S. M.; LeFevre, B., Jr.; and Borchardt, R.: Ultraviolet-Absorbing Components of Fossil and Modern Plants in Relation to Thermal Alteration of Lignins. Am. J. Sci., vol. 256, 1958, pp. 48–53.

Waber, J.; Williams, B. J.; Dubin, J.; and Siegel, S. M.: Changes Induced in Peroxidase Activity Under Simulated Hypogravity. Physiol. Plant., vol. 34, 1975, pp. 18–21.

Wardrop, A. B.; and Bland, D. E.: Process of Lignification in Woody Plants. Proc. 4th Intern. Congress Biochem., vol. 2, 1959, pp. 92–116.

Wise, Louis E., ed.: Wood Chemistry. Reinhold Publ. Co., N.Y., 1952.

Evolution of Man and Its Implications for General Principles of the Evolution of Intelligent Life

C. OWEN LOVEJOY

While "intelligent" animals may evolve on other planets by relational pathways similar to those seen on Earth, the phenomenon of "cognition" is quite distinct from that of intelligence and may well be quite rare.

Whether we are alone in the Universe is a perplexing problem admitting no easy solution. Unless another biological system is discovered, we have only one laboratory from which to generate hypotheses — our own biosphere. If one suggests that it is improbable that intelligent life exists elsewhere in the Universe, a common reply is "intelligent life *as we know it.*" This is nihilistic in the sense that what does not exist cannot be discussed; our knowledge is necessarily restricted to empirical experience and our capacity to imagine. What is known or imagined can be discussed and considered; what is not imagined is obviously beyond discussion.

Since we have no data on any exobiological system, we are also limited to a discussion of *process*, at the molecular, organismic, or ecosystemic level. It is reasonable to assume that organic evolution, as a process, obeys in exobiological systems the same fundamental strictures that define it on Earth, and that it consistently proceeds by differential reproductive success in variants of self-duplicating molecular bundles of one sort or another. In short, even if we cannot observe the structure of exobiological organisms, we can make quite reasonable inferences about the processes by which they might evolve.

Our existence on this planet provides the de facto basis for the conclusion that the evolution of intelligent life is possible. The critical point is estimating its probability. Such estimates have been frequently offered in recent years and range from 10^{-7} to 10^{-9} (Miller and Orgel, 1974) for our Galaxy. These estimates take into account such factors as the number of stars, the rate of star formation, the average number of planets per star, and so on. Any such estimate also requires a probability for the evolution of intelligent life. It is difficult to specify a basis for determining this particular probability. Usually it is simply stated and is clearly accompanied by several unstated assumptions about the evolutionary process, including the belief that biological systems, once initiated, tend to evolve toward higher levels of complexity, and that such complexity naturally leads to intelligent life. It is with these assumptions that I wish to deal here.

Before the fairly recent development of a synthetic theory of organic evolution, it was often held that organisms existed because of an *élan vital*. Huxley once commented that this explanation was about as effective as the invocation of an *élan locomotif* to account for the motion of a train. Early evolutionists believed in orthogenesis — that evolutionary progress was determined by some ultimate goal. Everything that we have learned about organic evolution contradicts such a view. Simpson (1949) concluded that only one universal trend characterizes evolution: "a tendency for life to expand, to fill in all the available spaces in the livable environments, including those created by the process of that expansion itself." Nothing that we know of the evolutionary process or of the factors that led to the appearance of man justify an *élan intelligent* or the existence of intellogenesis. If we wish to make estimates of the probability of intelligent life on suitable planets, then we must clearly identify the events and processes by which it appeared on this planet. This is the only method available short of pure fantasy. Are there fundamental evolutionary *trends* that led to the appearance of man, or is man to be viewed as a highly specific *event?* The question is far from trivial. If man represents the final expression of a recurrent process, then other communicating beings probably exist in the Universe. If, however, he is the historical consequence of a nonsystematic sequence of specific events, then the probability of other communicating beings depends directly on the likelihood of similar historical sequences. The question can be broken down into three specific questions:

1. Do biological systems naturally evolve toward more complex states?
2. Does increased intelligence favor survival and reproduction? That is, is it generally under positive selection?
3. Does increased intelligence generally evolve toward cognition?
The remainder of this paper will address each of these questions briefly.

EVOLUTION OF COMPLEXITY

An almost bewildering continuum of complexity exists in Earth's biosphere. It is natural to assume that this complexity, which is the product of evolution as interpreted from the fossil record, has evolved over a considerable time period and is the product of a tendency toward entropy reduction. An organic system, once established, gradually develops individual units (organisms) of greater complexity. Why should such a process occur? It is frequently assumed that complexity enhances the fitness of an organism — it increases its chances of survival and reproduction. This is, however, counterintuitive. A complex machine is inherently less reliable than a simple one. The principle of decreasing reliability with greater complexity is also evident from our knowledge of organic evolution: the survival of a mammal depends on the successful coordination and functioning of many highly sensitive organ systems.

One problem lies in a failure to distinguish between ecosystems and organisms. The principles governing these two kinds of antientropic devices are frequently quite distinct. Complex organisms are more likely to become extinct than simple ones, but complex ecosystems are more stable than simple ones. Organisms require some means of energy transformation or a direct supply of energy, and their energy resources are often limited. Efficiency in energy utilization militates against the development of backup systems, since such systems will only be required in unusual circumstances and are energy costly, reducing the chances of a particular organism's successful competition with conspecifics or other species. This process, constantly repeated in the evolutionary cycle, is known as specialization. The koala represents a pinnacle of adaptation with respect to finding, reaching, masticating, digesting, and excreting eucalyptus. It performs these functions with greater efficiency than any other mammal, but its systematic improvement of its ability in this area has reduced its capacity to respond to ecological change. Any serious interruption of eucalyptus supply could lead to its rapid extinction. Indeed, the great majority of mammal species have become or will become extinct because of specialization.

In a simple ecosystem, the loss of a single organism can be seriously disruptive. In a complex one, a loss is more likely to lead to replacement by a similarly functioning organism or minimal reactive shifts by a large number of participating organisms. In ecosystems, then, stability comes with complexity, and ecosystems tend toward complexity. These principles do not apply to organisms.

The fossil record shows quite clearly that Earth's most complex organism, the mammal, is recent (about 100–150 million years old) and is the end result of a trend toward more highly integrated organisms. This commonly held viewpoint is probably incorrect, as pointed out by Dobzhansky (1970, p. 392):

> . . . the transition from unicellular to multicellular organisms clearly involved a structural complication. Consider, however, the evolutionary sequence fish → amphibian → reptile → mammal → man. The sequence is usually taken to be progressive, yet "it would be a brave anatomist who would attempt to prove that recent man is more complicated than a Devonian ostracoderm." Although the development of a variety of sense organs is generally taken as progress, mammals and man lack certain kinds of senses present in other vertebrates — the lateral line organs of fishes, which perceive variations in pressure, or the directional receptors for heat radiation present in pit vipers.

If not a trend toward complexity, what then is the thematic characteristic of the sequence cited by Dobzhansky? The answer lies in the evolution of ecosystems and, more specifically, in trophic pyramids.

Once an entity is capable of energy transformation and utilization (whether by photosynthesis or perhaps by chemosynthesis), the stage is set for the development of "higher" organisms that can subsist on these auto-trophs. More complex mechanisms, especially those of food location (sense organs), locomotion, and digestion, must be evolved to ensure a constant supply of autotrophs. The so-called advanced adaptations of mammals are accountable almost entirely as specializations in food procurement and utilization. Vision, olfaction, rapid locomotion, and learning are all means of food location and capture. As each level of the trophic pyramid becomes established, the occupation of a higher level becomes possible, although the number of levels is strictly limited by the loss of about 90% of absorbed energy as heat at each level (Odum, 1962, 1971). The evolution of this trophic pyramid is obvious from the fossil record, and it clearly explains the different kinds of organisms in the chain cited above. Each presents a new form of exploitation of lower trophic levels, and the novel adaptations in each are directly related to this fact. Some obvious corrolaries of trophic evolution can be stated:

1. Evolution continued in each level, but there is no trend toward com-plexity. (Are angiosperms more "complex" than gymnosperms?) In fact, there are clearly recognizable trends toward simplification (e.g., mollusks).

2. There is a clear limitation and end point to the process of trophic pyramidization.

3. Diversity is reduced with each succeeding level added to the pyramid. There are about 2 million named species on Earth, including about 42,000 species of vertebrates and about 5000 species of mammals (Dobzhansky, 1970).

The above outline is subject to much variation and exception. All levels of the ecosystem are continuously reshuffled as evolution proceeds. Many shortcuts and new routes are established (herbivorous mammals feed directly on monocotyledinous angiosperms; mammals utilize bacterial fermentation in digestion), but the rules generally hold and explain the process of organic diversification much more satisfactorily than a supposed trend toward complexity.

Huxley once likened the process of evolution to the filling of a barrel. If we first put large rocks into a barrel, some (ecological/eutrophic) space remains; it can be filled with smaller rocks, then gravel, then sand, and so on. The important point is that the barrel does not expand, and that the remaining space is continually reduced. Once the barrel is filled, the system continues to reshuffle and adjust, but there is no inherent drive toward complexity. It is probable that the placental mammals represent the last major addition (in terms of trophic level) to Earth's ecosystem, and there is no basis to assume that, had man not evolved, some still more complex form of life would have appeared on this planet at some point.

EVOLUTION OF INTELLIGENCE

Our second question concerns the postulated tendency of complex organisms to develop progressively increased intelligence. Three orders of extant mammals — the Cetacea, Proboscidea, and Primates — demonstrate a greater degree of encephalization (brain volume allometrically adjusted for body mass) than other mammals. The question of intelligence in the Cetacea is still unresolved, but it is clear that much, if not most, of their unusual brain size is a direct consequence of their elaborate development of echolocation. They are social animals as well, and this also increases their need for association areas and pathways. It is difficult to judge the relative success of these mammals, inasmuch as they are the only major mammalian-grade fauna in the world's great ocean expanses, the pinnipeds (carnivora) being restricted to inland and coastal regions. The relatively small biomass of the cetaceans (compared to the remaining ocean biomass) is evidence of the limitations on the uppermost position of trophic pyramids.

The success of the Proboscidea is of interest because their levels of encephalization are also markedly higher than those of other mammals (Jerison, 1973). Yet they are clearly a relatively unsuccessful group, being represented today by only two living species, the remains of a much greater radiation that began to wither in the Miocene era (Maglio and Cooke, 1978). This order shows clearly that encephalization in no way guarantees the survival of a species or grade. The Proboscidea are survived by less-encephalized but equally large occupants of similar ecological zones.

In terms of understanding the role of intelligence in the survival and evolutionary progression of mammals, however, it is best to turn to the order Primates since our knowledge of the anatomy and fossil record of this group is relatively complete. In fact, the order provides a direct test of many of the claims often made for the selective advantages of "intelligence."

The primates can be traced to the earliest stage of mammal radiation at the beginning of the Cenozoic era (Simons, 1972). The origin of the adaptive complex that defines them can be related to the occupation of an arboreal econiche *and* a feeding strategy heavily reliant on insects and small verte-brate prey (Cartmill, 1975). Only this particular combination of selective factors can explain the grasping, prehensile chiridia of primates in contradis-tinction to the more effective clawed chiridia of arboreal insectivores. Sev-eral anatomical trends can be identified by a review of the comparative anatomy and fossil record of living and extinct primates. Among these are a tendency to retain many generalized ("primitive") characters, with the addi-tion of stereoscopic vision, olfactory reduction, and social behavior marked by intense parenting. Other trends include the bearing of single offspring, increased placental "efficiency," increased life span and maturation time, and finally an increase in the degree of specialization. In the search for trends toward complexity and intelligence, it should always be remembered that this order is distinguished by membership in the cohort unguiculata, the most primitive of the four mammalian cohorts (Colbert, 1969).

Encephalization in primates is a clear consequence of several anatomical complexes, including hand-eye coordination, keen vision, intense parenting, and complex social behavior. Ignoring for a moment the specific evolution-ary history of man, it is instructive to look at the sequence of events that can be read from the primate fossil record. All of the above trends can clearly be seen in the hominoids of the Early and Middle Miocene. This group was directly ancestral to the present-day great apes (gorilla, chimpanzee, orangu-tan) and man and inhabited most of the Old World. Enough is known to sug-gest that the primate trends described above were for the most part fully expressed in this group in a fairly advanced state, including considerable encephalization (Radinsky, 1974, 1975). Most of this radiation disappeared during the Late Miocene (Andrews and van Couvering, 1975), and its few extant descendants are relict species surviving only in small areas of the Old World in which few significant environmental changes have occurred (Lovejoy, 1981).

The Dryopithecinae are the best case of mammalian evolution in which to investigate the selective value of intelligence. Both the great apes and man are their direct descendants. Elements related to the evolution of cognition in man (prehensile forelimbs, social behavior, stereoscopic vision) were clearly present in the Dryopithecinae. Yet the descendants of this group (with the exception of man) are either extinct or relict and proceeding toward extinction. The primary reason for their disappearance is probably demographic and related to the reproductive process.

Two distinct strategies of reproductive behavior are often recognized in animal studies and referred to as "r" and "K" selection. While these terms represent ends of a continuum of reproductive adaptation, they are a useful heuristic device with which to discuss the dynamics of reproduction.

Only a portion of an organism's total energy budget can be used for reproduction. Other activities (predator avoidance, food searching, digestion, etc.) take up the remainder. An organism is faced with a "choice" of how its reproductive energy is to be employed in the most effective way. The energy may be spent in the production of a great number of offspring of which few survive for parenting; or, at the other end of the continuum, only one off-spring may be produced, with the remaining energy used in its protection and care. Cooperative effort may enhance the success of the latter strategy, and this is a prime selective factor favoring social behavior. The obvious terminus of the r–K continuum is the production of single offspring, a universal characteristic of higher primates.

There remains one further variable: the length of time devoted to the production and parenting of offspring (i.e., the length of gestation and infancy). Clearly, there is also a point at which parental investment exceeds reproductive yield. The development of the central nervous system, more than any other factor, determines the length of time required for development. Prenatally, the brain requires a constant, uninterrupted supply of oxygen and nourishment, and postnatally it would not serve any useful function were it not for a prolonged period of exercise and development (it may also be pointed out that the human brain does not mature histologically for almost 10 years). Intelligence can aid environmental understanding, but its development is at the same time an extreme liability. The vast majority of successful placental mammals display low levels of encephalization and correspondingly high levels of reproductive fitness and evolutionary success.

Unless intelligence enhances feeding strategy or locomotion, it may be expected to serve a minimal function in the survival of an organism. Intelligence may also serve in predator avoidance, but this is probably not crucial because equally successful but less costly avoidance mechanisms are normally available (coloration, olfaction, nocturnality, etc.). The encephalization found in primates (and cetaceans) is to a great extent directly related to feeding strategy and locomotion. It is a highly unusual anatomical complex, and it is unlikely to constitute a general trend in evolution.

To summarize this part of the discussion, an increased association capacity in the nervous system represents a reproductive liability both pre- and postnatally and may therefore be expected to undergo positive selection in only rare instances. Primates represent such an instance, because encephalization in this order can be accounted for directly by feeding strategy and locomotion at the first level and by reproductive cooperation at the second. No trend can be invoked toward encephalization in mammals using primates as an example since this order is unusually primitive in the majority of its mammalian traits. Within the evolution of the primates, a continuum may be seen in which a clearly defined limit on the degree of encephalization was reached sometime in the Miocene, and the expected endpoint of the K strategy was reached by hominoid primates, which were subsequently replaced by less encephalized, more reproductively successful cercopithecoids (Lovejoy, 1981). This clearly distinguishable end point on the r–K reproductive continuum serves as an ultimate "stop" in the evolution of "intelligent" organisms. There is no a priori advantage to intelligence, although it is a clear and unmistakable reproductive hazard.

EVOLUTION OF COGNITION

Taber (1973) defines intelligence as "the capacity to comprehend relationships." This definition is appropriate since it is practically synonymous with the functions of the association areas and pathways of the vertebrate cerebral cortex. All animals with the ability to encode and utilize relational experience may thus be called intelligent, and a partial anatomical indicator of intelligence can be obtained from Lashley's (1949) observation that portions of the brain not directly associated with somatic function (as judged by body mass) "seem to represent the amount of brain tissue in excess of that required for transmitting impulses to and from the integrative centers." Using this logic, Jerison and others have attempted to make estimates of "surplus" or "extra" neurons in mammalian brains by considering brain volume, neuron density, and body mass. Early hominids, which had brain/body mass ratios only 1/3 those of modern man, do not yield values greater than those of some other mammals. Yet these same hominids are known to have made tools (an indication of abstract symbolizing). The brain of modern man therefore should not be regarded as the product of a simple increase in neuron number, but rather as the result of a progressively enhanced capacity of an already (at least partially) structured cerebral cortex (Holloway, 1974,

1975, 1976, 1978). A computer analogy can be used to restate the argument as follows: A computer without a compiler or input/output device is useless, no matter what the capacity of its memory. Expansion of any computer to astronomical memory capacity would serve no purpose were it not accessible to at least one language. The human brain, then, is a highly specialized organ, as much so as the kidney of the desert-dwelling kangaroo rat or the salt gland of the albatross. Its mass is a product of selection for improved capacity but is not of itself responsible for that capacity (otherwise selection could not have acted). The critical structure of the human brain can thus be traced to an earlier evolutionary state in which it was not distinguished by its size, but rather by its structure. The antecedent selectional pathways of this structure are thus critical to an understanding of human evolution.

With reference to the computer analogy, a complex brain would not be accessible to selection in early hominids. Much of our ability to utilize symbols, formulate abstractions, and recognize the self depends on the phenomenon we call speech. The ability to speak is a neurophysiological character. It can be localized to specific areas of a single hemisphere and is associated with specialized histological structures in that hemisphere (Penfield and Roberts, 1959; Geschwind and Levitsky, 1968; Sperry, 1970). In short, the brain is preprogrammed for speech, including the necessary motocortical pathways to various peripheral anatomical structures such as the larynx, tongue, and lips. It is the capacity for complex symbolization, made possible by speech, that provides man the capacity for self-conception. Man is not an intelligent animal so much as a cognitive animal. The distinction is far from trivial. It is our ability to symbolize and not our ability to utilize relational experience that allows abstract communication. We are the only mammal with this ability, and the pathway by which this capacity evolved is critical for determining the probability of cognitive life elsewhere in the Universe.

It is important to point out that man did not develop the peripheral anatomical structures requisite for speech as a consequence of selection for speech. As duBrul (1958) established, the anatomical specializations that allow the formulation of voiced and unvoiced phonemes (including the control of the nasopharyngeal passage, the separation of the laryngeal and palatal aditi, the migration of the diagastrics from the hyoid, and the spreading of the oral diaphragm) are direct functional adaptations to the habitually erect posture induced by bipedal locomotion. Furthermore, the modifications of the oral cavity that permit consonant formation were primarily brought about by early hominid dietary specializations, including a reduction in mandibular length, loss of the sectorial canine complex and anterior tooth reduction, and elevation of the temporomandibular joint. These changes were induced by greater seasonality of food sources and a proliferation of Miocene mosaic environments; they were originally completely unrelated to speech capacity.

Speech as a communicating medium could not have arisen de novo. Pre-speech mechanisms of communication were almost certainly prerequisite and probably included a unique constellation of behavioral mechanisms such as manual gesturing, plasticity of facial expression, primitive vocalization, and intense parenting (Hewes, 1973; Holloway, 1969; Falk, 1978, 1980). No other order of mammals evinces the manual dexterity, facial plasticity, or parenting mechanisms of primates, and no other primate shows evidence of the dietary, locomotor, or sexual specializations of early hominids (Lovejoy, 1981). In short, man is not only a unique animal, but the end product of a completely unique evolutionary pathway, the elements of which are trace-able at least to the beginnings of the Cenozoic. We find, then, that the evolu-tion of cognition is the product of a variety of influences and preadaptive capacities, the absence of any one of which would have completely negated the process, and most of which are unique attributes of primates and/or hominids. Specific dietary shifts, bipedal locomotion, manual dexterity, con-trol of differentiated muscles of facial expression, vocalization, intense social and parenting behavior (of specific kinds), keen stereoscopic vision, and even specialized forms of sexual behavior, all qualify as irreplaceable elements. It is evident that the evolution of cognition is neither the result of an evolu-tionary trend nor an event of even the lowest calculable probability, but rather the consequence of a series of highly specific evolutionary events whose ultimate cause is traceable to selection for unrelated characters such as locomotion and diet.

CONCLUSIONS

The evidence bearing on the evolution of cognitive life on suitable planets is necessarily derived from our knowledge and understanding of the process and participants of organic evolution as it has occurred on this planet during the last 3.5 billion years. Three conclusions seem reasonable:

1. Planets on which complex autotrophs evolved may be expected to support pyramids of eutrophic organisms in ecosystems that tend toward complexity. There is no evidence, however, to support the belief that their participating organisms would demonstrate continually increasing complex-ity. The strictures of increasing cell/organ unreliability and the limitations of the r–K reproductive continuum would operate as effective "stops" in such complexity, as they appear to have done on this planet.

2. Intelligence is not the product of a general trend. Increased capacity of a central nervous system to encode relational experience is only useful in the selective sense if it improves the capacity of the organism's other activi-ties (locomotion, reproduction, feeding behavior). Intelligence also leads to

increased reproductive liability, and the most successfully adapted organisms (including mammals) on this planet exhibit lower brain/body ratios than most others. Only in rare instances where "intelligence" enhanced the locomotor or feeding strategies of mammals (echolocation, hand-eye coordination in arboreal environments, and small object feeding, etc.) was it favored on this planet.

3. While "intelligent" animals may evolve on other planets by relational pathways similar to those seen on Earth, the phenomenon of "cognition" is quite distinct from that of intelligence and may be expected to be exceedingly rare.

Thus I conclude that man is a highly specific, unique, and unduplicated species. If we wish to make probability estimates of the likelihood that cognitive (not intelligent) life has evolved on other suitable planets, the simplest and most direct question we may pose is: What is the probability that cognitive life would evolve on *this* planet, were not man already a constituent of its biosphere? From what we know of the human evolutionary pathway and of the critical elements that have directed it, the odds against its reexpression are indeed remote, if not astronomical. No other mammal even remotely shares the unique attribute complex that defines either man or his evolutionary pathway. Since *Homo sapiens* is a unique species, we may ask this same question in a slightly different way that allows greater objectivity: What is the probability that any named species, be it mammal, reptile, or mollusk, would evolve again on this planet? That is, what is the probability that the Bornean long-tailed porcupine, for example, would appear again were the evolutionary process to be reinstated on some imaginary planet identical to ours in every way save the last half billion years? I think it quite reasonable to suppose that despite the immensity of the known Universe, the specificity in the physiostructure of any organism is so great and its immensely complex pathway of progression so ancient that such probabilities are simply infinitesimal.

In any case it is reasonable to suggest that our understanding of the process of evolution responsible for the appearance of cognitive life will, for the foreseeable future, depend directly on our knowledge of the mechanisms that produced man, and that future predictions of exobiological cognition should be fully tempered by the knowledge of our own evolution.

REFERENCES

Andrews, P.; and van Couvering, J. A. H.: Palaeoenvironments in the East African Miocene. Contrib. Primatol., vol. 5, 1975, pp. 62–103.

Cartmill, M.: Primate Origins. Burgess, Minn., 1975.

Colbert, E. H.: Evolution of the Vertebrates, A History of the Backboned Animals Through Time. 2nd edition. John Wiley, N.Y., 1969.

deBrul, E. L.: Evolution of the Speech Apparatus. Thomas, Springfield, Ill., 1958.

Dobzhansky, T.: Genetics of the Evolutionary Process. Columbia Univ. Press, N.Y., 1970.

Falk, D.: Cerebral Asymmetry in Old World Monkeys. Acta Anatom., vol. 101, 1978, pp. 334–339.

Falk, D.: Language, Handedness and Primate Brains: Did the Australopithecines Sign? Am. Anthropol., vol. 82, 1980, pp. 72–78.

Geschwind, N.; and Levitsky, W.: Human Brain: Left-Right Asymmetries in Temporal Speech Region. Science, vol. 161, July 12, 1968, pp. 186–187.

Hewes, G. W.: Primate Communication and the Gestural Origin of Language. Current Anthropol., vol. 14, 1973, pp. 5–24.

Holloway, R. L., Jr.: Culture, a Human Domain. Current Anthropol., vol. 10, 1969, pp. 395–412.

Holloway, R. L., Jr.: On the Meaning of Brain Size. (Review of "Evolution and the Brain and Intelligence," by Harry J. Jerison. Academic Press, N.Y., 1973). Science, vol. 184, May 10, 1974, pp. 677–679.

Holloway, R. L., Jr.: Early Hominid Endocasts: Volumes, Morphology, and Significance for Hominid Evolution. In: Primate Functional Morphology and Evolution, R. Tuttle, ed. Mouton, The Hague, 1975.

Holloway, R. L., Jr.: Some Problems of Hominid Brain Endocast Reconstruction, Allometry and Neural Organization. In: Les Plus Anciens Hominides, P. V. Tobias and Y Coppens, eds., Centre National de la Recherche Scientifique, Paris, 1976.

Holloway, R. L., Jr.: The Relevance of Endocasts for Studying Primate Brain Evolution. In: Sensory Systems of Primates, C. R. Noback, ed., Plenum Press, N.Y., 1978, pp. 181–200.

Jerison, H. J.: Evolution of the Brain and Intelligence. Academic Press, N.Y., 1973.

Lashley, K. S.: Persistent Problems in the Evolution of Mind. Quart. Rev. Biol., vol. 24, 1949, pp. 28–42.

Lovejoy, C. O.: The Origin of Man. Science, vol. 211, 1981.

Maglio, V. J.; and Cooke, H. B., eds.: Evolution of African Mammals. Harvard Univ. Press, Cambridge, Mass., 1978.

Miller, S. L.; and Orgel, L.: The Origins of Life on the Earth. Prentice-Hall, Englewood Cliffs, N.J., 1974.

Odum, E. P.: Relationships Between Structure and Function in Ecosystems. Japanese J. Ecol., vol. 12, 1962, pp. 108–118.

Odum, E. P.: Fundamentals of Ecology, 3rd edition. W. B. Saunders, Philadelphia, Penn., 1971.

Penfield, Wilder; and Roberts, Lamal: Speech and Brain-Mechanisms. Princeton Univ. Press, Princeton, N.J., 1959.

Radinsky, L.: The Fossil Evidence of Anthropoid Brain Evolution. Amer. J. Phys. Anthropol., vol. 41, 1974, pp. 15–27.

Radinsky, L.: Primate Brain Evolution. Am. Scientist, vol. 63, 1975, pp. 656–663.

Simons, E. L.: Primate Evolution; An Introduction to Man's Place in Nature. Macmillan, N.Y., 1972.

Simpson, G. G.: The Meaning of Evolution; A Study of the History of Life and Its Significance for Man. Yale Univ. Press, New Haven, Conn., 1949.

Sperry, R. W.: Cerebral Dominance in Perception. In: Early Experience in Visual Information Processing in Perceptual and Reading Disorders, F. A. Young and D. B. Lindsley, eds., Washington, D.C., National Acad. Sciences, 1970.

Taber, C. W.: Taber's Cyclopedic Medical Dictionary, C. L. Thomas, ed., F. A. Davis, Philadelphia, Penn., 1973.

IV — Detectability of Technological Civilizations

The current consensus concerning the abundance of extraterrestrial life is that it exists in abundance in the Universe. We have heard the accumulating data to that effect in the previous sessions, along with some skepticism. Accepting this consensus, we then ask "how do we best proceed to detect these other civilizations?" An answer to this question has now existed for several decades: we should search for radio transmissions from these civilizations. But we accept that answer with a modicum of uncertainty and unease. We worry that we do not know enough about what the future of technology may hold for the typical civilizations. We worry that this answer has been unduly influenced by our booming expertise in radio technology. We worry that even if radio is the right answer, the radio transmissions of advanced civilizations may be in a form or at a radio frequency we have not thought to search for. All of these worries must be heeded, and we must constantly reexamine our thinking to ensure that we have done the best possible job in designing the searches for other civilizations. That is what this session is about.

Here we will ask questions such as: "what are the most promising stars to look for?" The numbers of stars and radio frequencies are so great that any guidance that would lead us to more promising stars or to eliminate some as candidates would greatly reduce the required search time and expense. We ask, once again, the important question: "are there other plausible technical activities of advanced civilizations that we should surely search for?" We review the evolving argument which still leads us to believe that radio searches are the most promising approach for our civilization. We hear the present wisdom leading to a plan for the first extensive, highly organized and developed search for other civilizations, the one being prepared jointly by NASA Centers. This approach is very promising, and its mere existence and the support it is receiving from important governmental figures and bodies is very encouraging. We also hear what our Earth might look like to a distant radio telescope. This gives us some guidance and insight as to the form in which signals might be falling on Earth.

Dr. Frank D. Drake
Professor of Astronomy
Cornell University
Director, National Astronomy and
Ionosphere Center at Arecibo, Puerto Rico

Identifiability of Suitable Stars

KENNETH JANES

In view of the uncertainties, a sine qua non for a targeted search is a catalog of the properties of all stars near the Sun. Constructing such a catalog would be useful for a variety of astrophysical problems in addition to the search for intelligent life elsewhere in the Universe.

We do not yet know what kinds of stars are most likely to be the homes of intelligent beings, or even what kinds of stars are most likely to have planets. It would seem reasonable to assume that very young stars (all those of spectral types O and B, for example), variables, some close binaries, red giants, or flare stars are rather inhospitable places, and the anthropomorphic point of view, of course, encourages us to consider most seriously the stars from spectral types F5 to K0. However, such assumptions could easily be wrong and, since we have no idea how common civilizations are, it may be necessary to search widely before we find any. In view of these considerable uncertainties, an important resource for any search program would be a catalog of the properties of all stars within some substantial distance of the Sun. In any type of search program, such a catalog would be useful, and for a targeted search it would be essential.

In this review, I want to describe the current state of our knowledge about the nearby stars, the methods used in learning about them, and the possibilities for a large-scale extension of present catalogs. Such a catalog would be useful for a variety of astrophysical problems in addition to a SETI program.

STELLAR PROPERTIES

We must first identify the stellar characteristics that we can hope to measure. Although the fundamental properties listed in the left-hand column of table 1 provide an essentially complete description of a star, several of these properties are not directly observable but must be inferred from a set of observable properties, such as those listed in the right-hand column. By combining observables with a theory of stellar evolution, one should be able to derive the quantities on the right-hand side. Not all of the quantities are of equal importance for either astrophysical purposes or for exploring the possible development of life, and in doing theoretical models most astronomers invoke the so-called Vogt-Russell theorem, which states that the physical properties of a star are determined entirely by its mass, age, and initial chemical composition. The equivalent observable parameters are temperature, luminosity, and surface chemical composition. The Vogt-Russell theorem is not entirely correct in that angular momentum, magnetic fields, and the presence of stellar companions can affect the life of a star and might be important for the evolution of life.

In theory, several (more than two) measurements of the precise position and spectral energy distribution would lead to a complete determination of the character of a star, but in practice a great deal of work is required to deduce even the approximate characteristics of a star. The complete set of properties listed in table 1 are known for only a small number of stars. The position, apparent magnitude, and proper motion of a star are the easiest to determine since they require only two photographs taken many years apart,

TABLE 1.– STELLAR PROPERTIES

Fundamental properties	Observables
Age Angular momentum Companions Initial composition Location Magnetic fields Mass Velocity	Apparent magnitude Companions Parallax Proper motion Radial velocity Right ascension and declination Rotation Spectral peculiarities Surface composition Surface gravity (luminosity) ⎫ Temperature ⎬ (spectral type) Variability ⎭

and its temperature can also be estimated relatively easily from either photographic or photoelectric measurements of the star's color or from low-resolution spectroscopy (spectral classification). A star's surface gravity (which is closely related to its luminosity) and its composition require higher spectral-resolution spectroscopy or narrow-band multicolor photometry and are therefore substantially more tedious. Parallaxes are even more time-consuming, requiring many photographs over a period of several years to obtain even an approximate value, although the distance to a star can be inferred roughly from its apparent magnitude and spectral class. Radial velocity and rotation require relatively high-dispersion spectroscopy, but variability, stellar companions, and some peculiarities are usually found in the course of observing the other properties. Finally, interstellar absorption should be mentioned, though it is not, strictly speaking, a stellar property, but is the dimming and reddening of starlight by dust between the star and the Solar System. This becomes increasingly important for stars more distant than 100 parsecs, but if the physical properties of a star are measured by a combination of photometry and spectroscopy, then the amount of interstellar reddening can usually be separated from the other properties.

For our present purpose, the important properties of a star are its spectral class (or temperature and surface gravity), position, composition, and distance. All of these can be determined approximately for large numbers of stars. In some cases, the presence of a close companion may profoundly affect the character of a star, and I have hidden a considerable variety of phenomena under the heading "spectral peculiarities"; but as a rule, when these properties are important, they are also easy to detect.

Finally, a star's age is a property of obvious importance for SETI. Although in a formal sense a star's age is a function of the other observable quantities, as a practical matter it is not generally possible to measure a star's properties with sufficient precision to permit a useful estimate of its age. It is possible to say something about the ages of certain spectral types (such as O stars or T-tauri stars) or stars in clusters, but generally one would have to say that at present the age of a star is an indeterminable quantity.

STAR CATALOGS

Astronomers have been compulsive catalogers, beginning in the 2nd century B.C. with Hipparchus, who compiled a catalog of the positions of 850 stars. The great age of astronomical catalogs was the last half of the 19th century and the beginning of the 20th. We have from this period the *Harvard Revised Photometry* of 9110 stars (which has been transformed into the *Catalogue of Bright Stars)*, the *Bonner Durchmusterung*, and the remarkable

Cape Photographic Durchmusterung, giving accurate positions and approximate magnitudes for 454,875 stars. The greatest of all these catalogs is, of course, the *Henry Draper* (HD) *Catalogue* (Harvard Observatory *Annals*, volumes 91–99, 1918–1924) published almost 60 years ago, which contains the spectral types of 225,300 stars.

All this work was done by hand, yet the HD catalog at least has not been surpassed, despite the enormous improvements of modern technology. More recently, the interest of astronomers has turned to other problems, and the catalogs of the last 50 or 60 years are somewhat specialized in nature (such as catalogs of proper motions or photometry) or are collections of data published separately by many astronomers. Table 2 is a partial listing of the major astronomical catalogs presently in use. Although some of the brightest stars are well studied, the overwhelming majority of the fainter stars appear in only one or two of the catalogs; that is to say, our knowledge of them is incomplete.

The lack of completeness and uniformity in catalogs of astrophysical data has been a continuing source of difficulty to astronomers, particularly since many of the catalogs in common use are simply compilations of work done in piecemeal fashion by several astronomers. This is one of the important aspects of the HD catalog: it is a uniform, systematic, and (nearly) complete tabulation of spectral types of stars brighter than magnitude 9.5 or so.

For our present purposes, ordinary catalogs of stellar data have another serious deficiency. They are generally magnitude-limited; that is, they are

TABLE 2.– CATALOGS OF STELLAR DATA

Catalog	Number of stars	Type of data
Henry Draper	225,300	Position, apparent magnitude, spectral type
SAO	260,000	Position, proper motion
G.C. Trigonometric Parallaxes	5,800	Trigonometric parallaxes
ADS/IDS	40,000	Information on visual binaries
G.C. Variable Stars	20,000	Information on variable stars
Radial Velocity Catalog	25,000	Radial velocities
Michigan Spectral Catalog	36,000[a]	Two-dimensional spectral classes
Bright Star Catalog	9,110	Properties of bright stars
Nearby Stars (Gliese, 1969)	1,890	Properties of stars within 20 parsecs
Nearby Stars (Woolley et al., 1970)	2,150	Properties of stars within 25 parsecs
C.S.I. (Strasbourg)	400,000	Cross-reference to all major catalogs

[a]Will eventually reach 225,300.

limited to stars brighter than some limit. It would be preferable to deal with a volume-limited sample, but the enormous range in the intrinsic luminosities of stars has made this an impractical goal. Consider, as an example, a volume of space 100 parsecs in radius centered on the Sun. There are within this volume approximately 250,000 stars brighter than absolute magnitude 14 (spectral type M5), but only 15,000 or so of these stars are in the HD catalog (Allen, 1973). Even in a volume 20 parsecs in radius around the Sun (as in the Gliese catalog in table 2), as many as 2/3 of all the stars may not yet have been noted in any catalogs (Wielan, 1974). It is true that the missing stars are extremely low-luminosity objects (many of them white dwarfs) and may not be friendly places for life, but we do not really know if that is a reasonable assumption.

Since the HD catalog was completed in 1924, there have been few systematic, large-scale surveys of stellar properties, although the basic astronomical data base has been enormously enlarged through a large number of limited studies.

Several projects are now underway to survey various stellar properties, and I will mention two projects to redo the HD catalog to give more detailed spectral classifications of the stars. Houk and colleagues are using objective prism spectra from the Schmidt telescope located at Cerro Tololo Observatory in Chile to reclassify the spectral types of the HD stars in the Southern Hemisphere. A traditional approach is being used in this project in the sense that the stars are being classified by visual inspection of the spectra by one person (Houk and Bidelman, 1979). In the Northern Hemisphere, a group of astronomers have set up a new observatory near Monterey, California, and plan to reobserve the northern half of the HD catalog (Overbye, 1979). They are taking a more modern approach and will be imaging the spectrum of one star at a time onto a solid-state, diode array detector, producing a spectrum in digital form that will be fed directly into a computer. These data should permit rapid, semiautomatic determination of the properties of the stars.

A MODERN HD CATALOG?

Present catalogs provide a somewhat limited search list for SETI. Is it practical or useful to undertake a large-scale extension of the HD catalog using available technology? A survey that reached magnitude 14 (i.e., a 5 magnitude increase over the HD limit) would contain about 90,000 stars within 100 parsecs, including virtually all stars of spectral type M0 or hotter. However, to identify the nearby stars, it would be necessary to observe all 15 million or so stars brighter than magnitude 14, virtually all of them luminous stars located farther than 100 parsecs from the Sun (Allen, 1973).

Although the more distant stars may not be relevant to SETI, such a massive catalog could be useful for a wide variety of astrophysical problems, particularly in helping to determine the structure and evolutionary history of the Galaxy.

Most of the catalogs of the past century have been based primarily on the measurement of photographs. It is possible, in fact, to photograph the entire sky in a relatively short time with a modest telescope. A photograph is also an incredibly efficient information-storage device (a single 35-mm picture can contain the equivalent of at least 10^7 bits of information), but the problem is one of obtaining rapid access to that information. So far, the best way to do it has been to make use of two other remarkable instruments, the human eye and brain, and this is why the HD catalog was so successful and why Houk is using the same approach 60 years later. Unfortunately, 15 million stars are too many to measure even for the most dedicated astronomer.

It is now possible to transmit data directly from the focus of a telescope to a computer for analysis, thereby saving the time and expense of intermediate storage in the form of a photograph. There are now several types of photoelectric array detectors, ranging from television systems to solid-state photodiode arrays, the most promising being the charge-coupled device (CCD), consisting of an array of small photodiodes on a single silicon chip. The quantum efficiency of the CCD is close to unity with spectral sensitivity from 4000 to 9000 Å. When extreme care is taken in their manufacture, CCDs can be made that will detect fewer than 10 photoelectrons per picture element (pixel). The chief limitation of the CCD is the relatively small number of pixels compared to a photograph. The largest experimental CCD arrays that have been manufactured are the 800×800 pixel arrays to be used in the space telescope.

Although perfect arrays of this size will always be difficult to produce, reasonably good, large arrays should become widely available in a few years, and many astronomers are already talking in terms of replacing photography with CCDs. While this may be a somewhat overoptimistic view, CCDs would appear to have tremendous potential.

A possible design for a survey system might consist of a meridian telescope (fixed to operate only on the celestial meridian) with a mosaic of CCD arrays at its focus. The telescope itself would be inexpensive; the system would acquire combined astrophysical and positional data at a rapid rate and be precise in measuring the positional data. By dividing the light into several beams to measure several bandpasses simultaneously, some spectral information could be acquired.

Such a system would operate by reading out the CCD arrays about 15 times per second and calculating the time and precise path of each star as it drifts across the field of view. If each pixel covered 0.5 arcsec, a star on the celestial equator would drift across an 800×800 array in a little under

30 sec. Its right ascension and declination could be calculated with a precision of about ±0.01 arcsec and the brightness could be recorded with a precision of about 0.02–0.03 magnitude in each bandpass.

By separating the light into 1000-Å bandpasses, it would be possible to record the magnitude of the star in five bands with sufficient precision to determine the basic astrophysical character of the star. A telescope with an aperture of 1.5 m or so would permit observations of stars as faint as magnitude 14. The possible characteristics of a meridian CCD telescope are given in table 3.

It is important to emphasize what this survey would *not* do. It would not yield highly accurate parallaxes or spectral types, nor would it detect planetary or even, in many cases, stellar companions. It *would* tell us the approximate temperature, distance, luminosity, and any unusual properties of 15 million or so stars. One could then abstract from this data set those stars of interest for a particular project. For SETI, one might want to separate out all nearby stars, which could then be studied in more detail using more conventional techniques.

TABLE 3.– CHARACTERISTICS OF A POSSIBLE STELLAR SURVEY TELESCOPE

Aperture	1.5 m
Focal ratio	f/6.67
Detector	9 CCDs
Array size	7200×800 pixels
Field of view	1°×0°.11
Pixel size	0.5 arcsec
Data rate	10^9 bits/sec
Average number of stars < mag 14 in field of view	200
Counts/readout/1000 Å (for magnitude 14 star)	150

REFERENCES

Allen, Clabon Walter: Astrophysical Quantities, 3rd ed. Athlone Press, London, 1973.

Gliese, W.: Catalogue of Nearby Stars. Veröff. Astron. Rechen-Inst. Heidelberg, no. 22, 1969.

Houk, N.; and Bidelman, W. P.: Reclassification of the HD Stars on the MK System: Current Status and Future Plans. Bull. Am. Astron. Soc., vol. 11, no. 2, 1979, p. 395.

Overbye, D.: Making It in Monterey. Sky & Telescope, vol. 57, 1979, pp. 223–230.

Wielan, R.: In: Highlights of Astronomy, G. Contopoulos, ed., D. Reidel, Dordrecht, vol. 3, 1974, p. 395.

Woolley, R.; Epps, E. A.; Penston, M. J.; and Pocock, S. B.: Catalogue of Stars Within Twenty-Five Parsecs of the Sun. Roy. Obs. Ann., no. 5, 1970.

Manifestations of Advanced Civilizations

RONALD N. BRACEWELL

All unorthodox suggestions warrant some consideration because future discoveries may not be obviously implied by what we know now. We should be on the alert for phenomena that might contribute to the detection of advanced civilizations elsewhere.

An advanced civilization somewhere in the Galaxy might be engaged in an effort to make its existence apparent to other civilizations (our own, for example) or, even if it is not trying, might nevertheless be manifesting its presence unintentionally. A good deal of discussion has centered on the idea that the other party might choose to advertise itself by means of a powerful radio beacon, and other specific means have also been suggested. Not so much consideration has been given to incidental manifestations such as the intense infrared radiation produced as a by-product of large-scale utilization of stellar energy. This latter idea was introduced by Freeman Dyson, who argued that in the asymptotic limit, where a civilization was approaching utilization of *all* the energy from its star, the full stellar luminosity would still be apparent to an outside observer, but a good part of the spectrum would be shifted to an infrared band corresponding to room temperature.

Thus two different categories exist into which search proposals for extraterrestrial civilizations might fit. Now I wish to view the problem in a different way by listing all the modes of information transfer that could conceivably transmit signs of an advanced civilization to us. The purpose of this approach is to set up an exhaustive framework for future discussion. In addition, a beginning will be made in assessing the possibilities, again for possible

343

future review. Each mode of information transfer must, of course, be considered under each of the two categories mentioned above. Here is the list of modes to be discussed:

A. Electromagnetic waves
B. Other waves
C. Matter transfer
D. Exotica

An apology is needed for mode D, which some readers might label science fiction, but there is a good reason for including items such as tunneling through black holes, tachyons (particles that travel faster than light), and telepathy. The reason is that the public is widely interested and that this interest is catered to by astronomers in the public eye.

Electromagnetic waves can be classified as radio, light, x, and gamma rays. In the radio spectrum, there is a universal optimum, not especially sharp, but nevertheless favoring the microwave band when the requirement is to transmit over great distances, conditions such as source power being held constant. The microwave band is defined broadly as the spectral region in which the waveguide technique is or was used — let us say from 1 mm to 1 m. Within this range and at longer wavelengths, say to a few meters, refined considerations (see Oliver, this volume) have revealed heavy handicaps for both long radio waves and light waves — in the advertising category.

The difficulty besetting long radio waves is the natural emission from the Galaxy itself. Energetic electrons executing helical orbits around magnetic field lines in interstellar space radiate radio waves with ever-increasing intensity down to frequencies below 1 MHz. This is known as synchrotron radiation because it also occurs in synchrotrons, where high-energy particles swerve through a magnetic field. At lower frequencies synchrotron self-absorption sets in, an effect that causes the interstellar medium to become effectively opaque. There are several reasons why we do not look for beacon transmissions anywhere in the wavelength range from a few meters to infinity. First, the natural competition from the Galaxy is too strong. Second, even if transmitter power is made high enough and bandwidth narrow enough that the natural background is dominated, a serious phenomenon of refraction sets in due to the presence of free electrons in interstellar space. That is, a transmission beamed in a given direction will be bent in accordance with electron density gradients, and especially so as the waves pass through planetary systems, where electron densities might be a thousand times higher than those in interstellar space. Moreover, these changes in direction vary with time. Such behavior is less than attractive from the standpoint of the designer of a point-to-point communication system. At frequencies lower than about 10 kHz, propagation cuts off in a medium with 1 electron/cm^3. Synchrotron self-absorption, which is important in determining the turnover frequency, has a negligible effect on propagation because the density of the high-energy synchrotron electrons is much less than the density of slow interstellar electrons.

The long-wavelength (low-frequency) side of the radio band is hopeless for beacons, but what about incidental radiation? When we turn to this category, considerations such as refraction lose weight because one would be perfectly happy to detect an advanced civilization anywhere in the sky, even if the apparent direction were not the true direction. Furthermore, even below 10 kHz, which is nominally cut off from propagation, there is the theoretical possibility of propagation in the presence of magnetic fields. That this suggestion is not merely academic is illustrated by the phenomenon of whistlers, radio pulses generated by lightning discharges, which in the general range of 5 kHz are able to penetrate Earth's ionosphere, whose cutoff frequency may be several megahertz. In fact, whistlers provide us with another sobering thought. Not only do they penetrate Earth's dense ionosphere with practically no attenuation, they may even *gain* energy as they traverse the magnetosphere, a fact revealed by the occurrence of long trains of echoes. Therefore, the detection of incidental low-frequency emissions from another civilization is by no means excluded a priori. Further study of this field is indicated.

Moving now to the other side of the radio spectrum, we reach the infrared, visible, and ultraviolet light waves. A determined case has been made favoring laser beacons, whose two distinctive features are high spectral energy density and narrow beamwidth. Special circumstances may exist in which lasers make sense. From the standpoint of a communications designer interested in detecting a weak signal against a noisy background and also in information rate, light waves suffer from concentrating a rather large amount of energy in each quantum. This means that, for given transmitter power, far fewer quanta can be transmitted as light than as microwaves. This consideration and others favoring microwaves weigh against laser beacons. Incidental light emission, however, is quite a different matter. Infrared as a by-product of energy utilization has already been mentioned, but probably no special followup is required because infrared astronomy, which is in vigorous condition, is likely to discover detectable Dyson sources in the normal course of events. Other incidental light might also turn up in ordinary astronomical observations. Of course, background light from an adjacent star works against detection of light produced by a civilization. Nevertheless, one should not forget factors facilitating detection. Examples would be characteristic time rates of change such as a nuclear explosion would exhibit, characteristic absorption spectra, and special emission spectra as from artificial auroras.

Passing now to x- and gamma rays, we see that the disadvantages of laser beacons apply even more strongly when we think about deliberate signals. Incidental signals offer more interesting grounds for speculation. There is, of course, the possibility of emission from bomb explosions. Sustained nuclear reactions for purposes of power generation do not seem a likely source because leakage would represent waste. There are, however, conceivable efforts of astroengineering on a colossal scale that could release x- and gamma rays. Generally speaking, light emission would accompany x-ray generation, but suppose that a galactic source were so distant that light was extinguished by interstellar dust. When we begin to talk about source distances on the order of 100,000 light years, we are encompassing a galactic habitat reaching far beyond the thousand or so light years that beacon theory contemplates. Somewhere in that vaster domain of the Galaxy there could be engineering projects entailing alteration of the orbits of planets and stars and arrangements for the collision or grazing approach of celestial bodies. The fundamental purposes of such activity would be to organize natural dispositions into a desired structure, to interfere with the degradation of energy in stellar processes, and to gain access to stellar matter. Ultimately, as when galactic evolution moves under intelligent control, the maneuvering and disassembly of stars becomes a goal, and the energies involved will liberate energetic rays. Such activities, if they exist, should be sought in external galaxies as well as our own, but we may expect the orderly development of survey instruments for x- and gamma-ray astronomy to reveal them.

Having covered the whole range of electromagnetic waves, we now turn to other waves. The only ones I can think of are shock waves, Alfvén waves, and gravitational waves. We have had considerable difficulty detecting gravitational waves of any kind, even from the strongest expected natural sources, galactic or extragalactic, so we may confidently forget them for now. Alfvén waves, or any mixture of magnetic-material wave motion, including shock waves, do, however, offer scope for thought. Magnetic field lines permeate the Galaxy and, if shaken at one place, will respond elsewhere in time. It is true that the direction of magnetic lines in space is contorted, but there are systematic patterns too. Near Earth, the magnetic field is also disturbed by solar phenomena, making it harder to see how information coming from outside could be detected. Even so, we now have the possibility of observing the magnetic field at great distances from the Sun, well out beyond Jupiter, and the records of the magnetic field will need to be scrutinized for wave energy arriving from outside the Solar System. Of course, the first detection of this kind will refer to some natural phenomenon, perhaps the dispersed echo of some ancient supernova, so for the time being we will just bear magnetohydrodynamic waves in mind.

Matter transfer between interstellar bodies must occur at a rather low rate, but it has not been negligible in totality, as we note from the fact that heavy elements of Earth come from the interior of stars other than the Sun. Currently, some comets may be engaged in transferring material from system to system. Solar wind and accretion of interstellar matter are factors. None of these modes of transport can be considered feasible for signaling. It is barely conceivable that unusual particles, molecules, or nuclides resulting from intelligent activity elsewhere in the Galaxy might filter into our Solar System and be recognizable as being of artificial origin. Of course, the source would be obscure but, one interesting thought, an extinct civilization that had flourished in our neighborhood billions of years ago might have left analyzable material traces. The idea of panspermia, the diffusion of spores or molecular precursors of life through space, originated with and continues to be entertained by distinguished people.

Deliberate transfer of matter brings us to space probes (Bracewell, 1979). The whole idea of interstellar travel by spaceship was severely criticized by Purcell, who relegated the notion firmly into the realm of science fiction. His discussion, however, assumed two-way travel to be completed within a human lifetime. With this restriction lifted, things change. I suggested in 1960 that, for distances around 100 light years, a one-way automatic space probe might be a good way for an advanced civilization, technically superior to our own, to establish first contact with us. Attention was focused on the 30–300 light year range because, for shorter distances say around 10 light years, contact could be established by radio, and for distances of the order of 1000 light years things looked gloomy because the expected average lifetime, if civilizations were so sparsely distributed, was too short to permit roundtrip communication. The idea that longevity and distance to the nearest civilization were interdependent was then new. The probe suggestion has attracted many criticisms, such as that the plan is too costly or takes too long, that probes are not durable enough, and that the plan calls for action by the other party but not by us. It is true that it is costly to launch a space probe to a neighboring star, and indeed the cumulative total, by the time success was achieved after sending out many inconclusive probes, might exceed that of a giant array of radiotelescopes such as was considered under Project Cyclops. However, the correct way to make the comparison is between alternatives open to the one party, not between actions that might be taken by different parties. Thus the cost of probes to the superior civilization might be compared with the cost to the same civilization of maintaining a beacon to achieve the same object. For example, if a 1000-MW beacon system was adopted, one item would be the cost of electricity over the time the beacon was kept going. This time might have to be comparable with a longevity of millions of years (in which case 10^{23} J would be radiated), or we might want to think in terms of substantial fraction of the generation time, say a billion years (10^{25} J).

The fact that a probe does not travel at the speed of light certainly increases the travel time spectacularly. However, when we understand that contact between civilizations is not exactly the same as contact between individuals, we realize that travel times comparable with the longevity of the civilizations rather than with the longevity of individuals are possibly acceptable. Of course, once first contact is achieved, communication thereafter may proceed at the speed of light.

When the cost and travel time have been worked out, as a function of distance to the target, the probabilities of successful detection of technological activity must also be factored in before comparison with an alternative approach becomes feasible. Much remains to be done here because of lack of basic data such as the weight and cost of advanced propulsion systems. Nuclear electric rocketry is a familiar idea but may have been superseded by new ideas, of which laser pellet propulsion is one. In addition, the design of electronics for long-term durability in space is an unknown and in itself constitutes technology on which work needs to be done for more immediate purposes of space science. Generally speaking, it would not seem that the durability of semiconductor electronics at the low temperatures in space would present insuperable obstacles.

Finally, the probe idea presents the psychological complication that no action seems to be called for on our part. Thinking further about this, it seems to me that some principle of equipartition of effort ought to be invoked. If the launching agent goes to the trouble to select us as a target, we are more attractive to him if we make our own contribution to the contact effort. Thus at the least we need to maintain radio receivers. But we could go further and maintain a space watch that would relieve the probe of the necessity of slowing down and stationing itself within the Solar System. It was already clear that if a probe failed to detect active radio communication on approach to Earth, then it would be more economical to send a second probe to a second likely target than to wait around for technology to evolve, which might take forever. Following this reasoning, I now think that it would pay the designer of the probe to shoot for civilizations that could detect and communicate with a fast flyby rather than to send off a smaller number of the slower and more elaborate probes that would be needed to kick into orbit in the habitable zone of the target star. If the equipartition strategy makes sense, we might expect to commit a fraction of the terrestrial GNP that is comparable with the effort of the other party. At present this seems to me to imply an intercept capability not unlike what is under discussion for a rendezvous with a comet.

To complete the discussion of deliberate matter transfer, we should look at particle beams and viruses. Beams of high-energy charged particles must negotiate magnetic fields, with a net effect over interstellar distances that will resemble diffusion, which is slow; so it is sufficient to consider neutral molecular beams, neutron beams, and neutrinos. Molecules may be propelled by laser beams and may be subjected to midcourse trajectory correction and refocusing. There may be other clever ideas to be discovered.

Neutrinos are costly to produce but may be collimated rather well, and they have the penetrating power to reach behind the dust clouds that lie in front of most of the stars of our Galaxy as seen from our neighborhood. A good many avenues need to be explored. The virus idea is that a message may be coded microscopically onto a long molecule by the methods of molecular biology. Such molecules could be propelled into space and in effect would be small probes. How to intercept and decode such viruses has not received much attention, but in principle condensing of messages to the molecular scale is possible, and there is nothing to say that an interstellar spore could not be germinated and grown to maturity. It might then make an announcement.

The first item on my list of exotica is tunneling through black holes, a subject that has clearly caught the public interest. We turn now to Sagan (1973) for the most authoritative exposition. His conjecture is that space-craft might use black holes for space travel. One would travel to the nearest black hole, typically 20 light years away, he estimates, taking 21 years for the trip. The spacecraft would then plunge down the black hole. Sagan states that he does not know whether it is possible to reach the point of emergence faster by going down a black hole than it would take by the direct route. However, he thinks black holes may be the transportation con-duits of advanced technological civilizations, the pneumatic tubes of the Galaxy. He imagines great civilizations growing up near the black holes, just as urban development clusters around rapid transit stations, and he pictures vehicles being rapidly routed through an interlaced network to the black holes nearest their destinations. In my opinion, great danger would set in while one was still at a considerable distance from a black hole, much as it would be dangerous to be in a panic-stricken crowd of football spectators all starting to rush down a manhole in the center of the 50-yard line. Further-more, it would not be possible for a spacecraft to emerge from a black hole and negotiate its way up through the infalling debris.

Tachyon is a term which, tangible though it may look to the public, resembles the "square circle" in having no referent in the physical world. However, there has been speculation that there may be faster-than-light enti-ties that cannot be decelerated below the speed of light, just as the particles we know of cannot be accelerated to speeds above that of light. Should such entities be discovered, it would clearly be interesting to speculate on their use as information carriers. The same applies to other phenomena that have not been discovered yet, such as telepathy.

Although there are no generally agreed facts relating to telepathy and related words such as telekinesis, there is a body of lore. For example, among believers in telepathy it is widely held that thought-transference telepathy does not suffer from attenuation by matter. If this proved to be so, there would be a serious signal-to-noise problem in separating the wanted signal from all the other thoughts from all over the Universe simultaneously clamoring for attention. This may be why telepathy is so hard to exhibit on demand!

To my mind, all unorthodox suggestions warrant consideration, even though one might choose not to devote personal effort to following them up, because the future undoubtedly holds discoveries that are not obviously implied by the present state of knowledge, and we should be on the alert for any phenomena that might contribute to the detection of advanced civilizations elsewhere.

REFERENCES

Bracewell, Ronald N.: The Galactic Club: Intelligent Life in Outer Space. W. W. Norton, N.Y., 1979.

Sagan, Carl: The Cosmic Connection. Doubleday, N.Y., 1973, pp. 248 and 266.

Search Strategies

BERNARD M. OLIVER

If intelligent life is common in the Universe, many species may have attempted to discover their neighbors. We are beginning to make our first attempts; our search may have to grow by orders of magnitude before we succeed.

If intelligent life is common in the Universe, it seems unlikely that the millions (or billions) of advanced cultures in our Galaxy will all go through their entire histories totally isolated from one another. Surely, as their convictions grow that theirs is not the only advanced society, many will attempt to discover their neighbors. In fact, many may have done so already. We are beginning to make our first attempts; our search may have to grow by orders of magnitude before we succeed.

How shall we best proceed? How can we have the greatest chance of success at each stage for the time, energy, and effort expended? We present here some of the constraints imposed by the magnitude of the problem and by physical law. It turns out that enormous times or energies are required for interstellar travel; unless we do it properly, large energies are needed for contact by radiated signals. The principle of least energy expenditure, and therefore of least cost, leads us to a preferred part of the radio spectrum, but several alternative modes of examining this region remain. Many of these modes cannot be ignored, but because we cannot do everything, we will be forced, as in any search, to give our highest priorities to those modes deemed to have the highest a priori probability of success based on all we know. We can never design a search strategy based on what we do not know.

351

ENERGETICS OF INTERSTELLAR CONTACT

Decades of science fiction have lulled many of us into accepting inter-stellar spaceflight as a reality for "them" and a near reality for us. In fact, interstellar travel requires such enormous expenditures of time or energy that it may not exist, or it may be attempted only *in extremis.* Since the nearest star is over 4 light years from us, any round-trip interstellar flight completed in a human working lifetime (τ) of, say, 40 years requires a ship speed (v) at least 1/5 that of light (c). Such a ship is far beyond our present technology. But let us assume it is not beyond theirs and ask how much energy an ideal rocket would require. We will express speeds as fractions of the speed of light ($\beta = v/c$) and energies as fractions of the energy equivalent of the rocket payload mass ($\rho = E/m_p c^2$).

The most efficient rocket is one whose motor uses all the available fuel energy to accelerate a propellant to an exhaust speed equal to the sum of all speed increments experienced by the ship since launch. (An auxiliary initial launch vehicle or a period during which the exhaust speed is greater than the ship speed is needed to avoid an infinite mass ratio.) If all speed increments are in the same direction, such a rocket leaves its exhaust stationary in the original rest frame; all the fuel energy is converted to kinetic energy of the payload. So far as the rocket is concerned, a retrofiring is the same as a forward-firing, and we may calculate the energy expenditure as the kinetic energy the ship would have in the original rest frame if all firings had been additive.

Adding β to β relativistically we find that, for a one-way trip,

$$\beta_2 = \frac{\beta + \beta}{1 + \beta\beta} = \frac{2\beta}{1 + \beta^2} \tag{1}$$

and

$$\rho_2 = \frac{1}{\sqrt{1 - \beta_2{}^2}} - 1 = \frac{2\beta^2}{1 - \beta^2} \tag{2}$$

while for a round trip,

$$\beta_4 = \frac{2\beta_2}{1 + \beta_2} = \frac{4\beta + 4\beta^3}{1 + 6\beta^2 + \beta^4} \tag{3}$$

and

$$\rho_4 = \frac{1}{\sqrt{1 - \beta_4{}^2}} - 1 = \frac{8\beta^2}{(1 - \beta^2)^2} \tag{4}$$

Finally, we define k as the ratio of elapsed ship time to the light time:

$$k = \frac{\sqrt{1-\beta^2}}{\beta} \tag{5}$$

The elapsed ship time is k years per light year traveled. Figure 1 shows the behavior of k, β_2, ρ_2, β_4, and ρ_4 as functions of β.

In a search for extraterrestrial intelligence by starship, the strategy would presumably be to search all likely stars out to some radius R before

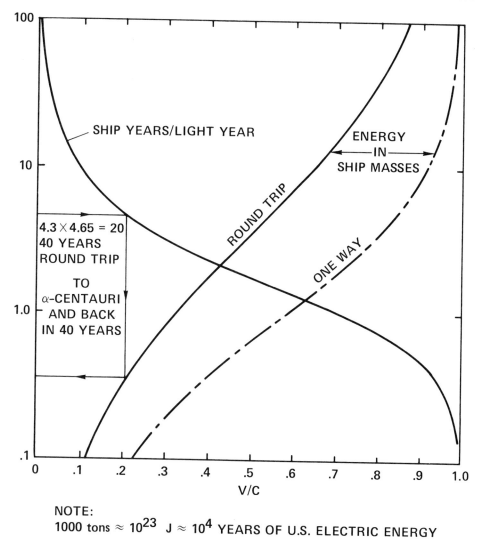

NOTE:
1000 tons $\approx 10^{23}$ J $\approx 10^4$ YEARS OF U.S. ELECTRIC ENERGY

Figure 1. *Ideal rocket performance. The curve that descends to the right shows the ship's time required to travel 1 light year while the ascending curves give the energy expended in making a one-way trip and a round trip at the velocity shown by the abscissa.*

searching any at a greater distance. Of course, if a humanly habitable planet were found that was devoid of intelligent life, a colony might be established and (much later) a new search started from there. We will ignore this serendipitous complication.

Solving equation (5) for β^2 and substituting in (4), we get

$$\rho_4 = 8 \left(\frac{1}{k^2}\right) + \left(\frac{1}{k^4}\right)$$
(6)

If a round trip to a range r must be completed in a ship's time τ, then $k \leqslant \tau/2r$. But for minimum energy we choose k as large as we can: $k = \tau/2r$. Then equation (6) becomes

$$\rho_4 = \frac{32r^2}{\tau^2} + \frac{128r^4}{\tau^4}$$
(7)

The number of likely stars out to several hundred light years is given very nearly by

$$n = \left(\frac{r}{r_0}\right)^3$$
(8)

where $r_0 \approx 8$ light years. The total energy needed to search all likely stars out to radius R is

$$\rho_S = \int_0^R \rho_4(r) \frac{dn}{dr} \, dr = \frac{96R^5}{5\tau^2 r_0{}^3} + \frac{384R^7}{7\tau^4 r_0{}^3}$$

$$= \frac{96R^5}{\tau^2 r_0{}^3} \left(\frac{1}{5} + \frac{4}{7} \frac{R^2}{\tau^2}\right)$$
(9)

Figure 2 shows ρ_S as a function of R, taking $r_0 = 8$ light years and $\tau = 40$ years.

In assessing the significance of figure 2, one should realize that with all the necessary engines, auxiliary power systems, control, communication, and guidance equipment, repair shops, crew quarters, and life-support systems, it is inconceivable that the payload would weigh less than 1000 tons. Thus $\rho_S = 1$ corresponds to at least 10^{23} J or 1000 years of total energy consumption by the United States. The ordinate can thus be labeled "millennia of U.S. energy."

Doubling τ implies two generations, which doubles the living space needed and adds nursery and educational facilities, so that the payload is

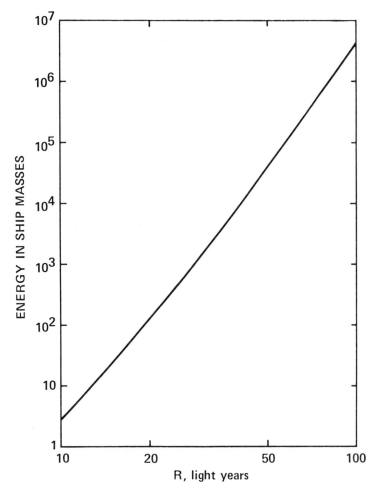

Figure 2. *Energy expended in searching by spaceship all the good suns out to the range indicated by the abscissa if the ship's mass is 1000 tons. The ordinate could be labeled millenia of U.S. energy consumption.*

increased. The longer time also increases the risk of disaffection or actual mutiny by the crew: the parents were presumably screened for psychological stability; the children are not. For $R > 40$ light years, the results of the search are denied to those who launched it. SETI by spaceship is a prohibitively expensive and risky undertaking. (We have said nothing about impacting interstellar debris at near-optic velocity.)

The costs drop drastically if we replace the spaceships with automatic probes. Payloads on the order of 10 tons seem reasonable, one-way journeys suffice, and τ can easily exceed a working lifetime. In fact, all the probes can be designed for the same β, regardless of their range. From equations (2) and (8) the energy needed would be

$$\rho_p = \left(\frac{R}{r_0}\right)^3 \frac{2\beta^2}{1-\beta^2} \qquad (10)$$

We have considerable latitude in our choice of β, so the energy required can vary over a wide range. It might, however, be politically impossible to get funding for results that will not appear for centuries. If $R = 100$ light years and $\beta = 1/2$, the most distant returns will take 3 centuries to come in. With these values, $\rho_p = 1300$. With a 10-ton probe this represents 13,000 years of U.S. energy rather than 4.5 billion years. We might (optimistically) estimate the cost of each launch at $1 billion, or $2 trillion for the entire fleet.

How do we retrieve the data from these probes? By radio, of course! If each probe carries a 10,000 W X-band transmitter and beams the signal at us with a 10-m antenna, its effective isotropic radiated power will be 10^{10} W and at 100 light years this will produce an intensity of 10^{-27} W/m^2. About 20 dishes each 100 m in diameter would be needed to receive the signal at a low error rate (10^{-3}) at 1 bit/sec from each probe at or near maximum range. Since there would be 1000 probes at 80 light years or more, we would need more than a thousand 100-m dishes, if all are to be monitored continuously.

Thus we would require an antenna system for the probe search that is at least as large as any yet proposed for a radio search. Yet the cost of a radio search is only a small fraction of the cost of the probes. Why, then, bother with the probes? Why not first listen for more powerful transmitters that may have been transmitting for centuries? This brings us to a radio search as the logical way to begin a SETI program. Because we are very uncertain of the strength of the signals we might receive, the radio search should start with modest, even with existing, antennas before a large dedicated facility is built.

But before we jump to radio, are there other signaling means we should consider? Signals across empty space must be carried by some kind of physical particle. To qualify, the particle should:

1. require a minimum amount of energy to exceed the natural background,
2. travel at, or close to, the speed of light,
3. not be deflected by galactic or stellar fields,
4. be easy to generate, detect, and beam, and
5. not be absorbed by the interstellar medium or by planetary atmospheres and ionospheres.

Requirement 3 excludes charged particles. Requirements 1 and 2 exclude all particles except those with zero rest mass; an electron traveling at half the speed of light has a kinetic energy a hundred billion times that of a photon at the best part of the radio spectrum. Of the zero-rest-mass particles, gravitons and neutrinos fail requirement 4. There is hardly anything harder to generate than gravity waves, hardly anything harder to detect than a neutrino (except perhaps a quark!).

Only photons survive all the requirements, and so electromagnetic waves of some frequency are the only known suitable signal. Later we will use requirements 1 and 5 to find the best frequencies.

The energy required to contact other intelligent life forms by radio depends greatly on the assumptions one makes. To correspond more or less with the spaceship and probe cases, let us assume that we have built a 1000-element array of 100-m dishes and have failed to detect any signals from stars within 100 light years, and that we then elect to beam signals at the 2000 likely stars within that range. Assume that we beam the same power at each star, and that this power is sufficient to be detected at 100 light years with another 100-m dish. Then the average power required is about 100 MW. If we operate the beacons for 30 years (with occasional interruptions to listen to stars within 15 light years), the total energy consumed will be about 10^{17} J. We might then listen for responses over the next 200 years, so that the total time is much like the probe case (see table 1).

TABLE 1.– ENERGY REQUIREMENTS FOR
THREE MODES OF SETI

Mode	Energy, J	= U.S. consumption for
Spaceships	$\sim 4 \times 10^{29}$	~ 4 billion yr
Probes	$\sim 10^{24}$	10,000 yr
Beacons	$\sim 10^{17}$	9 hr

A rational radio search does not begin by establishing beacons. It begins with the assumption that others have already been radiating for a long time, for if that is true, we can discover the signals as soon as our receiver sensitivity is high enough. We need not wait out the round-trip light time. Further, the listening program begins with existing antennas and grows to larger antennas or arrays only if earlier searches fail.

The energy cost of this listening phase is miniscule; even the hardware costs are reasonable. A significant listening program can be conducted at a cost that is small compared to other space missions.

BEST PART OF THE SPECTRUM

The energy argument, which brings us to electromagnetic waves, can be continued to find the part of the spectrum where the least energy is needed for detection. Since we must receive at least one photon, and actually some number *n*, to be sure we have a signal, and since the energy per photon is proportional to frequency, we should use the lowest frequency allowed by other considerations. This would also reduce the interstellar absorption. Below about 60 GHz, the cosmic background radiation enters the picture, and we now require not *nhv* but *nkT* (joules) to be sure of our signal, where *k* is Boltzmann's constant and *T* = 2.76 K. Finally, below about 1 GHz, *T* rises rapidly because of the synchrotron radiation of electrons in the galactic magnetic field. Figure 3 shows the sky "noise temperature" (the blackbody temperature needed to produce the observed noise) as seen from space, as a function of frequency and galactic latitude. Quantum effects are included, so that the necessary received energy per bit is everywhere proportional to the ordinate. The low-temperature (and hence low-noise/low-received-energy) valley from about 1 to 60 GHz is the so-called "free-space microwave window."

Notice that there is nothing geocentric about this window. Radio observers anywhere in the Galaxy would see substantially the same window and would conclude that this was the best part of the spectrum for interstellar communication.

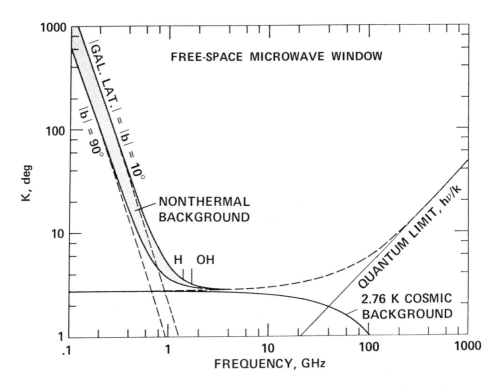

Figure 3.   *Free-space microwave window, in which the basic noises that limit radio communication over interstellar distances are least disruptive.*

There are several reasons to prefer the low end of the window and few, other than existing radio-frequency interference, to make us prefer the high end for search purposes. At the low end of the window:

1. Antenna surface tolerances are greater, so collecting area is cheaper;

2. For a given collecting area, antenna beams are broader (more sky is searched per pointing direction);

3. Frequency drift rate from Doppler effects and other causes is less, making the detection of monochromatic signals easier;

4. Power densities in transmitters and waveguides are less, allowing higher power transmitters; and

5. Atmospheric attenuation and noise are less, as is receiver noise.

Figure 4 shows the microwave window as deteriorated by atmospheric oxygen and water vapor. The upper end of the window is ruined on any Earth-like planet, but because of points 1 through 4 above, we do not mind going to the lower end to achieve point 5. There the atmospheric penalty is about 3–4 K. We conclude that for search purposes the optimum part of the spectrum is in the range 1–2 GHz.

By an eerie coincidence, right in the middle of this optimum region we find the spectral lines of hydrogen (1420 MHz) and hydroxyl (1.612, 1.615,

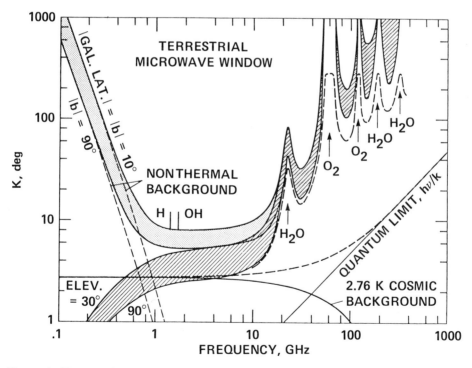

Figure 4. *Terrestrial microwave window. Atmospheric water vapor and oxygen degrade the upper end of the microwave window for receivers on Earth's surface and raise the temperature in the lower portion of the window.*

1.667, and 1.720 MHz). Cocconi and Morrison (1959) were the first to point out the suitability of the microwave window for interstellar contact and the significance of the hydrogen line as a signpost. In 1971, the Cyclops team pointed out the significance of OH as the other disassociation product of life-giving water, and suggested that the "waterhole" between these signposts be considered the prime spectral region where intelligent species might meet. The waterhole is probably a preferred region deserving intensive, but not exclusive, attention.

ANTENNA AND TRANSMISSION FORMULAS

To appreciate some of the constraints and tradeoffs in a radio search, we need to know the basic factors determining gain, directivity, and signal range. (Those familiar with the subject may want to skip this section.)

If a transmitter radiates a power P_i isotropically, then at a range R this power will be spread uniformly over a sphere of area $4\pi R^2$. An antenna of area A will collect a fraction $A/4\pi R^2$ and thus receive a power

$$P_r = \frac{A}{4\pi R^2} P_i \tag{11}$$

If the transmitting antenna radiates only into a solid angle Ω, the effective isotropic radiated power (EIRP) for receivers in the beam will be greater than the actual transmitted power, P_t, by a factor $4\pi/\Omega$; this is the power gain g of the antenna:

$$g = \frac{P_i}{P_t} = \frac{4\pi}{\Omega} \tag{12}$$

The effective area of an isotropic antenna can be shown to be

$$A_0 = \frac{\lambda^2}{4\pi} \tag{13}$$

where λ is the wavelength of the transmitted radiation. This is the area of a circle one wavelength in circumference. The gain of a receiving antenna is proportional to its area and, since the gain of an isotropic antenna is unity,

$$g = \frac{A}{A_0} = \frac{4\pi A}{\lambda^2} = \left(\frac{\pi d}{\lambda}\right)^2 \tag{14}$$

where the latter equality holds if A is a circle of diameter d.

When these relations are combined with equation (11), the ratio of received-to-transmitted power can be expressed in several ways:

$$\frac{P_r}{P_t} = \frac{A_t A_r R}{\lambda^2 R^2} = \frac{g_t A_r}{4\pi R^2} = \frac{A_t g_r}{4\pi R^2} = \left(\frac{\lambda}{4\pi R}\right)^2 g_t g_r \tag{15}$$

For single-unit antennas, the maximum *area* tends to be proportional to λ, so λ^2 disappears from the bottom of the first expression; for arrays the *gain* tends to be independent of λ. Equation (15) recommends low frequencies for long-distance transmission provided we have no fixed-size constraint on our antennas.

By reciprocity, equations (12), (13), and (14) apply to both transmitting and receiving antennas. If n is the number of directions in which an antenna must be pointed to cover the sky, we see from equation (12) that (ideally)

$$n = g \tag{16}$$

Directivity and gain are equal; both are proportional to the antenna area measured in square wavelengths. Unless we use multiple-feed horns on a single antenna or multiple beams in a phased array and associate a receiving system with each horn or beam, we can only get broader sky coverage per pointing direction at the expense of sensitivity.

The nominal range limit is the distance R at which the received power equals the receiver noise power, $N = kTb$, where T is the system noise temperature and b is the resolved bandwidth. From equation (11),

$$R = \sqrt{\frac{A_r P_i}{4\pi N}} = \frac{d}{4}\sqrt{\frac{P_i}{N}} \tag{17}$$

where, again, the latter relation holds for circular antennas.

From a SETI standpoint, the only variables under our control are d, the antenna diameter, and N, the receiver noise in the signal channel. The effective isotropic radiated power is up to "them." To increase d is expensive; and while this may ultimately be necessary, we should first minimize N by using the lowest noise receivers, feeds, and antennas and by using optimum detection methods.

MATCHED GATES AND FILTERS

If $s(t)$ is a signal of arbitrary shape but finite duration, one way to detect the signal in the presence of noise is to multiply the received waveform by a gating waveform $g(t)$ that is zero (gate closed) if and only if $f(t)$ is zero, and to integrate the resulting waveform. It can be shown that the best possible signal-to-noise ratio in the output is obtained if the amplitude spectrum, $G(\nu)$, of the gate is given by

$$G(\nu)\,\overline{F(\nu)} = \gamma\,\frac{S(\nu)}{\Psi(\nu)} \tag{18}$$

where $\overline{F(\nu)}$ is the conjugate of the transmission of any preselection filter, γ is a constant, $S(\nu)$ is the signal amplitude spectrum, and $\Psi(\nu)$ is the noise power spectrum. For our purposes, we set $\Psi(\nu) = kT = $ constant. Two limiting cases are of interest:

(a) *No preselection, $F(\nu) = $ constant.* Then $G(\nu) = $ constant $\times F(\nu)$, and therefore

$$g(t) = \text{constant} \times f(t) \tag{19}$$

The ideal gate is called a matched gate; in the white-noise case it has the shape of the signal itself and thus weights each *instant of time* in proportion to the expected signal amplitude.

(b) *The gate is u δ-function.* Then $G(\nu) = $ constant and

$$F(\nu) = \text{constant} \times S(\nu) \tag{20}$$

The ideal filter is called a "matched" filter; in the white-noise case it has a transmission proportional to the amplitude of the signal spectrum and thus weights each *frequency* in proportion to the expected signal amplitude there. The conjugacy aligns all frequency components to peak at $t = 0$ and thus produces the highest peak at that time.

For a matched gate or matched filter or any optimum combination as specified by equation (18), the signal-to-noise power ratio in the output is

$$SNR = \frac{W}{\Psi/2} \tag{21}$$

where W is the signal energy.

If the signal is oscillatory, the matched gate will be oscillatory also and will serve as a homodyne detector. But if we do not know the phase of the oscillation in advance, we must use both quadrature phases and take the quadratic sum. This is the same as using a square-law detector following a matched filter and degrades the signal-to-noise ratio to at best

$$SNR = \frac{W}{\Psi} \tag{22}$$

The signal-to-noise ratio out of a square-law detector is

$$SNR = \frac{(P/N)^2}{1 + 2(P/N)} \tag{23}$$

where P is the signal power and N, the noise power. Thus the degradation is 3 dB (2 to 1 in power) as indicated by comparing equation (22) to (21) when P/N is large. For lower input signal-to-noise ratios, the degradation is more.

Note in equations (21) and (22) that the signal-to-noise ratio depends only on the total signal energy received and not at all on how that energy is distributed in time or frequency. To design a matched filter in the first place, we need to know that distribution, and this is not possible in SETI. We can only approximate a matched filter during the search phase by making a variety of plausible assumptions as to the signal distribution in the time-frequency plane and testing each assumption.

PUTATIVE SIGNALS

We can expect to find two kinds of ETI signals: those broadcast or beamed for the use of the senders, which we merely intercept, and those intended for our receivers, which we will call beacons.

It seems clear that we can only hope to detect leakage signals if, like ours, they contain strong monochromatic components or carrier waves. Some argue that as civilizations advance they make their transmissions more efficient by eliminating carriers. In their next breath, these same people expect advanced races to become Kardashev type II cultures with a substantial fraction of the energy of their star at their disposal. It seems likely that many ETI signals will contain carriers for the same reason ours do: to save complexity and cost in both the transmitter and receiver. Even if these carriers are present, however, we will not detect them unless we build a large receiving array or unless their signals are much more powerful than ours. Early searches using existing radio telescopes will depend heavily for their success on the presence of beacons or very powerful sources such as orbiting solar power stations with microwave downlinks.

What characteristics might we expect of a beacon signal? We may assume that it will be as cost-effective as possible; that is, it should, with the least transmitted energy, announce its nature unmistakably to the least expensive receiver in the shortest time. Thus the following characteristics are likely:

1. It will be located at the best part of the spectrum, in or near the waterhole.

2. It will not resemble natural signals such as spectral lines or pulsars.

3. The effective duty cycle will be high.

4. It will be designed to minimize the difficulty of detection. Thus, for example, it will be circularly polarized since this reduces the ambiguity: only two receivers are needed to cover the possibilities.

5. Any information-bearing modulation will not destroy the detectability or its distinctiveness as an artifact.

A strong case can be made for a simple monochromatic (or nearly monochromatic) signal. It is nonnatural and always there. Binary-coded modulation can be transmitted by occasional polarization reversals at regularly spaced times. Once found, it can be homodyne-detected for improved SNR. If generated monochromatically, the diurnal and annual Doppler shift will identify it as originating on a planet and give the local day and year lengths. If Doppler-corrected in our direction, the receiver bandwidth (or binwidth in a multichannel spectrum analyzer) can be made very narrow indeed, and the coherent observing time can be made correspondingly long. An upper limit on coherence time appears to be set by multipath phenomena in the interstellar medium. This effect decreases with increasing frequency and is one of the few reasons, perhaps the only reason, favoring the high end of the microwave window. At the waterhole, coherence bandwidths on the order of 1 mHz are possible out to 250 light years, which is beyond our present range limit.

Because of the finite observing time, even a monochromatic signal is received as a pulse. The matched filter is the transform of this pulse, and this is, in fact, the filter shape provided by each bin of a multichannel spectrum analyzer (MCSA). Thus the MCSA provides a matched filter for the (arbitrary) observing time used, provided the signal is at the center of a bin.

To receive a coherent beacon efficiently, then, we are forced to divide the spectrum into bins whose width is determined by the reciprocal of the observing time or by the Doppler drift rate, whichever is limiting. A signal drifting at a rate $\dot{\nu}$ will cross a band b in time $t = b/\dot{\nu}$. The response time of the filter is $\tau \approx 1/b$. If $t > \tau$, the bandwidth is larger than needed; if $t < \tau$, only a partial response will occur. In either case, the SNR suffers, so for best results we choose $t = \tau$ and find that

$$b = \dot{\nu}^{1/2} \tag{24}$$

The peak diurnal Doppler drift rate is

$$\dot{\nu} = \left(\frac{r}{c}\omega^2 \cos\theta\right)\nu \tag{25}$$

where r is the planet radius, ω its angular velocity, θ the latitude, and ν the operating frequency. For Earth, near the equator, $\dot{\nu}/\nu \approx 10^{-10}$/sec or $\dot{\nu} \approx 0.15$ Hz/sec if $\nu = 1.5$ GHz. Thus for uncorrected Earth Doppler, we would choose $b \approx 0.4$ Hz. However, we have no way of knowing whether Earth is typical. We might not expect to find r more than twice or less than half that of Earth, but ω is very uncertain. Of the planets in our Solar System, Mercury and Venus barely rotate while Earth and Mars both have 24-hr periods. But the Moon, an unusually large satellite for a planet of our size, may have slowed Earth considerably. It would not be surprising to find Earth-like planets with periods of a few hours.

There does not appear to be any natural value for b; in fact, we may wish to provide a wide range of resolution bandwidths, perhaps from 0.001 to 1000 Hz, the wider bandwidths being reserved for pulse detection. For early searches of nearby stars, we would assume that beacons are apt to be aimed at us, and if monochromatic, that they are corrected for their Doppler drift. We can correct for ours.

To observe 100 MHz of spectrum at once with millihertz resolution requires an MCSA with 10^{11} channels, a formidable data processor even with very large-scale integration. One reason to consider pulses as likely beacon signals is that the detection job is made easier.

Now suppose we replace the isotropic antenna on a monochromatic omnidirectional beacon (which cannot be built anyway) with a rotating antenna having a fan beam. The beacon is still omnidirectional, but now pulses are received in each direction each time the beam sweeps by. Each pulse contains as much energy as would have been received from the original beacon in one rotation period. The pulse duration can be arbitrarily short, but the period between pulses should be short compared with a typical observing time. If this is true, the pulses will not be missed, and the *effective* duty cycle is not reduced.

Such a beacon would be just as detectable as a CW beacon, and the MCSA would be far less complex, because to receive millisecond pulses, for example, 10^5 channels would suffice. A series of regularly spaced flashes would be every bit as conspicuous as a monochromatic signal, perhaps more so. That is the way we build lighthouses. Polarization reversals between successive flashes could carry the information.

In a beamed beacon, pulses could still be used to advantage. So long as the average power is held constant, the detectability is not reduced; in fact, because of square-law detector performance, it is improved. Ultimately, peak power sets a limit to reducing the duty cycle, but by then the peak power may be 100 times the CW value and the SNR is correspondingly improved.

Although pulses introduce the new dimension of modulation type (i.e., pulse length, period, etc.), it appears possible to cover a wide range of alternatives with a properly designed MCSA. Pulses will not be so short as to be smeared by dispersion in the interstellar medium. Because of their advantages, we should be prepared to detect pulses as well as steady beacons.

FILTER–SIGNAL MISMATCH LOSS

We cannot hope, during the search phase, to match precisely the receiver filter to the incoming signal. The best we can do is to provide a variety of filters that approximately match a wide range of putative signals. The mismatch loss will then depend on the degree of mismatch, the nature of the signal, and how we combine the outputs of various filters.

Assume first that the signal pulse has the shape

$$f(t) = \frac{\sin(\pi t/\tau)}{\pi t/\tau} \tag{26}$$

Its spectrum is then

$$F(f) = \begin{cases} \tau, & |f| \leqslant 1/2\tau \\ 0, & |f| > 1/2\tau \end{cases} \tag{27}$$

and the pulse energy is

$$W = \int_{-\infty}^{\infty} f^2(t)dt = \int_{-\infty}^{\infty} F^2(f)df = \tau \tag{28}$$

The matched filter has unity transmission out to $|f| = 1/2\tau$ and is zero thereafter. The noise power will be

$$N = kTf_0 \tag{29}$$

If $f_0 > 1/2\tau$, the pulse is unaffected, so the peak height $h = f = 1$. Thus

$$SNR = \frac{h^2}{N} = \frac{1}{kTf_0} = \frac{2W}{kT} \frac{1}{m} \tag{30}$$

where $m = 2f_0\tau > 1$ is the mismatch factor. If $f_0 < 1/2\tau$, the signal becomes

$$h(t) = 2f_0\tau \frac{\sin(2\pi f_0 t)}{2\pi f_0 t} \tag{31}$$

which has the peak height $h = 2f_0\tau$. Thus

$$SNR = \frac{h^2}{kTF_0} = \frac{4f_0\tau^2}{kT} = \frac{2W}{kT}\frac{1}{m} \tag{32}$$

where now $m = 1/2f_0\tau > 1$ is the mismatch factor. Exactly the same result holds for the case of a rectangular pulse and a $(\sin x)/x$ spectrum. The mismatch loss (in dB) is therefore

$$L = -10 \log m \tag{33}$$

as shown by the lowest curve in figure 5.

If the signal is a Gaussian pulse

$$f(t) = e^{-t^2/2\tau^2} \tag{34}$$

and the filter has the response

$$K(\omega) = e^{-\omega^2/2\omega_0{}^2} \tag{35}$$

the same sort of analysis shows the mismatch loss to be

$$L = 10 \log \frac{m + 1/m}{2} \tag{36}$$

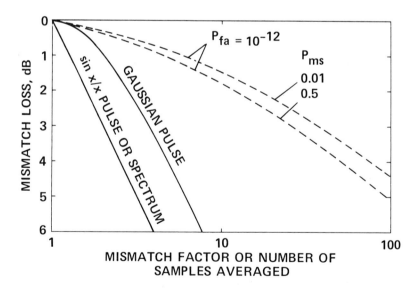

Figure 5. *Mismatch loss. When a $(\sin x)/x$ pulse is received by a filter half or twice the width it should be, there is a 3-dB loss in signal-to-noise ratio. The loss is only 1 dB for a Gaussian pulse, and 0.3–0.4 dB for a pulse detected by an MCSA if the powers in two consecutive samples or two adjacent bins are added, as the case may be.*

where $m = \omega_0 \tau$ or $1/\omega_0 \tau$ is the mismatch factor. This gives the next higher curve in figure 5.

These curves show the degradation in *output* SNR with filter mismatch. The effect on the *input* SNR required for a given quality of threshold detection is much less severe. Project Cyclops (Oliver and Billingham, 1973) proposed adding many spectra out of an optical MCSA with various offsets sample-to-sample to align drifting CW signals into the same final cells. An exact analysis of the detection statistics led to figure 11–14 in the Cyclops report, reproduced here as figure 6. It shows the input ratio of signal power to noise power per bin needed to give probabilities of signal detection of 0.5 and 0.99, as a function of the number of samples averaged, when the false alarm probability per cell is 10^{-12}. Taking the solid curve for $p_m = 0.5$, we see that for a single sample ($n = 1$) the input signal-to-noise ratio must be

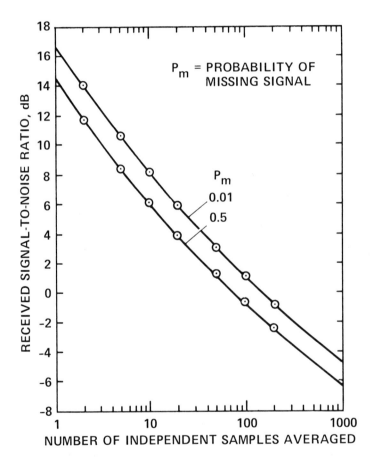

Figure 6. *Statistics of square-law signal detection. If the threshold is adjusted so that the probability of a false alarm is 10^{-12}, then the input signal-to-noise ratio indicated by the ordinate is needed to obtain the probabilities of missing the signal shown on the curves.*

14.34 dB, while for 100 samples we need –0.57 dB, a sensitivity improvement of 14.91 dB. But bins 1/100 the width would have given 20-dB improvement. Hence the "mismatch" loss of averaging 100 successive samples from bins 100 times too wide is 20 – 14.91 = 5.09 dB as shown by the lower dashed curve in figure 5. So long as the signal energy is the same and the noises are independent, the same curves should apply whether we add successive time samples or adjacent bins. If we average 2 and 4 successive samples as well as 2 and 4 adjacent bins, the mismatch loss will be at most 0.86 dB and each MCSA output will cover a 16:1 range. Two MCSA outputs having bin widths in the ratio 32:1 will cover a 512:1 range, while three will cover a frequency range of 16,384:1. The effective bin widths of a typical system of this sort are shown in table 2. In order to center on the times of occurrence and on the pulse spectra, the time and frequency averages should be *running* averages; that is, each new average should be formed by dropping the oldest sample and including the latest, or by dropping the lowest bin in the sum and including the next highest.

TABLE 2.– EFFECTIVE BIN WIDTHS IN A SYSTEM
WITH SAMPLING

Number of samples added			Direct	Number of adjacent bins	
32	4	2	1	2	4
1/1024	1/128	1/64	1/32	1/16	1/8
	1/4	1/2	1	2	4
	8	16	32	64	128

If, instead of adding the powers of successive or adjacent samples, one adds the complex amplitudes, a variety of new filters can be synthesized. Suppose the complex amplitudes of two successive time samples are added. Then, for a signal centered in the bin, the amplitudes will be in phase and the total amplitude will double. But for a signal at the bin edge, the amplitudes will cancel. We have in effect, an MCSA with twice the resolution but with every other bin missing. To recover the missing bins, we must also *subtract* the successive samples. This is similar to doing a two-point transform; in fact, if the samples are added with phase shift of ±i, this is a two-point transform.

The effective filter response for an MCSA when the time window is rectangular (input applied at constant level for a time τ) is

$$F_1 = \frac{\sin \pi x}{\pi x} \tag{37}$$

where $x = 2\tau(f - f_0)$ and f_0 is the center frequency of the bin. If two adjacent bins are added with the weighting $\pi/4$, the equivalent filter response is

$$F_2 = \frac{\cos \pi x}{1 - 4x^2} \tag{38}$$

which matches a pulse having the envelope $\cos 2\pi t/\tau$, $|t| < \tau/2$. Or if three adjacent bins are added with weightings $1/2$, 1, and $1/2$, we obtain

$$F_3 = \frac{\sin \pi x}{\pi x (1 - x^2)} \tag{39}$$

which matches a pulse of the form $(1 + \cos 4\pi t/\tau)/2$, $|t| < \tau/2$. As the number of bins added is increased, the matching pulse gets narrower and narrower and the MCSA becomes more and more insensitive to the pulse if it occurs near the ends of the observing period. To provide outputs sensitive to pulses at n adjacent times in the interval τ, the n bins must be added with n different relative phase shifts. This amounts to doing an inverse n-point transform.

All these variations appear to require more additions than merely adding the bin powers as suggested earlier. However, further study might reveal tricks similar to the fast Fourier transform that would simplify the whole process.

SKY SURVEYS

Many surveys of the radio sky have been carried out by astronomers, but their low sensitivity, restricted frequency ranges, and limited data processing have excluded the possibility of detecting ETI signals. Let us see what kind of sensitivity and frequency coverage is achievable in a SETI sky survey.

If t_s is the time allotted for a complete scan of the sky for one frequency band, then from equation (16) the observing time per direction is

$$\tau = \frac{t_s}{n} = \frac{t_s}{g} \tag{40}$$

The energy received during one observation from a CW beacon delivering a flux ϕ is

$$W = \phi A \tau = \phi \frac{\lambda^2}{4\pi} g\tau = \phi \frac{\lambda^2}{4\pi} t_s \tag{41}$$

which is the energy that would be received by an isotropic antenna in the search time t_s. To avoid too many false alarms, we need to set a threshold at $W = mkt$, where $m \approx 30$. Thus the sensitivity of an all-sky survey is

$$S = \frac{1}{\phi} = \frac{\lambda^2 t_s}{4\pi mkT} \tag{42}$$

We note that, for a given search time and threshold, the sensitivity is proportional to λ^2 because the longer the wavelength the broader the beam in both dimensions. *Antenna area does not affect the sensitivity.* However, from equation (40) we see that the MCSA matched bandwidth is

$$f = \frac{1}{\tau} = \frac{g}{t_s} = \frac{4\pi A}{\lambda^2 t_s} \tag{43}$$

The total bandwidth per scan is $B = Nf$, where N is the number of MCSA channels. A good figure of merit for a survey system is the product of sensitivity and bandwidth:

$$M = BS = \frac{AN}{mkT} \tag{44}$$

which depends only on the antenna size, the data processor power, and the noise temperature. Search time is a variable that can be used to trade bandwidth and sensitivity for fixed antenna and MCSA.

With our present technology, only a small fraction of the microwave window can be received at one time. Several scans of width B are needed to survey all frequencies within the window. The time required to do this depends on our choice of sensitivity as a function of frequency. Let t_0 be the time devoted to the lowest frequency scan, centered at f_0.

If we keep the time per scan constant, then from equation (43) the sensitivity will fall off as λ^{-2}. The antenna diameter must be proportional to λ^{-1} (area proportional to λ^{-2}) to keep the beamwidth constant. The time required for s scans covering the total bandwidth sB is simply

$$T_0 = st_0 \tag{45}$$

If we make the time per scan proportional to band-center frequency and make the antenna diameter proportional to $\lambda^{-1/2}$ (area proportional to

λ^{-1}), the sensitivity will be proportional to λ and the time t_R for the scan with center frequency $f_0 + kB$ is

$$t_R = t_0 \left(1 + k \frac{B}{f_0}\right) \tag{46}$$

The total time is

$$T_1 = \sum_{k=1}^{S} t_0 \left(1 + k \frac{B}{f_0}\right) = t_0 \left[s + \frac{B}{f_0} \frac{s(s-1)}{2}\right] \tag{47}$$

If we make the time per scan proportional to the square of the band-center frequency and keep the antenna diameter and sensitivity constant, then

$$t_R = t_0 \left(1 + k \frac{B}{f_0}\right)^2 \tag{48}$$

and

$$T_2 = \sum_{k=1}^{S} t_0 \left(1 + k \frac{B}{f_0}\right)^2 = t_0 \left[s + \frac{B}{f_0} s(s-1) + \frac{B^2}{f_0^2} \frac{(s-1)s(2s-1)}{6}\right] \tag{49}$$

Let us assume a 100-m-diameter antenna. At 1 GHz its gain is 1.45×10^6, so we have this many directions. If we allow 1 sec per direction, then $t_0 = 17$ days, and $b = 1$ Hz. If $B = 300$ MHz, $N = 3 \times 10^8$, which is a formidable MCSA, but feasible. Now $B/f_0 = 0.2609$. The time to execute s scans for the three cases is shown in figure 7. To cover the microwave window from 1 to 10 GHz requires 30 scans and times $T_0 = 30\,t_0 = 1.4$ yr, $T_1 = 143.5\,t_0 = 6.7$ yr, and $T_2 = 839\,t_0 = 39$ yr.

Taking $m = 30$, $T = 10$ K, and $\lambda = 0.2609$ m, we find from equation (42) that $\phi = 5 \times 10^{-25}$ W/m^2. Figure 8 shows sensitivity vs frequency and the frequency coverage of surveys made with the above system with constant sensitivity and with sensitivity proportional to λ and to λ^2. Only the latter allows full coverage of the terrestrial microwave window.

From figure 9 we see that, at its most sensitive, the above sky survey would detect a 1 MW beacon at 1.5 GHz beamed at us with a 64-m antenna out to a range of about 40 light years. Such beacons, if located near likely stars, would be detected by a targeted search to much greater range. Beyond 100 light years we appear to them as only one of thousands of likely stars, so it is difficult to see why any beacon farther away than this would be

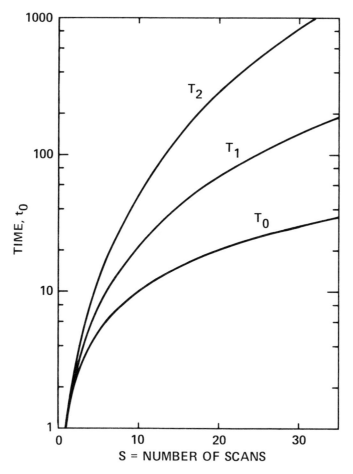

Figure 7. *Total time required to conduct the number of scans shown by the abscissa when the time to conduct the lowest frequency scan is* t_0.

beamed at us. At this range an omnidirectional beacon would have to radiate 5×10^{12} W (i.e., the output of 5000 nuclear plants) to be detectable by this sky survey. We can improve the sensitivity by observing for a longer time in each direction, but this will reduce the spectrum coverage proportionally.

TARGETED SEARCHES

The orthodox approach to SETI is to concentrate the search on "good" stars, that is, main-sequence F, G, and K stars beginning with the closest and progressing outward to greater and greater ranges. Most of the effort to date

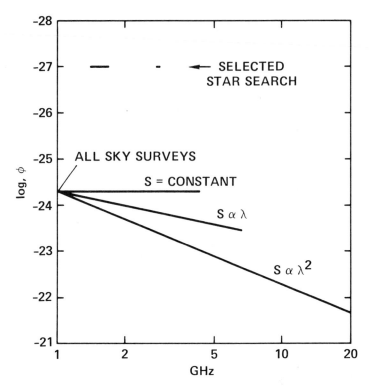

Figure 8. *Frequency coverage possible in a given time for sky surveys and targeted searches.*

has been aimed at detecting monochromatic or slowly drifting CW signals. In view of the foregoing discussion, pulse signal detection probably also belongs in a targeted search.

The search system for the targeted mode is identical with that for the sky survey in most respects. Both involve a large antenna, a low-noise receiver, and an MCSA. The difference is that for the targeted search the MCSA should contain a variety of bin widths from wide bins (>100 Hz) to look for pulses to very narrow bins (<0.1 Hz) to look for monochromatic signals for times much longer than allowed in a sky survey.

A targeted search does not posit any bizarre life form or any technological prowess that far transcends our own. Essentially, it assumes that intelligent life is most likely to be found on terrestrial planets circling Sun-type stars and that the powers they will devote to beacons are consistent with our technology and economic values.

The minimum detectable flux for a targeted search is simply

$$\phi = \frac{mkTb}{A} \tag{50}$$

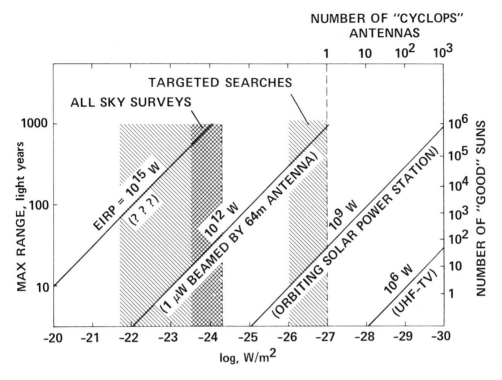

Figure 9. *SETI system capabilities.*

where *m* is to be found from figure 6. If we choose $b = 1/32$ Hz and integrate 32 samples, the observing time per star is 1024 sec and, from figure 6, we will have a 50% chance of detecting a signal if $m \leqslant 1.8$, that is, if the threshold corresponds to an input signal-to-noise ratio of 2.62 dB. For a 100-m antenna, $A = 2500\,\pi$. Taking $T = 10$ K we find

$$\phi = 10^{-27} \ \text{W/m}^2$$

With this sensitivity we see from figure 9 that we can just detect an EIRP of 10^{10} W (10 kW at 1.5 GHz beamed by a 64-m antenna) at a range of 100 light years. Thus we can detect beacons of modest power around any of the 2000 good stars within that range.

The time to scan all these stars is 2×10^6 sec or 24 days. In 3 years, 45 different frequency bands could be searched. Assuming the MCSA again has 3×10^8 bins, 10 MHz would be searched per scan, for a total of 450 MHz. This is enough to cover the waterhole and other cardinal frequency bands.

The Cyclops concept was to start with a single 100-m antenna and add others in a phased array, thus increasing the sensitivity for each successive search. We see from figure 8 that several hundred elements would be needed before leakage signals like our own UHF-TV signals became detectable. Early SETI searches using existing antennas will have a sensitivity range of about 10^{-26} to 10^{-27} W/m^2 (as shown in fig. 9). These searches are heavily dependent for their success on the existence of beamed or powerful beacons.

CONCLUSIONS

Given present technology, or any we can foresee, the only practical way to search for intelligence outside our Solar System is to listen for (and possibly later to radiate) coherent signals in the decimeter range of the radio spectrum. Current low-noise receivers have only about 2 K noise temperature, which raises the total to perhaps 10 K in the best part of the spectrum. Present receivers are therefore only 1 dB short of perfect. Present data-processing technology makes multichannel spectrum analyzers with perhaps 10^8 bins affordable. Computer hardware costs (memory and microprocessors) are dropping rapidly. Very powerful MCSAs (10^9 bins) will soon cost under $1 million. It is timely to begin tests using state-of-the-art hardware and existing antennas so the proper data processors for SETI can be developed as the art matures.

REFERENCES

Cocconi, Giuseppe; and Morrison, Philip: Searching for Interstellar Communications. Nature, vol. 184, 1959, pp. 844–846.

Oliver, Bernard M.; and Billingham, John, eds.: Project Cyclops, a Design Study of a System for Detecting Extraterrestrial Intelligent Life. NASA CR-114445, revised edition, 1973.

Eavesdropping Mode and Radio Leakage from Earth

WOODRUFF T. SULLIVAN III

Earth is even now advertising itself splendidly to the Universe. We cannot know whether the first artificial nonhuman signal detected would place us in the role of intended recipient or of eavesdropper, but in our explorations we should allow for both possibilities.

One can make either of two basic assumptions about our first contact with an extraterrestrial civilization: (1) that it will arise through a purposeful attempt, perhaps through the use of an interstellar radio beacon or (2) that a civilization will be detected through no special efforts of its own. The latter hypothesis, often called eavesdropping, is concerned with the extent to which a civilization can be unknowingly detected through the by-products of its daily activities. While much thought has gone into the idea of purposeful contact, eavesdropping has been somewhat neglected; we will argue that it deserves more attention.

The overall likelihood of contact through eavesdropping depends on the nature and intensity of the civilization's "leakage," as well as on how long that leakage continues. Very general arguments (Oliver and Billingham, 1973, and the other papers in this volume) show that radio waves provide the most economical and reliable means of contact at interstellar distances. This is true not only for intentional contact, but probably for eavesdropping as well. In any case, there can be no argument with the fact, first discussed in print by Webb (1963) and Shklovskii and Sagan (1966), that the presence of humans can already be detected at interstellar distances as a result of the complex communications and transportation network now spread over our globe. Of course, we do not know how applicable our present situation is to the more general case of all galactic civilizations over all time. It may be that our present "leaky" state will soon be terminated by advancing technology, but on the other hand it may continue for a very long time, perhaps even longer than any period in which we might have the perseverance to send out purposeful messages. If we are at all typical, then we should perhaps also be looking at least as diligently for unintentional signals from others as for intentional ones. To explore the principles involved, we might start by asking about emissions from Earth as detected at interstellar distances.

LEAKING RADIATION

Detailed consideration of all parts of the electromagnetic spectrum reveals that *radio* waves represent by far the most important "leakage" from Earth (see Bracewell, this volume). For instance, nothing that we do with visible light, not even exploding a hydrogen bomb, compares in the least with the Sun's output. But at wavelengths from 1 cm to 30 km, our society has organized a host of activities that give our planet an unnatural "radio signature": television and radio broadcasting, radars used for weather, navigational, and military purposes, "short-wave" communications ("hams," CBs, taxis, police), satellite communications, and so forth.

We now want to put ourselves in the "shoes" of an extraterrestrial radio astronomer on a planet revolving about a star far from our Sun. Which of these radio services would be best for our eavesdropper to tune in on? Which is detectable to the greatest distances? Which potentially carries the most information of use to the eavesdropper? To answer these questions one must study many factors, including the power of each service's transmitters, the frequencies and bandwidths involved, the types of antennas used, and the fraction of time spent transmitting. (Readers desiring technical details on these and other matters should consult Sullivan et al., 1978). One example of these factors is the general tradeoff between the information content (TV picture, spoken words, Morse code) of a transmitted signal and the farthest distance at which it can be detected. This can be understood by noting that one gets more range by concentrating transmitted power at the fewest number of frequencies possible. But the information content of a signal is contained in the arrangement of its power among a number of neighboring frequencies and increases as we spread the power over a greater bandwidth.

Three other important criteria in the evaluation of each radio service are: (1) that the signal should be exactly the same from day to day, (2) that the amount of sky illuminated by the transmitting antennas should be large, and (3) that the number of transmitters on Earth should be large. Regarding (2), remember that the radio waves from even a stationary antenna can sweep out a large portion of the sky as a result of Earth's rotation. Furthermore, each antenna has a characteristic beam into which the transmitter power is directed. If an antenna is designed so that the power is concentrated into a relatively small region of the sky, the range of detection for the signal increases, but at the expense of excluding many potential listeners.

ACQUISITION AND INFORMATION SIGNALS

Keeping the above factors in mind, an examination of all radio services reveals two categories of strong signals escaping Earth that might be of interest to an extraterrestrial observer.

An *acquisition signal* merely announces our presence over a large region of space by its very existence but is not generally useful for careful study because it fails to meet one or more of the criteria given above. An *information signal*, however, satisfies all three criteria. At the present time on Earth, some of the most important acquisition signals originate from a half-dozen or so U.S. military radars (and their presumed Soviet counterparts). These Ballistic Missile Early Warning System (BMEWS) radars sweep out a large fraction of the local horizon with extraordinarily powerful transmitters. The result is that this "radio service" provides by far the most intense signals that leak from our planet to a large fraction of the sky.

While BMEWS radars pass criterion (2) above, they fail (3) and partially fail (1) because they are so few and often change their frequency of operation to avoid being jammed. Nevertheless, if an external observer used equipment comparable to the most sensitive radio telescope on Earth (the 305-m-diameter dish at Arecibo, Puerto Rico), we calculate that a BMEWS-type radar could be detected as far away as 15 light years. This distance includes only about 40 stars, but, of course, it is possible that our eavesdropper possesses a much more sensitive radio telescope than we do. If "he" had something like the largest one ever proposed for Earth, namely, the array of a thousand 100-m dishes called for by Project Cyclops (Oliver and Billingham, 1973), he could detect a BMEWS-type radar at a distance of 250 light years. In this case at least 100,000 stars are possible candidates for such an eavesdropper's location. But note that radio waves travel at the finite speed of 1 light year per year, and thus it will take until the 23rd century, or 250 years from now, before all these stars have had a chance to be bathed in the radiation of our defense system radars!

After picking up a BMEWS (or other) acquisition signal, the observer needs at least 100 times more sensitivity in his equipment to reach the rich lode of information signals emanating from Earth. It turns out that television broadcast antennas (or stations) are the most intense sources of such signals.

All other services either have their transmitter power spread over too broad a frequency band (for instance, FM broadcasting and most radars) or do not transmit continuously (ham radio operators) or from the same location on Earth each day (taxis, aircraft). Many signals, such as medium-wave AM broadcasting and almost all short-wave communications, never even penetrate the ionosphere, the reflective layer of charged particles that surrounds Earth. We thus concentrate on TV broadcasting; all other services that leak from Earth are less intense and merely add to the background noise a distant observer would measure in the direction of our Sun as seen in his sky.

But again note that TV broadcasting from Earth has been in existence for only 40 years. Figure 1 illustrates the phenomenal growth in intensity of the resultant ever-expanding "power bubble." On a cosmically infinitesimal time scale, Earth has indeed become a very bright planet, outshining the Sun by orders of magnitude in certain narrow frequency ranges.

TV BROADCASTING SIGNALS AND ANTENNAS

To understand why television is so valuable to the eavesdropper as an information signal, we should discuss some of the characteristics of TV broadcasting signals. Perhaps the most important facts are that there is a large number of very powerful TV stations on Earth (fig. 2) and that about half a station's broadcast power resides in an extremely narrow band of frequencies, only about 0.1 Hz wide, called the *video carrier signal.* The other half of the power contains the picture information and is spread out in a complex manner over a far larger frequency range of about 5 MHz. Nowhere in this broader region is the power per Hertz even a thousandth that at the video carrier frequency. It would therefore be much more difficult (a factor of 2×10^4 is a good estimate) for the eavesdropper to receive full program material than to simply detect the presence of the carrier signal. (Given the quality of most TV programs, we find this fact very reassuring.) An observer near Barnard's star, third closest to the Sun at a distance of 6 light years from Earth, is thus about to receive television signals originating from the 1974 House Judiciary Committee hearings, but he probably cannot find out if Nixon was impeached. In the discussion below, we assume that only the video carrier signals of stations, *not* program material, are detected.

The combination of reasonably high power and small bandwidth means that the most powerful TV carrier signals can be detected (at optimum frequencies of 500–600 MHz) from distances as large as 1/10 those discussed for the BMEWS radars. The narrow-band nature of the signal also enables the observer to measure extremely accurate Doppler shifts in the frequency of the carrier signal, allowing a determination of the relative speed with which

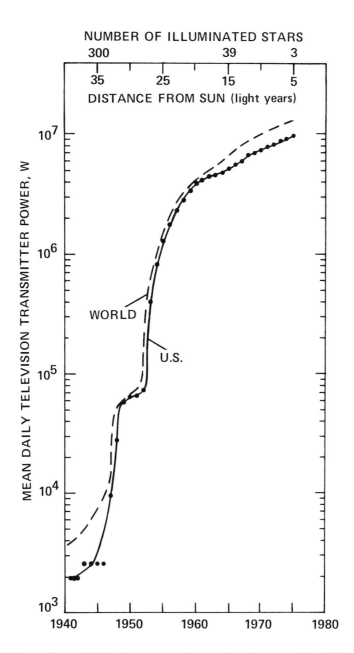

Figure 1. *Estimated growth in time-averaged transmitted power since TV broadcasting began. The solid curve for the United States is reasonably accurate since starting dates for all stations are available, but it was still necessary to use a model for the growth of transmitter power and daily broadcast hours for each station. The dashed curve for the entire world is correct for 1975 and only estimated for earlier dates. The increasing number of stars bathed by the expanding power bubble (as of 1980) is indicated at the top.*

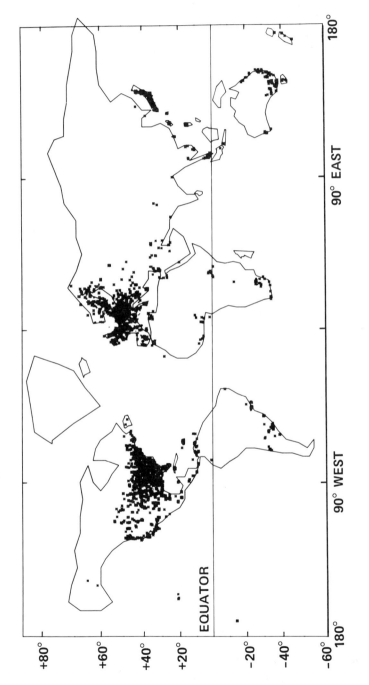

Figure 2. Map of Earth showing the 2200 most powerful TV transmitters, possessing about 97% of the world's total TV power. Note the absence of stations in the Southern Hemisphere and the marked concentrations in North America (46% of all stations), Europe (35%), Japan, and Australia. Full information for stations in the Soviet Union and China was not available, but estimates indicate that there is only a minor amount of TV power in these countries.

each station is moving to an accuracy of about 0.1 m/sec. Each station's signal thus contains information concerning the myriad motions in which its broadcast antenna participates while anchored to the rotating and revolving Earth (see fig. 3). Note that stations on a common channel will not fall precisely on top of each other's frequency because the combined effects of engineering sloppiness, deliberate frequency offsets, and Doppler motions all shift a station's video carrier frequency by much more than its width (fig. 4). This means that our hypothetical observer could not obtain a more favorable signal-to-noise ratio by trying to receive simultaneously all the Channel 5s,

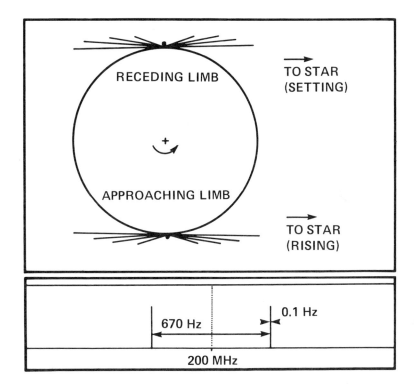

Figure 3. Top: *Sketch of two TV broadcasting antennas as seen from above Earth's pole (also applies to a single station as it would be seen at 12-hr intervals). The length of a particular line of radiation indicates the relative amount of power "beamed" in that direction. As seen from a distant star located to the right, both stations are at maximum intensity, but one is just coming into view, the other just disappearing. From the point of view of the stations, one sees this star rising and the other sees it setting.*

Bottom: *Radio spectrum of the two stations' video carriers as measured at the distant star. Both stations are assumed to radiate from Earth with the same rest frequency (dotted line); the observed frequencies differ as a result of the Doppler effect arising from Earth's rotation. The numbers given are for stations taken to radiate at 200 MHz (approximately U.S. Channel 11) on the equator, and are typical of those for most stations.*

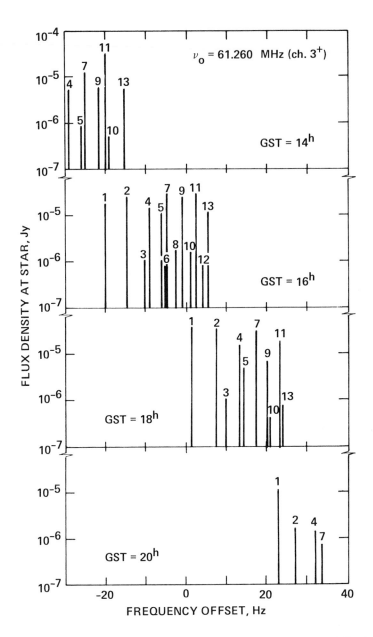

Figure 4. *One of the most crowded regions of the radio signature of Earth: calculated spectra of 0.1-Hz-wide video carriers for Channel 3 (61.260 MHz) as observed from the star Kruger 60 at a distance of 12.9 light years. The four Greenwich sidereal times (GST) cover the period during which the star scrapes the northern horizon at lower culmination as seen from the United States. The illustrated frequency offsets arise solely from the Doppler shift due to the rotation of Earth; in reality, the degree of crowding is much less since the frequencies of video carriers often wander by as much as ±200 Hz from their assigned frequencies. Individual stations are labeled by numbers.*

for example. The problem of detecting radio leakage from Earth as a whole is thus essentially identical to the problem of detecting its single strongest transmitter.

The beam patterns into which TV broadcast antennas radiate are important in such an analysis. It turns out that these antennas (whose purpose, after all, is to broadcast to Earth and not to the stars) confine the transmitter power to within a few degrees of the horizon, but distribute it about equally in all compass directions. Those radio waves directed above the horizon completely escape Earth's atmosphere, and even about half those below the horizon manage to escape by bouncing off the ground. (Only a negligible portion ever reach any TV set.) Since most of the power is broadcast near the horizon, only when a star is rising or setting, that is, when it is on the horizon as seen from a given antenna location, will it be illuminated with radio power (fig. 3).

After his initial discovery of these radio waves from the direction of our Sun, our eavesdropper would undoubtedly first ask, "Is this some kind of strange natural radio emission or has some form of civilization produced it?" It would seem that the narrow-band nature of the signals would be one of the best clues that the signal is artificial in nature, since no astrophysical process known to us can channel comparable amounts of energy into such small-frequency intervals. Other clues, such as polarization of the signals, also exist. And yet, who knows? Perhaps the theorists of another planet are clever enough to come up with a substance whose emission spectrum matches that of the observed radio waves! Clever theorists notwithstanding, for this discussion we assume that the signals from Earth would be recognized as artificial.

As shown in figure 3, when a star is near the horizon and thus illuminated by a particular station, the station must be near the edge of Earth as seen from the direction of the star. The result is that Earth has a very "bright" edge or *limb,* when observed with a receiver for TV frequencies (40 to 800 MHz), but the great distance to our eavesdropper's radio telescope means that he is unable to discern the disk of Earth. Nevertheless, the Doppler shift of each station due to the rotation of our planet can tell him not only whether the station is on the approaching or receding side of Earth, but also whether its latitude is near the fast-spinning equator or the more slowly moving polar regions. Furthermore, he could discover a station's longitude from the *times* of the twice-daily appearance of the signal from each station. Thus he could construct a map (just like fig. 2, but of course without the outlines of the continents) of all detected stations, each located to an accuracy of a few kilometers.

Because of the extremely nonuniform distribution of stations on Earth, the total number of stations visible at any one time to an outsider will vary with a period of 24 hr (sidereal). The situation as it would be measured from

Barnard's star is shown in figure 5. The peaks correspond to the times when population centers with concentrations of television transmitters are on Earth's limb. By combining data on these intensity variations and Doppler shifts in a straightforward fashion, any eavesdropper could deduce his position relative to our equator (we would say his *declination*), the radius of Earth (6000 km), and the rotational velocity at the equator (0.5 km/sec).

With this information in hand, the observer is likely to suspect that he is dealing with a planetlike body. His next step might be to study Earth's annual motion about the Sun (at a rate of 30 km/sec), which causes very large Doppler shifts in the signals of all the individual stations. By tracking these shifts over a year or more, the Earth-Sun system can then be investigated exactly as astronomers here study what they call single-line spectro-

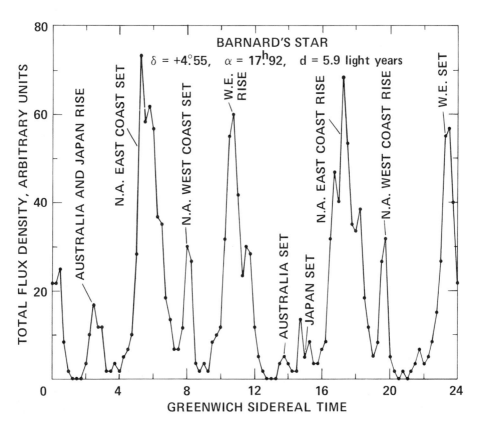

Figure 5. *Calculated flux density (summed over all frequencies) of TV radiation that would be measured over a sidereal day by an observer located in the direction of Barnard's star, third closest to the Sun and by chance near our celestial equator. The origin of the various peaks is indicated; "rise" and "set" refer, respectively, to the appearance at the western limb and disappearance at the eastern limb of a particular region on the rotating Earth (N.A., North America; W.E., Western Europe).*

scopic binaries. In such a system, two bodies (usually two stars) are orbiting about each other, but only the Doppler shifts in the spectral lines of one member (usually the brighter of the two) can be measured. In the present case, the "spectral lines" are the TV carrier signals and the "bright" member is Earth, far outshining the Sun at the radio frequencies we are discussing. It can be shown that radio observations of Earth, together with standard optical observations sufficient to give an estimate of the mass of the associated G2 dwarf star, which we call our Sun, would yield all the vital orbital data for Earth — its orbital period, its eccentricity, the true Sun-Earth distance, etc. The eavesdropper would then be able to provide his colleagues in the Exobiology Department with a good estimate for Earth's surface temperature, allowing them to place constraints on the possible forms of life responsible for the radio signals. It also turns out that from the duration of each station's daily appearances, one can readily deduce the beam size and thus the dimensions of the transmitting antennas (typically 15 to 20 m), yielding vital clues to the size scales of terrestrial engineering.

There are also more subtle effects contained in the TV carrier signals, effects which may or may not remain ambiguous to the observer. For instance, seasonal variations in vegetation, weather, and the ionosphere will leave their mark in each station's signal. Vegetation has an influence on the amount of power reflected from the surface, as does the choppiness of the sea for coastal stations. Weather and ionosphere affect the direction and intensity of the radiated power, either through winds flexing the antenna structure or through our upper atmosphere bending and absorbing the radio waves on their way out. These conditions will cause the observed power levels and times of station appearance to vary slightly and, at first, inexplicably from those predicted. Detailed study may nevertheless allow a few basic conclusions; for example, the presence of an ionized gas around the planet might be deduced from the clue that the lowest frequency stations are much more affected than those at higher frequencies.

A second type of complexity results from such factors as a station's daily sign-off hour and the specific frequency and antenna conventions that it follows. These generally vary from one country to another, but can be the same even for countries that are widely separated but otherwise cooperative in trade, politics, or technology. For example, frequency assignments and other conventions are very similar in Japan and the United States. We can interpret these diverse patterns with our detailed cultural and historical knowledge, whereas the extraterrestrial probably cannot, unless his social theory is advanced far beyond our own. The overall problem is not unlike that confronting an archeologist trying to understand an ancient city with a knowledge of only its street plan. It can only be hoped that the many unsolved puzzles would not hinder the eavesdropper from understanding the more regular and straightforward features of Earth's radio spectrum.

OBSERVATIONAL TEST

In order to sample the radio signature of Earth from an external site and thus test whether TV broadcasting is in fact the principal component, S. H. Knowles and the author used the Moon as a handy and objective reflector of Earth's leakage. Using the 305-m Arecibo radio telescope, we scanned a wide range of frequencies between 100 and 400 MHz and found that, once local interference was eliminated (using an on-Moon, off-Moon technique), the frequencies of most observed signals could indeed be identified with the video carriers of various nationalities (fig. 6). Not all of the countries mentioned in figure 6 were on the limb of Earth as seen from the Moon, however, illustrating that some power leaks also from relatively high elevation angles at the transmitters, although of lesser intensity.

SHOULD WE TRY TO EAVESDROP?

The above discussion is, of course, relevant to the larger issue of our own attempts to contact extraterrestrial civilizations. We note that the one civilization we know something about (our own) has sent out virtually no purposeful signals, yet has been leaking radiation for several decades. How typical this situation will prove to be in our own future or for other galactic civilizations is impossible to say. Cable television may replace the present system of broadcasting antennas, but new forms of radio leakage may appear as well. For example, even the slightest bit of back-lobe leakage from the giant transmitting antennas proposed for the solar power satellite would create extremely intense, repeatable, narrow-band microwave signals. It is true that the range of detection of a purposeful beacon is probably much larger than for leaking signals, but another civilization might be leaking prodigious amounts compared to us, or they might have even set up powerful navigational beacons for interstellar travel. Furthermore, purposeful signals require decoding of any received message, while we have seen that unintentional signals, so long as they are narrow-band and periodic, yield a great deal of information using only standard astronomical techniques. Not only that, but in a sense the information gained may be a more accurate reflection of the society's major concerns. At least this seems to be true for the case of our own civilization with its military and TV leakage, although we might not wish to admit it.

In terms of actual search strategies at the radio telescope, the eavesdropping hypothesis does not come into direct conflict with that of purposeful signals, but rather it suggests that a well-designed system must allow for

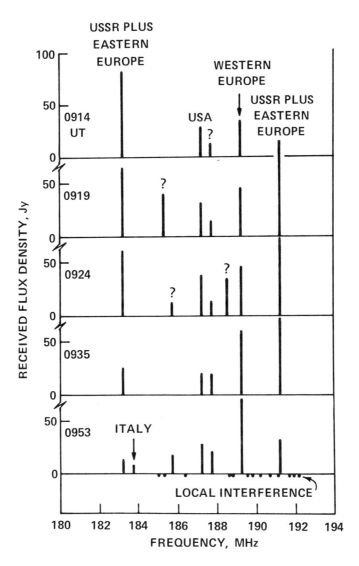

Figure 6. *Radio signature of Earth as revealed in a synoptic series of spectra (10-kHz resolution) in one of the very high-frequency bands assigned to TV broadcasting on December 21, 1978. The Arecibo telescope was pointed alternately on and off the Moon; reflected signals were taken to be only those absent in the off spectrum. Probable identifications, based on frequency allocations for video carriers, are indicated.*

more possibilities. In particular, extremely narrow-band signals drifting in frequency and variable in intensity must not be automatically filtered out. And, of course, the rationale behind special "natural" frequencies fails; on the other hand, leakage will occur at such frequencies as well as at any others, so they should still be used, although perhaps not exclusively.

In summary, then, we should keep two possibilities in mind when searching for extraterrestrial signals. We cannot know whether the most likely signals to be detected would place us in the role of intended recipient or of eavesdropper.

REFERENCES

Oliver, Bernard M.; and Billingham, John, eds.: Project Cyclops: a Design Study of a System for Detecting Extraterrestrial Intelligent Life. NASA CR-114445, revised edition, 1973.

Shklovskii, I. S.; and Sagan, C., eds.: Intelligent Life in the Universe. Delta, N.Y., 1966, pp. 255–257.

Sullivan, W. T., III; Brown, S.; and Wetherill, C.: Eavesdropping: The Radio Signature of the Earth. Science, vol. 199, Jan. 27, 1978, pp. 377–388 (related correspondence in Science, vol. 202, p. 374 ff.).

Webb, J. A.: Detection of Intelligent Signals from Space. In: Interstellar Communication, A. G. W. Cameron, ed., W. A. Benjamin, N.Y., 1963, p. 178.

SETI — The Search for Extraterrestrial Intelligence: Plans and Rationale

J. H. WOLFE, R. E. EDELSON, J. BILLINGHAM, R. B. CROW,
S. GULKIS, E. T. OLSEN, B. M. OLIVER, A. M. PETERSON,
C. L. SEEGER, and J. C. TARTER

A moderate but wide ranging exploratory program is described which would use existing radio telescopes and advanced electronic systems with the objective of trying to detect the presence of just one signal generated by another intelligent species.

As presently envisaged, the SETI effort involves a 10-year program which will search a well-defined volume of a multidimensional search space for microwave signals of extraterrestrial intelligent origin, using existing antennas and special purpose data acquisition and analysis systems having very high throughput. The entire sky will be surveyed in two polarizations between 1.2 and 10 GHz with resolution binwidths down to 32 Hz. More than 700 nearby solar type stars and other selected interesting directions will also be searched in two polarizations between 1.2 and 3 GHz with resolution binwidths down to 1 Hz. Particular emphasis will be placed on those solar type stars that are within approximately 20 light years of Earth. An analysis system will detect the presence of a wide range of pulses, carriers, and complex or drifting signals. The sky survey will be about 300 times more sensitive and will cover 20,000 times more frequency space than previous surveys. The targeted search will extend the type of signals being sought, the number of targets by a factor of 4, and the range of frequencies covered by a factor of 3×10^6.

391

INTRODUCTION

The only practical way we know now to test the idea that life exists beyond our Solar System depends on an intelligent fraction of that life providing an electromagnetic signature we can recognize. The physical laws of the Universe, a relatively mature microwave technology, recent digital solid-state achievements, and a minimum number of ad hoc assumptions have permitted the development of a promising set of exploratory strategies for detecting a range of possible electromagnetic signals of extraterrestrial intelligent (ETI) origin. As a result, Ames Research Center (ARC) and the Jet Propulsion Laboratory (JPL) are proposing a moderate research and development exploratory program using existing radio telescopes and advanced electronic systems with the objective of trying to *detect* the presence of just one signal generated by another intelligent species, if such exists.

In the scientific community, the notion that intelligent life might exist elsewhere is a widely held hypothesis with decades of increasing confidence in its essential validity in the presence of increasing controversy. It follows directly from two other hypotheses: given a suitable and sufficiently enduring environment, life is a normal, natural consequence of the long-term application of the basic physical and chemical processes of our Universe; and, as all human experience seems to attest, once a physical process has been found to occur, it can be found occurring elsewhere. But the notion is still an extrapolation based entirely on indirect and mostly Earthbound evidence gathered from many disciplines by a single example of life, intelligent by self-definition and ignorant of any prohibitory physical law. To some skeptics, Martin Rees once retorted, "Absence of evidence is not evidence of absence!" (Oliver and Billingham, 1973). But, of course, neither is it evidence of presence. Despite the many elaborate and largely anthropocentric discussions to date, there is no foreseeable way to test such an interesting hypothesis other than to suitably explore the outer Universe.

BACKGROUND

Much of the extensive history of the Life in the Universe concept is covered under "Additional Reading" at the end of this paper. The essence of the scientific arguments developed in these papers and in many others is as follows. Modern astrophysical and astronomical theory predicts that planets are the rule rather than the exception and are likely to number in the hundreds of billions in our Galaxy alone. Given a suitable location and environment for any single planet, current theories of chemical evolution and the origin of life predict that life will begin. Once life has been established and given a period of billions of years of comparative stability on the planetary surface, it is argued, life will sometimes evolve intelligence. In some cases, the next step may be the emergence of a technological civilization. While we believe ourselves to be an example that this complex path was followed at least once, there is no broad agreement as to how many other technological civilizations might currently exist in our Galaxy. A joint discussion of this topic during the 1979 General Assembly of the International Astronomical Union in Montreal, Canada, showed the optimists and pessimists to be separated by at least six orders of magnitude in their estimates. On that occasion, as he had done 20 years earlier in his pioneering paper, Morrison urged that a search be conducted to experimentally determine (or at least bound) the number of other technological civilizations, as it cannot be calculated from first principles.

In parallel with the scientific arguments for the existence of ETI, there has been a rapid growth in the techniques and technology used in radio-astronomy. In recent years, there has been an increasing acceptance of the hypothesis that the most effective way now to detect the existence of other civilizations is to listen for their signals in the microwave region of the spectrum. Indeed, there have been numerous separate searches over the past 20 years. In all cases, these pioneering SETI observations have been pursued with small budgets and with comparatively primitive data processing equipment. It seems clear that reasonable chances of detecting an ETI signal can come about only from more thorough observational procedures using more sensitive and sophisticated data processing systems.

A series of SETI science workshops, chaired by Philip Morrison (Massachusetts Institute of Technology) and supported by the NASA Office of Space Science, was conducted as part of a two-year feasibility study. The results of the science workshops were published as NASA SP-419, "The Search for Extraterrestrial Intelligence" (Morrison et al., 1977). The conclusions of the workshops were:

1. It is both timely and feasible to begin a serious search for extraterrestrial intelligence.

2. A significant SETI program with substantial potential secondary benefits can be undertaken with only modest resources.

3. Large systems of great capability can be built if needed.

4. SETI is intrinsically an international endeavor in which the United States can take a lead.

The proposed SETI effort is an integrated program incorporating the proposals generated by the science workshops under conclusion 2. The plan recognizes the timeliness of conclusion 1, not only in terms of available technology, but in terms of the very serious problem of man-made interference at radio frequencies. For a SETI program to require only modest resources, it must be ground-based yet still have access to those portions of the frequency spectrum where the search for potentially weak signals is to be accomplished. The most recent allocations of the microwave spectrum to numerous users worldwide serve to emphasize the need to proceed with a SETI exploration as soon as possible.

In accord with conclusion 2 above, a significant SETI program can be carried out without new radiotelescopes by equipping existing telescopes with instrumentation of enormous capability, unavailable to previous searchers. Recent electronic developments offer the opportunity to conduct more efficient searches with higher sensitivity and with broad sky and frequency coverage. Key aspects are the ability to process simultaneously more than 10^7 frequency channels, ultra-low noise cryogenic receivers having wide bandwidth and tunability, and sophisticated on-line and off-line signal processing and identification systems.

Starting with Project Ozma in 1959, there have been some 22 separate radio searches for ETI signals, many of them still continuing. Though they represent an impressive effort, they have covered only a very small fraction of the parameter space in which an ETI signal might be expected. The program described here should provide at least a 10-million-fold increase in search space coverage, compared to the sum of all previous searches.

In summary, the underlying rationale for SETI is based on the confluence of major developments in science and technology over the last two decades, leading to both increased interest and increased capability. Controversy over the probability of success can only be resolved by conducting the search.

GOAL AND OBJECTIVES

SETI has as its goal the detection of evidence of the existence of extraterrestrial intelligence. To approach this goal, a series of specific objectives, covering a 10-year period, has been established:

1. Develop a sound scientific and technological search strategy, considering the laws of physics as currently understood, the present state of technology, and the opinions and expertise of the engineering and scientific community.

2. Systematically test for the presence or absence of a wide variety of radio signals.

3. Explore and evaluate other approaches where a theoretical basis suggests a promising search regime.

4. During the search, gather the crucial environmental and engineering data required to better understand the natural and man-made constraints on the observations and simultaneously acquire scientifically interesting astronomical data.

5. Maximize the utility of technology developed for SETI to other objectives such as deep space communication and information management.

SEARCH STRATEGY

A comprehensive search should examine the basic dimensions of signal space: source location, transmission frequency and power, signal modulation, and polarization. But, clearly, it is not possible to search for all kinds of signals at all frequencies from all directions to the lowest flux level, even though weak signals may be more likely than strong signals. There must be a tradeoff between sensitivity and signal character, and between spatial and frequency coverage and the constraints imposed by limits of time and resources.

The very first choice made in the proposed strategy is to limit the search to the electromagnetic spectrum. The use of communication carriers other than electromagnetic waves for signaling over interstellar distances is either prohibited by the nature of the interstellar medium (absorption, scattering, or magnetic bending) or involve exotic particles whose detection is not possible with present-day technology.

Astrophysical sources produce large quantities of electromagnetic radiation throughout the spectrum, forming a background noise that impedes detection of any weak signal of extraterrestrial intelligent origin. Therefore, it makes sense to search first for signals wherever this cosmic pollution is least. This leads directly to the second strategic choice: to limit the search in frequency space to the relatively quiet microwave region of the spectrum. State-of-the-art technology, cost, and time constraints further restrict the portion of the microwave region which may be explored effectively.

The strong general preference for a microwave search, as the most promising near-term strategy, first appeared in the science workshop report "SETI" (Morrison et al., 1977). There have been continuing discussions of this position, much of it pertaining to the advisability of conducting a search in the far infrared region of the spectrum; but it remains the conclusion of the discussants that, for now, a search of the microwave region is the preferred strategy.

It is useful to attempt to estimate the volume of the search space which may need to be explored to detect a microwave ETI signal. A comparison can then be made between the portion of this total volume which has been searched to date by individual researchers and the fraction of the total volume to be covered by the program being considered here.

Figure 1 is an attempt to give a three-dimensional graphic representation of a multidimensional search space that has been named the "cosmic haystack." To accomplish this dimensional compression, it is assumed that the search is conducted in two orthogonal polarizations and that the signal is present for a significant portion of any randomly chosen observing time, and that any modulation present does not give the signal such extreme complexity as to render it noiselike to the detector. Two of the three spatial dimensions can be represented on one axis as either the number of directions examined or the number of telescope beams needed to tessellate the sky. The frequency axis covers the entire microwave region of the spectrum from 300 MHz to 300 GHz. The remaining axis combines both the third spatial dimension and the unknown equivalent isotropic radiated power (EIRP) of the transmitter; this axis is the sensitivity of the search (measured in Wm^{-2}) received within the narrowest channel of whatever detector is being used. The boundaries of this last parameter are the most arbitrary. The low sensitivity end has been set at 10^{-20} Wm^{-2}, which corresponds to 1 jansky (10^{-26} Wm^{-2} Hz^{-1}) over 1 MHz of bandwidth and is roughly the level at which previous radio astronomical surveys of the sky might have detected a signal if such existed at the frequencies of these surveys. The high sensitivity limit is what would be required to detect an Arecibo planetary radar transmission (EIRP = 10^{13} W) if the transmitter were located 30 kpc away on the far side of the Galaxy. The ceiling to the cosmic haystack slopes because it has been drawn as the number of directions on the sky in which a 213-m telescope (equivalent to Arecibo) would need to be pointed to conduct an all-sky survey; this number increases as the square of the observing frequency.

Figure 2 shows the portions of parameter space explored by SETI observations reported during the past 20 years. The total volume encompassed by these searches is about 10^{-18} of the volume of the cosmic haystack in figure 1.

The strategy proposed here is simply to detect as wide a range of types of ETI signal as possible while expanding the boundaries of the parameter space searched. It is also a systematic exploration of the microwave regime which will determine physical limitations on SETI strategies resulting from background radiation, and will necessarily develop methods to deal with man-made radio frequency interference (RFI). The proposed strategy surveys the entire celestial sphere over a broad range of the spectrum at limited sensitivity and examines specific directions at higher sensitivity but over a smaller portion of the spectrum. The observation in specific directions is intended to permit detection of weaker signals originating from the neighborhood of nearby stars selected *a priori* to present especially promising pos-

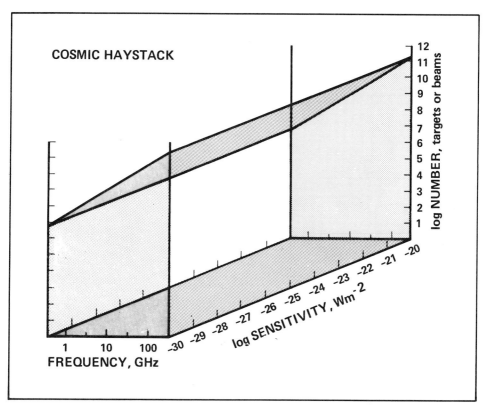

Figure 1. *Cosmic haystack — a three-dimensional graphic representation of multidimensional search space. To accomplish this dimensional compression, it is assumed that the search is conducted in two orthogonal polarizations and that the signal is present for a significant portion of any randomly chosen observing time, and that any modulation present does not give the signal such extreme complexity as to render it noiselike to the detector. Two of the three spatial dimensions can be represented on one axis as either the number of directions examined or the number of telescope beams needed to tessellate the sky. The frequency axis covers the entire microwave region of the spectrum from 300 MHz to 300 GHz. The remaining axis combines both the third spatial dimension and the unknown equivalent isotropic radiated power (EIRP) of the transmitter; this axis is the sensitivity of the search measured (in Wm^{-2}) received within the narrowest channel of whatever detector is being used.*

sibilities or of stronger signals from particularly interesting aggregates of stars at greater distances. This mixed strategy ensures that all possible life sites are surveyed to a significantly low flux level.

It is assumed that ETI transmissions will be strongly polarized, and that circular polarization is very probable in the case of an intentional interstellar transmission because its polarization remains unaltered during propagation through the interstellar medium. We further recognize that signals may be relatively narrow-band ones, pulsed or continuously present. Signals of this

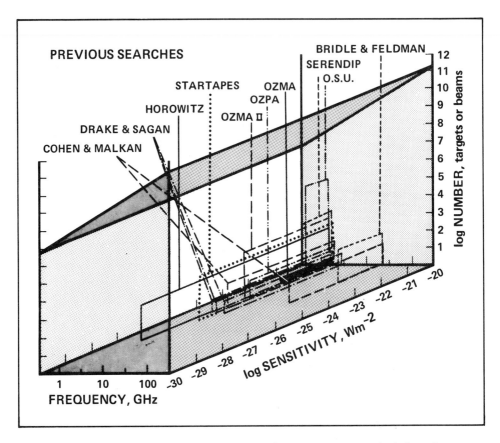

Figure 2. *Cosmic haystack plus those portions of parameter space which have been covered by previous searches.*

nature are much easier to distinguish from natural astrophysical sources and noise than are complex broadband ones. They are also more economical for the transmitting society and, for a given EIRP, can be detected at a much greater range. Thus the search will be carried out in two circular polarizations (although the receiver will be adjustable to permit matching any polarization), using dual solid-state multichannel spectrum analyzers (MCSAs) capable of resolving the broad instantaneous bandpass into many narrowband channels on a real-time basis.

Three general classes of signals can be envisaged:

I. Signals that are compressed in frequency; an obvious example is a CW signal or carrier (drifting or nondrifting).

II. Signals that are compressed in the time domain; the simplest example being a regularly pulsed signal (drifting or nondrifting).

III. Signals so complex as to exhibit little or no fine structure over the temporal and spectral windows of the detector; many kinds of intercepted transmissions may belong in this class.

The characteristics of the proposed SETI systems should allow detection of the exemplary signals of classes I and II to a power level of order one and ten times the mean noise power per channel, respectively. For the third class of signals, or other examples of the first two classes, the sensitivity achieved will depend on the data analysis schemes implemented and therefore is indefinite at present. In the limit of broadband, featureless signals, detection reduces to a total power mode with the sensitivity set by the long-term gain stability of the receiving system.

To assure that sensitivity is not seriously degraded through a mismatch between the resolution of the MCSA and the spectral or temporal structure of a possible ETI signal, in one mode of observing, on-line power thresholding and accumulation will be carried out over a frequency resolution range of 4096:0.25 Hz in steps of factors of 2. Complex signal outputs are planned at resolutions of 1, 32, and 1024 Hz. For pulsed signals, successive additions of power spectra from two or four adjacent channels or two or four sequential time samples provide spectra having sensitivities nearly equivalent to matched filter detectors at all intermediate steps in the frequency resolution hierarchy.

OBSERVATIONAL PLAN

The large number of unknown factors inherent in SETI has led to the adoption of a strategy that examines the entire sky while also concentrating on certain selected targets, in order that a significant portion of search parameter space be covered. This requires the use of both northern and southern latitude telescopes. The specific goal and objectives of the target and sky survey searches have been identified and now must be translated into a realistic and achievable observing plan.

The observational plan assumes the use only of existing radio telescopes; modifying or upgrading of facilities is possible, but not constructing new telescopes. Therefore, telescope availability is a key factor that directly impacts the objectives of the microwave observing program and the resources required to achieve them.

A. Targeted Mode

The primary candidates for particular observational attention will be the solar type stars (spectral types F, G, and K, luminosity class V) which have been identified within 25 parsecs of the Sun. The frequency range to be covered will be $1.2 \leqslant \nu \leqslant 3.0$ GHz, and as many spot bands between 3 and 25 GHz as time and resources permit.

This range of continuous frequency coverage roughly corresponds to a 1-dB decrement in the free space spectral utility function relative to its maximum value near the "waterhole" (1.4 to 1.7 GHz). The free space spectral utility function (fig. 3) is just an inversion of the more familiar temperature bandwidth index (see fig. 3, p. 71, in Morrison et al., 1977) which represents the effective noise in a receiver whose channel widths are optimized for any constant Doppler drift rate.

For a continuously present signal, sensitivity depends mainly on which telescope is used since the integration time required to reach a given sensitivity depends on the fourth power of the effective diameter of the aperture. Only 243 solar type stars are visible to the 213-m-diameter* facility at

*213 m diameter is that portion of the 305-m-diameter reflector which can be illuminated with a line feed having a 1-dB bandwidth of 30 MHz.

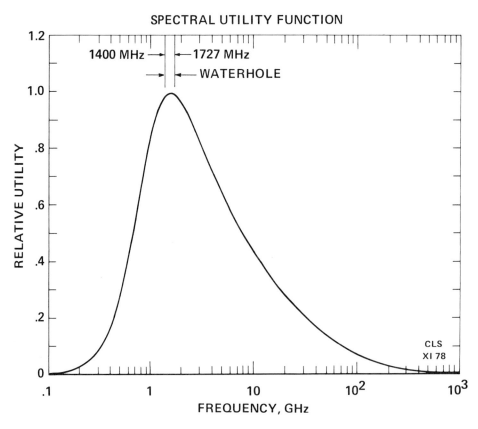

Figure 3. *Spectral utility function; an inversion of the more familiar temperature-bandwidth index which represents the effective noise in a receiver whose channel widths are optimized for any constant Doppler drift rate. The minimum in the temperature-bandwidth index represents the frequency of maximum utility for interstellar communication with ultranarrow-band signals.*

Arecibo; the remaining candidate stars will have to be observed using antennas at least a factor of 2 smaller in diameter. The resulting increase in integration time required to achieve the Arecibo sensitivity, a factor of approximately 20, cannot be contemplated for a program extending over a reasonable period of time. Instead, at all other sites, stars will be observed for approximately the same integration duration per frequency band as at Arecibo. That integration time is determined by the type of signal being sought and by our best estimate of the time likely to be available for SETI.

The observing program allows for additional or alternate options, besides the systematic search of nearby identifiable solar type stars. If telescope time or the number of SETI systems constructed permit, it is desirable to dwell for a longer time on a larger subset of the candidate stars than currently planned, in an attempt to pick up weaker signals or those having a lower effective duty cycle which might be more characteristic of leakage radiation or rotating beacons. Aggregates of stars may also be selected as targets even though they are more distant. These aggregates may contain a distribution of transmitted signal powers, many or most of them too faint to detect; however, the most luminous transmitter at the tail of the unknown distribution may, in fact, be observable. As one example, a small number of directions near the galactic plane, towards the galactic center, contains the vast majority of all stars in our Galaxy.

B. Sky Survey Mode

This microwave search mode will include 100% of the celestial sphere. The frequency range to be covered will be $1.2 \leqslant \nu \leqslant 10$ GHz and as many spot bands between 10 and 25 GHz as time permits. This range of continuous frequency coverage spans the flat minimum of the terrestrial microwave window, and represents the highest frequency to which a sky survey can be made at reasonable sensitivities, in reasonable times. As in the targeted mode, the lower frequency limit is determined by engineering costs for feed construction.

To gain a better understanding of the constraints on time and sensitivity for this sky survey, it is necessary to consider a specific mode of observation. It is proposed that the telescope primary beam be swept across the celestial sphere at a constant rate. This is not the only choice possible, but is, in fact, a good compromise between search sensitivity and duration.

The reason for the choice of constant scan rate is the tradeoff of sensitivity against time required. A constant angular tracking rate ensures that sensitivity varies slowly with frequency ($\sim \nu^{1/2}$), while the time required to cover the frequency range required is not overwhelming ($\sim \nu^2$). If the rate were chosen so that the sensitivity would not vary with frequency, the time required to complete the entire survey would vary as v^3. On the other hand, maintaining a constant beamwidth at all frequencies would require a time to complete the survey that varies as v, but it would result in a sensitivity that varies as v^2. The constant tracking rate approach is an attractive intermediate case and one which is operationally tractable.

C. Search Space Coverage

It is clear that, all other things being equal, sensitivity to narrow-band signals is increased as the individual channel width decreases. The sensitivity remains constant beyond the resolution at which the ETI signal is broader than the channel. The true physical limit to the narrowness of a signal is not to be found in the hardware of the transmitting society or the receiving one; it is due to the dispersive medium through which the signal must pass. For distances $\geqslant 1000$ light years, the interstellar medium will probably degrade any transmitted signal to a bandwidth of $\geqslant 10^{-3}$ Hz. This bandwidth limit may be small compared to uncertainties in providing compensation for stellar relative motion. Finally, modulation of the signal will further widen the received spectrum. We conclude that 0.1 Hz represents the minimum resolution applicable.

It is also clear that the search duration decreases as the product of number of channels and channel width, Nb, increases. A design choice (based on funding and state-of-the-art technology) has been made to limit the number of channels to $N = 7,864,230$. The maximum instantaneous resolution, b, will be 32 Hz for the sky survey and 1 Hz for observing targets. Thus the instantaneous bandpass of the sky survey receiver will be 252 MHz and that of the targeted survey receiver will be 7.9 MHz. Figure 4 is an attempt to display the volume of the cosmic haystack which could be searched, employing the foregoing observing strategy. This volume of search space represents an improvement of a factor of $\sim 10^7$ over the coverage shown in figure 2.

Figure 5 shows the range of sensitivities that may be realized during the observing program in both the sky survey and the targeted observations. For comparison, sloping lines representing a range of transmitter EIRP have been included so that the sensitivity required to detect classes of transmitters having the indicated EIRPs can be deduced as a function of distance to the transmitter. The numerical limits of sky survey and target sensitivity are estimated, based on reasonable values for available collecting area, system noise temperatures, integration time, and acceptable false alarm rates which are considered to be achievable in the observing program.

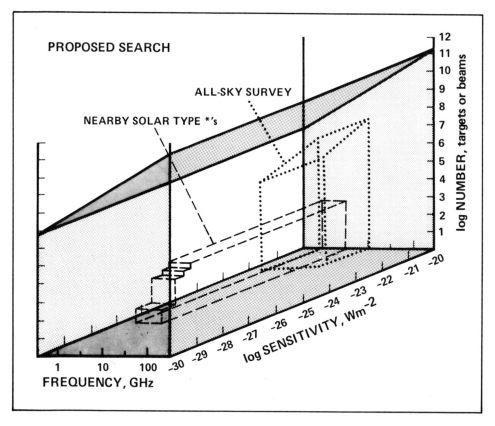

Figure 4. *Cosmic haystack plus the volume of search space which will be covered by the proposed search strategy.*

Neither of the search modes will be capable of detecting leakage radiation at power levels comparable to our own TV transmitters. However, the planetary radar transmitter at Arecibo could be detected at a distance of 400 light years with the *minimum* sensitivity of the targeted search or out to about 18 light years at the lowest frequency of the sky survey.

INSTRUMENTATION

This section describes the approach and current concepts used in developing the SETI instrument system. Figure 6 is a functional block diagram of the receiving system. There are three major elements. First, a wide-band dual polarization feed and low-noise amplifier system. Second, a digital spectrum analyzer constructed with modules that can be configured as appropriate for the capabilities of the site being used and for the specific search strategy

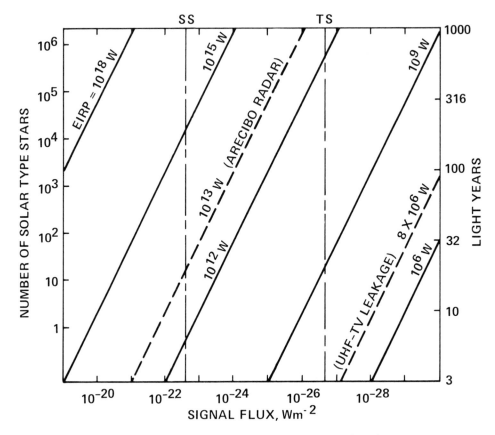

Figure 5. *Maximum sensitivities that may be realized during the observing program in both the sky survey (SS) and targeted observations (TS). For comparison, sloping lines representing a range of transmitter EIRP have been included so that the sensitivity required to detect classes of transmitters having the indicated EIRPs can be deduced as a function of distance to the transmitter.*

being undertaken. Third, a signal processing and control element to examine a reduced data set (preselected by the high-speed computational capabilities of the spectrum analyzer) to determine which information should be kept for further study and/or archiving, and to control the real-time actions required to collect data on interesting signals detected.

It is important to note that we are describing technological aspects of the SETI program while we are simultaneously engaged in a very active design phase. Therefore, although no fundamental philosophical change in the nature of the instrumentation is expected, the precise manner in which the functional requirements are met is still under consideration. With this caveat, table 1 gives the functional requirements on the instrument system as dictated by the search strategy.

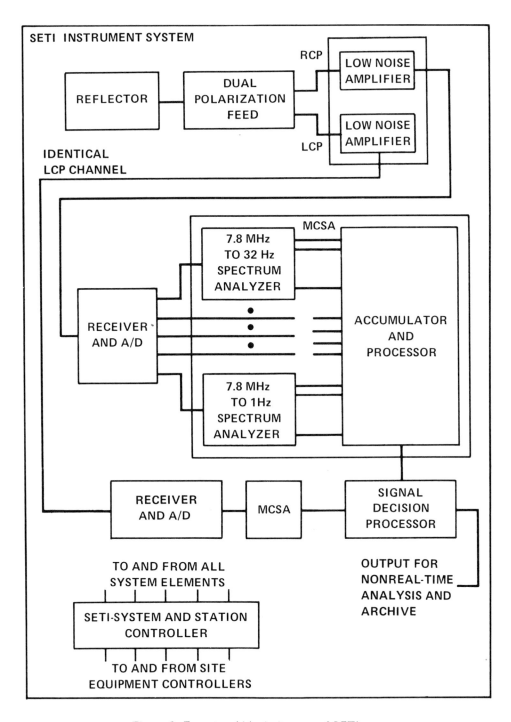

Figure 6. *Functional block diagram of SETI system.*

TABLE 1.– OVERVIEW OF INSTRUMENT SYSTEM FUNCTIONAL
REQUIREMENTS

Specification	Requirement
RF tuning range	1.2 to 25 GHz, in 9 or 10 bands
Instantaneous RF bandpass	300 MHz or to antenna capability
Polarization	Simultaneous RCP and LCP, adjustable to match any signal polarizations
Gain stability	$\Delta G/G \sim 10^{-3}$
Gain calibration	Noise adding diode to measure gain and gain stability as function of frequency across passband(s)
Frequency stability	$\Delta \nu/\nu \leqslant 10^{-13}$ over 10^3 sec; all local oscillators (LOs) locked to single-frequency standard
Equivalent system input noise temperature	State of the art consistent with resources
Computer control of	Antenna aspect and motion; LOs, gains, polarizations, integration times, passbands and resolutions, signal analysis and memory functions, peripheral equipment, data flow, etc.
MCSA bins per polarization	*Narrow band,* 7.9 MHz; 7,864,320 1-Hz bins; 245,760 32-Hz bins; 7680 1024-Hz bins; pseudo bins of 0.25, 0.5, 2, 4, 8, 16, 64, 128, 256, 512, 2048, 4096 Hz
	Broadband, 252 MHz; 7,864,320 32-Hz bins; 245,760 1024-Hz bins; 4096 61-kHz bins
Available digital MCSA output signals	Power and unprocessed complex voltages at 7.9 MHz, 61.4 kHz, 1024 Hz, 32 Hz, 1 Hz
Signal processor; instant replay; auxiliary tests	To be determined

A. Low-Noise Amplifiers and Feeds

The low noise amplifier system envisioned for the SETI system is based on an existing design which JPL constructed for the National Radio Astronomy Observatory. It is a tunable cryogenic microwave maser operating in the range 19 to 25 GHz with an instantaneous bandwidth of better than 300 MHz. (A newer design with a center frequency of 32 GHz may prove to be the ultimate choice.) By preceding this device with a cryogenically cooled, parametric upconverter, the large instantaneous bandwidth of the maser can be moved quite far down in frequency while still maintaining a very low-noise temperature. Systems such as this have now been operated in the laboratory at an input RF frequency of about 2.3 GHz and with an equivalent noise temperature of 4 K. To obtain the full frequency range required for SETI requires eight or nine upconverters.

The baseline plan envisions the implementation of two upconverter-maser devices in a single cryostat. Each device is used for the amplification of one sense of polarization from a dual polarized feed. The maser system is a technically optimum choice because of its exceedingly low-noise temperature. However, current technology in cooled FET (field effect transistor) amplifiers, especially at the lower frequencies, may provide a less costly alternative without too great a sacrifice in performance. Hence a study will be undertaken to evaluate the tradeoffs between these two approaches.

The antenna feed structure depends on the reflector system it illuminates. The use of corrugated feed horns with large paraboloidal reflectors will permit the full bandwidth capability of the masers to be used. In clear dry weather, a zenith system noise temperature of about 15 K can be expected at 1.2 GHz and about 35 K at 10 GHz. With other than the paraboloidal reflectors, special feed designs will be required. For example, the 1000-ft Arecibo spherical reflector currently requires the use of a line feed, with an instantaneous bandwidth of about 30 MHz and a tuning range of only 80–100 MHz, regardless of center frequency.

B. Multichannel Spectrum Analyzers (MCSAs)

Four basic spectral analysis techniques have been examined for their suitability in SETI: classical RLC comb filter technology, surface acoustic wave technology, optical power spectrum technology, and digital techniques. The studies to date show that the digital approach is far superior in terms of capability, flexibility, reliability, and cost.

The search strategy requires two different configurations of the basic MCSA modules. The all-sky, wide-frequency range survey needs MCSAs with large instantaneous bandwidths. The targeted search needs MCSAs with higher frequency resolution and a wider range of time resolution. After studies of various competing digital techniques, it has been found both technically and economically desirable to construct both types of MCSAs with a high degree of commonality down to the board level.

The basic MCSA architecture is as follows. A real instantaneous intermediate frequency (IF) bandwidth of 7.9 MHz is down-converted to two 3.95-MHz base bands by a quadrature mixer (fig. 7). After low-pass filtering, the two signals are digitized at a 25% oversampled rate of 9.83 MHz. The two outputs, I and Q, are then processed together by two stages of finite impulse response (FIR) filters, which produce 7680 parallel channels, each 1024 Hz wide.

The use of digital filters, rather than, say, a single pipeline FFT approach, has a number of important advantages. At insignificant added cost, linear phase, low ripple passbands can be formed with channel to channel crossovers at the 0.1-dB down points and 60 or more dB attenuation

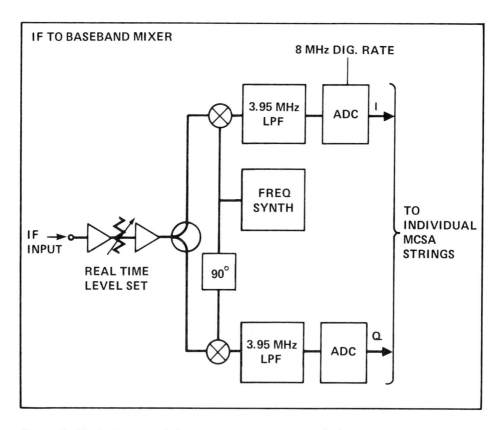

Figure 7. *Block diagram of the intermediate frequency (IF) to baseband mixer. A real instantaneous IF bandwidth of 7.9 MHz is converted to two 3.95-MHz bands by a quadrature mixer. After low-pass filtering, they are each digitized at a slightly oversampled rate of 9.83 MHz, producing two outputs, I and Q.*

by the middle of the adjacent channels (and in all other channels). This provides a desirable freedom from aliasing effects when observing in the presence of even quite strong interference in one or more channels. At the crossover frequencies, there is no loss of sensitivity. The fact that a signal can appear in two passbands is not a significant problem. The rapid development of parallel channels enhances system reliability. The 128-channel filter can be built on one quickly replaceable plug-in board, and the bulk of the components are downstream. The parallel strings, though working on separate data, are controlled in parallel. Furthermore, a wide range of additional spectrum manipulation can be carried out on line by software changes alone, without altering the wiring and parts of the preceding stages as is so often necessary with pipeline FFT systems. Note, however, that the final decision on the MCSA architecture (BPF/DFT or pipeline FFT) will not be made until extensive comparison testing of breadboard hardware has been completed.

Figure 8 is a block diagram of one of 32 parallel spectrum analyzers, which together make up the 7.9×10^6 channel broadband MCSA for the all-sky wide-band survey. The total instantaneous bandwidth is 252 MHz, and the frequency resolution is 32 Hz. The transition from the 1024-Hz level to the 32-Hz level is carried out by microprocessor-controlled DFT algorithms. The microprocessors are controlled by software and plug-in firmware, so that, for example, dedrifting or dedispersion transforms can be implemented if desired. Further details can be found in Peterson et al. (1980).

Power spectra and unprocessed complex voltages are available at the outputs of each stage of the MCSA. Accumulation before or after thresholding can be accomplished within the MCSA at each stage in the resolution string. If the analysis of a suspected signal requires it, higher resolution over a more limited bandwidth can be obtained in either of two ways: by clocking down the MCSA or by adding one or more of the 1024-Hz to 1-Hz DFT boards used in the narrow-band MCSA (discussed below).

Figure 9 shows the configuration of the narrow-band MCSA which consists of one or two strings of the broadband MCSA plus some additions. In parallel with the 32-Hz DFT boards are 1024-Hz to 1-Hz DFT boards to provide 7.9×10^6 1-Hz output bins. Power spectra and unprocessed complex voltages are available at three resolution levels. Accumulation before or after thresholding can be accomplished at each resolution bandwidth. Most important, though, is the construction of a series of real and pseudo resolution bandwidths in powers of two which are thresholded for pulses.

The strings of different resolutions are formed in the following way. At each of the real resolution levels, 1, 32, and 1024 Hz, two and four adjacent frequency bins are added together to give pseudo binwidths of twice and four times the basic binwidth; and two and four successive individual bin outputs are added together to give 1/2 and 1/4 the resolution of the basic

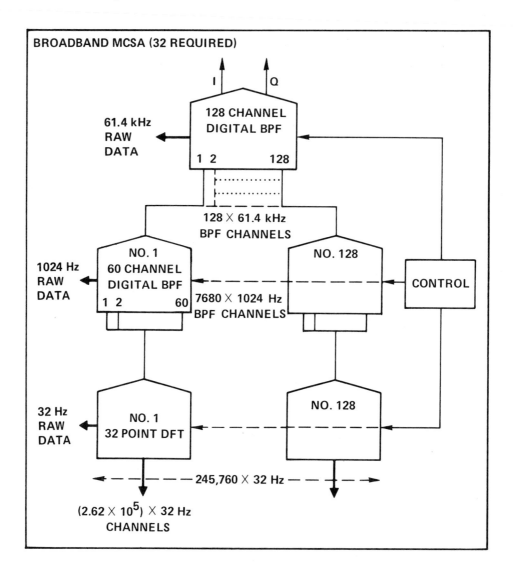

Figure 8. *Block diagram of one of 32 parallel spectrum analyzers which together make up the 7.9×10⁶ channel MCSA for the all-sky wide-band survey.*

binwidth. This pseudo binwidth construction process introduces, at most, a 3/4-dB signal loss, compared to direct DFT computation. As a result, one may search for a wide range of relatively narrow-band energetic pulses with durations increasing in powers of two from 0.25 msec to 4 sec. Longer pulses are handled at a later stage in the data manipulation, as is the storage and analysis of pulses from all the different spectral bands.

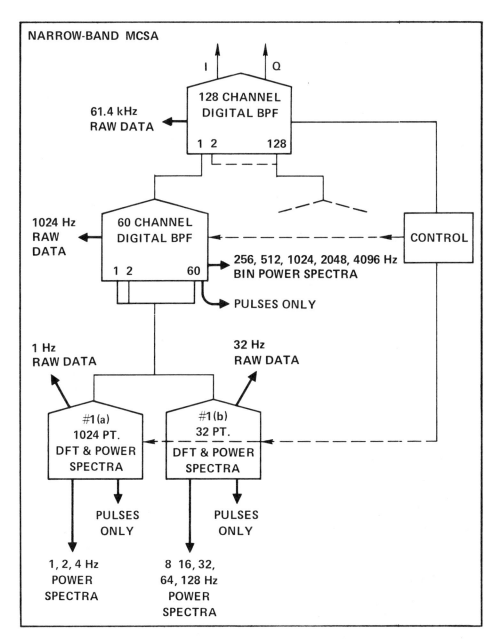

Figure 9. *Narrow-band MCSA consists of one or two strings of the broadband MCSA plus some additions. In parallel with the 32-Hz DFT boards are 1024-Hz to 1-Hz DFT boards to provide 7.9×10⁶ 1-Hz output bins. Power spectra and unprocessed complex voltage are available at three resolution levels.*

C. Signal Processor

The design of the signal processor is just beginning, as is the complete definition of the auxiliary signal identification tests, on line and off line. But a few remarks are useful at this time. For the case of the sky survey, spectral data from the broadband MCSA can appear with frequency resolutions from 32 to 1024 Hz, in powers of two, by suitably programming the microprocessor DFT operations in the third resolution stage of the MCSA. Successive spectra will be accumulated and read out four or five times per full half-power beamwidth scanned on the sky. Maximum sensitivity is preserved if adjacent scans are saved and reduced together. There is the option to threshold the accumulated spectra on line or at some later time, or both. Still to be determined is just how much of what data is to be archived. Early experience testing prototype units under real observing conditions will be important in this determination.

Thresholding on line will be employed to determine if any signals are present. If there are any, a series of automatic tests will be performed to identify their nature and origin.

For the targeted search, 10 to 32 complete 1-Hz spectra from the narrow-band MCSA will be saved and searched for drifting continuous signals in one or more channels, and for pulses of duration longer than 4 sec which show some kind of coherence in the time-frequency-polarization domain — for example, three or more pulses equally spaced in time, or pulses alternating in frequency and/or in polarization. This process will be repeated for successive groups of 10 to 32 spectra and, at the end of observing in a particular direction, the accumulated threshold surviving residues from the successive spectral batches will be analyzed for signals.

Pulses appearing in the 0.25- to 4096-Hz resolution spectra will have been thresholded at a level that produces, for example, an *average* of 10^3 to 10^4 false alarms per spectrum. These will be saved for a period of time and scanned for regularities. This procedure permits thresholding at appreciably lower S/N, thus increasing system sensitivity to pulsed signals. As with the broadband system, the appearance of a possible signal will initiate auxiliary tests. How much of the 1-Hz spectral data is to be preserved for off-line study, distribution, or archiving is yet to be determined.

Studies are planned to see if there is a practical way to optimize sensitivity to an even wider range of signal types, using the frequency-time-polarization data in the 32-spectrum matrix. In addition, there will be studies to see if there is a practical way to employ predetection integration. In principle, for classes of signals containing enduring, coherent frequency components (e.g., carriers or subcarriers), sensitivity increases in proportion to time with predetection integration; sensitivity only increases with the square root of time for post-detection integration.

An "instant replay" facility will be provided so that the relevant raw (complex) data can be preserved for off-line analysis when the real-time system is unable to analyze the nature and origin of a signal satisfactorily. One attractive hardware solution to this task is the Mark III recorder being developed by NASA-Goddard and the Haystack Observatory for VLBI Astronomy and Geodesy.

The signal processor will store data on tape in two batches for the permanent archives — one for data relevant to SETI, the other for data deemed to be of long-lasting significance, based on prior consultation with the scientific community at large. The bulk of the incoming data is noise of negligible interest and will not be saved.

SIGNAL IDENTIFICATION

The data rates expected, on the order of a gigabit per second at every real frequency resolution, require that most of the data be examined automatically on line and immediately discarded if no interesting patterns are detected. Only that very small fraction deemed significant in that it contains a possible ETI signal or important astronomical data would be recorded for posterity and off-line analysis.

The simplest class of signal is a continually present, nondrifting narrowband pattern that appears in one or more channels. The signal processor is required to integrate power spectrum samples over a period of time and to test at the end of that period whether one or more channels display a sufficient excess power level over the mean noise power level so that one may confidently conclude that a signal has been received. The signal processor requires an accumulator vector array that will not overflow in the integration time, and a software/hardware package that carries out the statistical analysis.

A more difficult class of signal is a pattern that is pulsed, may be drifting, and may appear in one or more channels. A threshold test is made during each sample of the power spectrum, with the level set so that the false alarm rate is about 10^{-4}. Thus, each sample yields about 10^3 or more data points, which must be stored to permit time-frequency-polarization series analyses. Data so stored over the duration of the observation in one direction must be analyzed before the succeeding observational set is ready for analysis. The analysis must search over frequency and polarization space for drifts and discrete, systematic shifts, and in the time dimension for periodicity. One is looking for pulse repetition periods from about twice the reciprocal binwidth up to about 1/4 the observing time in a single direction.

A class of signal still more difficult to detect is a continuously present, slowly drifting pattern. No threshold test is applied to such a signal until it is dedrifted. The entire power spectrum is stored sample by sample. Data so stored over some appreciable time must then be analyzed in the time required to take the sample set. Dedrifting is memory-intensive. Ideally, one would like to save perhaps 1000 spectra at 1-Hz resolution, dedrifting in increments of 1 mHz/sec. An attractive compromise would appear to be as follows: decrease memory by a factor of about 32 to 1, save and analyze spectrum samples in groups of 32, and preserve only the residues from the 32 sample sets.

Once a potentially interesting signal pattern is flagged by normal observing procedures, a hierarchy of signal identification protocols will be brought on line; for examples, see the partially ordered listing in table 2. To

TABLE 2.– CANDIDATE SIGNAL IDENTIFICATION TESTS

Apparently Continuous Signals:

1. Signal in RF, IF, BASE, or IMAGE BANDS?

2. Does signal correlate with output from "omni" RFI receiver?

3. Consult catalog of RFI and natural signals in disk memory.

4. Check harmonic radiation possibilities.

5. Are other signals being received simultaneously?

6. Optimize resolution bandwidth and polarization.

7. Any frequency/phase/polarization/time structure?

8. Can local oscillator be phase-locked to signal?

9. Any recognizable modulation scheme?

10. An observable frequency drift rate? Terrestrial? Other?

11. How does signal change with antenna pointing?

12. Source direction apparently fixed on celestial sphere?

13. Signal receivable at another site? Any parallax?

14. Does signal show a frequency-drift pattern cyclic in time?

15. Further tests for proper motion and parallax: Are interferometric observations possible? VLBI? Reobserve at many later times.

16. Off-line study of frequency assignments.

17. Off-line analysis of signal, using recorded raw data. Search for coherent structure in signal. Search for a suitable "matched filter" for signal.

Intermittent or Pulsed Signals:

18. Look for coherent patterns in and among these parameters: Frequency, phase, polarization, and time; in the short term and in longer terms.

19. Same tests as for apparently continuous signals.

conserve scarce telescope time, a major effort will be devoted to designing automated procedures to quickly identify the usual types of RFI. For instance, tests for items 1 through 5 in the table will be designed to be carried out in seconds or less, if there is enough signal, and should be able to identify a large percentage of the RFI commonly experienced at radio observatories.

CONCLUSIONS

The total effort envisaged for this program will require about 10 years — 5 years for the SETI instrumentation development followed by 5 years for observations. To accomplish this, several narrow-band SETI systems would need to be replicated so that simultaneous northern and southern hemisphere targeted observations could be performed during the period in which the sky survey was being conducted with the wide-band SETI system. The present program calls for the use of the NASA Deep Space Network antennas and facilities as well as non-NASA observatories as SETI observing sites. Although the procedural details for vitally needed SETI involvement by non-NASA observatories and by the general scientific community have not yet been determined, it is anticipated that this will be accomplished through scientific proposals to NASA. In addition, alternative approaches are expected to be forthcoming and will be encouraged.

REFERENCES

Morrison, Philip; Billingham, John; and Wolfe, John, eds.: The Search for Extraterrestrial Intelligence, SETI. NASA SP-419, 1977 (reprinted, with minor revisions, by Dover Publications, New York, 1979).

Oliver, Bernard M.; and Billingham, John: Project Cyclops, a Design Study of a System for Detecting Extraterrestrial Intelligent Life, NASA CR-114445, revised edition, 1973.

Peterson, A. M.; Narasimha, M.; and Narayan, S.: System Design for a Million Channel Spectrum Analyzer (MCSA). Conference Record, 13th Asilomar Conference on Circuits, Systems, and Computers; IEEE Catalog No. 79CH1468-8C, 1980, pp. 14–17.

ADDITIONAL READING

These references cover some of the extensive history leading to the Life in the Universe concept.

Cameron, A. G. W., ed.: Interstellar Communication. Benjamin Press, N.Y., 1963.

Cocconi, G; and Morrison, P.: Searching for Interstellar Communications. Nature, vol. 184, 1959, p. 844.

Drake, Frank D.: Project Ozma. Physics Today, vol. 14, 1961, p. 40.

Kaplan, S. D., ed.: Extraterrestrial Civilizations: Problems of Interstellar Communication. NASA TT F-631, 1971, translated from Russian.

Mallove, E. F.; Connors, M. M.; Forward, R. L.; and Paprotny, Z.: A Bibliography on the Search for Extraterrestrial Intelligence. NASA RP-1021, 1978.

Oliver, Bernard M.: Strategy for SETI Through Radio Waves: An Introduction. In Strategies for the Search for Life in the Universe, Proc. of Special IAU General Assembly Session in Montreal, Canada, Aug. 1971, M. D. Papagiannis, ed., vol. 83, pp. 79–80, D. Reidel Pub. Co., 1980.

Ponnamperuma, C.; and Cameron, A. G. W., eds.: Interstellar Communication: Scientific Perspective. Houghton Mifflin, N.Y., 1974.

Sagan, C., ed.: Communication with Extraterrestrial Intelligence. M.I.T. Press, Cambridge, MA, 1973.

Sagan, C.; and Drake, F.: The Search for Extraterrestrial Intelligence. Scientific American, vol. 232, 1975, pp. 80–89.

Seeger, C. L.: The Recognition of Extraterrestrial Artificial Signals. Conference Record, 13th Asilomar Conference on Circuits, Systems, and Computers; IEEE Catalog No. 79CH1468-8C, 1980, pp. 18–22.

Shklovskii, I. S.; and Sagan, C.: Intelligent Life in the Universe. Holden-Day, San Francisco, 1966.

Special issue on CETI (Communication with Extraterrestrial Intelligence). Acta Astronautica, vol. 6, no. 1-2, 1979.

Zuckerman, B.; and Tarter, Jill: Microwave Searches in the U.S.A. and Canada. In Strategies for the Search for Life in the Universe, Proc. of Special IAU General Assembly Session in Montreal, Canada, Aug. 1971, M. D. Papagiannis, ed., vol. 83, pp. 81–92, D. Reidel Pub. Co., 1980.

Reflections

Reflections

PHILIP MORRISON

It would be very hard after these two days to say something original. The most I can hope to do is to lead a ceremony that may lend some perspective to what we have heard.

Looking over the schedule of these interesting and varied papers, I think it is fair to say this has been not only the largest but the most diversified meeting ever held in this general domain. In a sense it represents a milestone for SETI; but that is not its most important aspect. To isolate its importance, I want to make a few remarks about the history of the subject and about the context of the American scientific community in which it is embedded. A number of scientific points will doubtless creep in.

The first session began with the broad sweep of cosmic evolution, which has become the myth of our time for the scientist. When I say myth, of course I don't mean falsehood nor do I mean a mere tale. I mean a world view embodying substantial values of deep cultural significance, centered around efforts to explain origins. That is clearly what we begin with. Our myth today differs from those of the past in that we have a much wider series of tests for its acceptance. We expect it to transcend past experience, to go beyond language. That, in a way, distinguishes post-Renaissance myth-making, especially at the cosmogonic level, from what went before. I would hasten to say that I think a good deal of what we have put in the myth will fall before the test, just as would fall the Ramayana if we asked to see the stance from which Hanuman leapt in order to get to Ceylon. We won't find that. We might find something like it, but we won't find the actual footprints.

Our myth is a grand cosmic myth. But there is also a style of great *personal* myth: a hero, his life and times, how he was chosen, what he did in

421

childhood, the difficulties he went through, the triumphs he had, the disasters he endured. The myth becomes over time a set piece, perfect in its symmetry, brooking no changes. In India, if you go to see a dramatic performance or a dance, or hear a song, 8 out of 10 times it is from the Ramayana. You know what is going to happen, but that is not a problem: novelty is not a positive value here. We now have such a personal myth growing in our culture, a very important myth that is only about a century old. Its presence has hovered over all the papers at this meeting, although we have almost never mentioned it explicitly. That is, of course, the life of Charles Darwin.

The story of evolution is the Ramayana of the scientist, and Charles Darwin is Rama, or as close as we can come in our impious and divided times. There is nothing we have done that does not have somewhere behind it the evolutionary model. We have transferred that model from self-replicating and mutating systems to a whole variety of cookery, but in the end it is the myth we are aiming at. And the outlines of the myth were not imaginable even 150–200 years ago.

I won't tell the story of the Darwin Ramayana, though it is a nice one to tell, but I will mention a few events that everybody should know about. There was, for example, the voyage of the Beagle, and the letter written by Uncle Josiah to get Charles on the Beagle. Charles's father, who was a self-assured, well-to-do man, did not like to be crossed in his opinions. He conceived that it was a very poor idea for a young man who was going to become a clergyman to go off on a trip with all those sailors for several years, to waste his time when he could be sitting in the country trying to avoid conducting too many funeral services. He told his son in no uncertain terms that he had half a dozen reasons for opposing the trip. He said: (a) It is a waste of time; it doesn't lead to your vocation. (b) It was offered to somebody else first; you're second choice; what good is that? (c) Being off with all those rude sailors can be very bad for your settling down in the future. And so on. Darwin was crushed. Here was a tremendous opportunity that had opened for him in a letter from his professor at Cambridge. Darwin said to his father, "Well, can I appeal in any way? Is there anything I can do to change your mind?" His father was a reasonable man. He was firm, but he believed he was controlled by reasonable persuasion, so he said, "Okay, find any thoughtful man of affairs, who knows the world, who agrees with you and not with me and I'll reconsider the whole thing." So Darwin went off 10–15 miles to his uncle's house, his Uncle Josiah Wedgwood, who made the pottery. (He was the second or third generation of the pottery owners. By the way, it was Josiah's daughter who became Mrs. Charles Darwin a couple of years after he came back from the Beagle. So he was obviously pretty friendly in that house.) He came over there, age 22, and talked to Uncle Josiah. Uncle Josiah wrote a wonderful letter back to the father explaining every point. "Of course, now, it's true sailors are unsettled, but how often do you hear the stories of the sailor who comes back and settles, runs a little farm, and never wants to leave home again. That's an equally likely consequence of going off to sea."

Now suppose that Charles hadn't gotten Uncle Josiah to write the letter. He would not have gone on the Beagle and would not have seen the Galápagos Islands. He would instead have been a country clergyman and would have written a brilliant journal of ornithology about the birds of some little parish. I am sure that is quite true. Of course, someone else, some Wallace, would have developed the theory of evolution in a decade or two anyhow. Because A. R. Wallace did write the first paper offered for publication which correctly described the evolution of species by the operation of natural selection. As you know, he sent this to Darwin to find help in getting it published. Darwin was disturbed, because after 10 or 12 years of closely filling his big notebooks with data on this point, here was a man he knew only slightly who had anticipated the entire story, but without any strong evidence, just a very clear logical statement of the whole story. Wallace himself was away in Indonesia, where he worked as a professional collector of animal skins, birds, eggs, and so. He traveled about the tropics on commission and sent back to the museums and to wealthy private collectors in England the things that he could get. He had malaria in Ternate, Indonesia, which took him out of the field for a few months. During that time he wrote his paper. Darwin was the only naturalist he knew who had been a traveler and was therefore likely to be sympathetic, not a bookish Cambridge professor with dusty disdain for people who traveled about far from libraries. Wallace was not a university man. He wrote to Darwin, "I think you would be interested in this general idea; I know you have some concern for the species problem." And that was the story. It is, of course, full of myth. I imagine that if we last a thousand years (and I think we will), this story will be told in movies and film, and heaven knows what three-dimensional holographic schemes I can't even foresee, in one way or another, as the great myths of the past have told about Rama, and Sita and the deceiving deer, and the jumping of the monkey army across the sea.

I tell this story to indicate that even in human events there is a kind of convergence. This, of course, argues against one of the objections to the SETI program: that evolution can never get to the same point twice. It is true that an exact retracing of a path is impossible. You cannot come from where you slept last night to this auditorium another time by exactly the same path; certainly not to the resolution of a micron, and I think not even to the resolution of a yard. When you cross the highway, for example, the traffic pattern will be different. But it does not matter if the path is not the same provided the end is the same. This is what is easily forgotten. It is improbable that you will find any second path that is close, but it does not seem so improbable that you will come to the same end, by a very different path in that complex interweaving of world lines that is both human history and the history of biological structures. I am not going to talk anymore about the principle of convergence, but I think it is worthwhile to bring it forward sharply because some of the papers we have heard took a very strong view. They argued, quite correctly, that many conditions of preadaptation and so on need to exist for any big evolutionary change. When you look at those preconditions in any particular case, you may find them unrelated to the end result, and therefore you say, "How could it be that, since the essential preadaptation has nothing to do with the end result, that same end result could nevertheless happen?" Of course, the answer is just the same. Were the end result to happen, by my hypothesis, along a very different route, you would look back and say, "Aha, this is preadapted in such and such a way and therefore it could never happen any other way."

The particular case I would like to mention here, the preadaptation of human beings for speech, was probably quite real. But I would like to focus on something that has moved me very much in the past few years. One of the great new events in the human sciences relates to the development of sign language. It is now generally recognized by linguistic scholars that the large and randomly scattered deaf community has developed, without any assistance, a flexible, vigorous language of signs and gestures. This community is comprised of totally deaf persons in every country, and in our country in particular — persons who at a very early age before the onset of speech have been made totally deaf. It has taken more than a hundred years and the contributions of millions of people, but a gesture language has been developed which you can now see repeated very rapidly along with the television news. It is gradually becoming salient in the American scene. This language, when examined by the methods of grammatical analysis that, since the second World War, have revolutionized our view of languages, seems to show not only a great unity between the Chinese version and the American version (which have little historical connection, though the French and American have), but also a grammar, highly adaptive and flexible, which violates some of the principal canons of human speech. If this new work is sustained, it will demonstrate that human communication can be achieved on a complex and subtle but distinct level that shares many of the characteristics of normal speech, such as plays on words, words in bad taste, poetry, and song. All of these things have been identified and verified in American sign language in the last few years. There are lots of dirty jokes going around; teenagers giggle over them in the corner, but they won't do them in public. In the same way there are slang expressions that are not good form, but are fast. And there is an analog to song, that is, rhythmic exaggerated gestures that have the quality of music; there is also poetry marked by intricate word play, gesture play in which puns and imitations, one gesture for another, are very important. All this is gradually becoming understood by the academics.

I remark on this because it is, of course, a completely different style of language. It may not connect to Broca's area or to any other of the principal compiler regions of the human cortex, where there exists our facile and rapid instrumentation for speech. One of the papers today said that it was difficult to see how speech could have evolved. Of course, I am not saying that it is just as good to do gestures as to have auditory communications; I honestly do not know. It might turn out to be better in the end. But, in any case, if you evolve with one mode and develop the other as a substitute, this is a tribute to the flexibility of the human mind, to a talent not evolutionarily built in at all; it shows that there is more going on than we imagine. It is another example of the convergence that I have talked about and tried to illuminate by the Darwin-Wallace story.

One of the most important occurrences over the 15 or 20 years that people have been thinking much about SETI has been talked about implicitly here, but not a great deal explicitly. Outside this rather rarefied atmosphere, though, it has had a big effect on public imagination, something we should not forget. The event we thought might happen, but did not, was that another form of life would be discovered on another planet. But what has occurred that we did not foresee, but that supports our general view while complicating our detailed quantitative understanding, is the finding of complex polyatomic carbon compounds in a wide variety of cosmic contexts. From studies of the molecular species in the great condensed galactic clouds and from the putative analysis of some asteroids and maybe satellites (in which we find some indication that carbonaceous compounds are important), we have derived a renewed interest in the carbonaceous chondrites themselves. As was noted today, this may represent a physical sequence or simply a set of analogies. But it also demonstrates a point on which biochemists were already very clear 50 years ago: the flexibility and peculiar subtlety of carbon compounds, as well as the high abundance of carbon, makes them preeminently the source of complex chemistry in the Universe.

I am led to another little story that I think will be familiar to many but perhaps not to all. When we seek the input of the physical sciences — astronomy (dynamical astronomy especially), geology, climatology, aeronomy — we know that their practitioners are masters of a powerful deductive structure, with quantitative possibilities. Of course, they are beset by the necessary complexity of their models. They must try to pin down from some a priori model just what the first 6 or 7 hundred million years of Earth history were like. They search for a necessary prelude for the biologists, something essential to the total picture of cosmic evolution. On the other hand, it was explicitly stated by someone at this meeting that it can be the other way around. If the biologists were to explain why they needed certain conditions, perhaps it could be determined whether these conditions were at all plausible under some existing model. It is clear to me that strong interchange must go on in this domain. We clearly have to learn a more interdisciplinary way of facing such problems.

A famous 19th-century interdisciplinary dispute illustrates this point. There was a lot of dispute, even full conflict, but no resolution at all; logic was clearly on one side, but it turned out to be wrong. This is a charming story, resolved in print by the distinguished geologist T. C. Chamberlain, of the University of Chicago, about the turn of the century. It was also resolved in a lecture by Ernest Rutherford in the early years of the 20th century at the Royal Society. Many would guess what I am talking about: the famous problem of the time scale available for Darwinian evolution.

The absence of substantial observed changes in speciation in the natural world, as compared with the swift changes produced by domestic hybridization during the course of history, is a strong argument for the slowness of natural speciation. This argument was made even stronger by the fact that paleontology showed change clearly; you had to say that the time available in the geological record was very large. Darwin, while he was a man of extraordinary logical ability and in very simple ways a brilliant experimenter, often cutting right to the heart of the matter, was a bad mathematician. He could not calculate anything. He had a touching faith, however, in instrumental methods. His son writes that he discovered his father making measurements with an old paper ruler and writing these down to high accuracy. The son commented, "Well, you know that ruler probably is not right." So he got a better ruler; sure enough, his paper ruler was stretched and deficient. Darwin just fell into depression; the notion that a calibrated ruler, a thing you trust to measure with, might not be right was a breach of faith he could hardly accept from his ruler-making colleagues! That was his style. He kept saying that his study of geological records suggested that there was a great deal of time indeed. He liked to put it that biology required almost infinite time. He didn't, I think, mean what infinity means mathematically. But what he meant was very sensible. He meant a time quite long compared with all times that had been suggested so far. That is what we usually mean physically by *infinite*. You don't literally mean infinite, you mean a willingness to neglect the reciprocal; that was what he was prepared to do.

Now comes Lord Kelvin, armed with the most powerful physics of the 19th century, who was able to show with hammer blows, one after another, that the time available for Darwin's evolution could not be 100 million years, it could not be 80 million years, it could hardly be 60 million years, perhaps not even 10 million years. Because, if the Sun burns carbonaceous fuel or any similar chemical fuel, it can only last thousands of years — a palpably inadequate stretch of time. If it derives its energy from gravitation, we know its mass and we know its size, and we can show that it cannot last for more than a few tens of millions of years. The proposal that it catches and feeds on comets and meteors can also be excluded, though perhaps it can push us into the 50–100-million-year domain. Thus there was a direct conflict. Kelvin would come to the biologists and heckle them terribly by saying, "Tell me what time you want and I'll show if it can be; I can calculate everything." He calculated cooling and so on. Of course, they couldn't name a time. They simply said, "Well, we know you're wrong. We feel it in our bones, in our fossils, but we can't prove it." So the evolutionists were regarded as people without any quantitative basis for their science, though they were, of course, on to something profound.

That was noticed by T. C. Chamberlain immediately after the discovery of radioactivity about 1900. By 1905 or so, Rutherford himself gave a famous evening talk. It was a most distinguished and formal lecture to the Royal Society of London. As a young breaker of rules, a young discoverer, he noticed that in the front row sat Lord Kelvin himself, very elderly, but very stern, trying hard to stay awake and check on this young man who was going to talk about new forms of energy. (Kelvin didn't like radioactivity either, by the way.) But Rutherford tells us that he wondered how to avoid offending Lord Kelvin. Perhaps the famous man would get up and leave when Rutherford talked about the fact that nuclear energy can keep the Sun going for a good long time. Finally Rutherford thought of the right formulation. He said that he had been able to solve an old problem, whose magnitude and importance had been shown by Lord Kelvin himself, when he pointed out that there was no known source of energy capable of keeping the Sun going. At last by experiment they had been able to find a new source, bringing out fully what Kelvin had shown all those years ago. Says Rutherford, Kelvin went immediately to sleep and the whole session was a huge success.

Let us hope that some meeting of minds among physicists and biologists will come to relieve us too of our dilemmas. Of course, there are interesting traps: we don't really know whether we have to have a strongly reducing Earth environment, whether we can find special microenvironments, or whether we can cross one or another of the bridges set before us from the physical to the biological side.

Today, for the first time in the many symposia I have attended on this our subject, adequate attention has been devoted to that great span of time from the stromatolitic microorganisms, complex as they are, to the earliest trilobites. This multicellular phase of early evolution was given us in interesting detail, from several points of view. I found it an exciting thing. I hope that gradually more nonpaleontologists will try to understand and cope with it. We begin to grasp the remarkable progress that was made during that limited time at the dawn of the Cambrian, for we are now examining faunal assemblages from all over the world. We begin to see the richness of change emphasized by Professor Valentine. I find that a most fascinating account.

Of course, there goes with this progress some remarkable inventions, the inventions of life. It is dangerous to pin our view of what happened wholly on the success of this or that invention. It is very easy to do and might even be true. But it also might have been that another invention would have occurred had "invention #1" not won the day. That is a question you must always ask. Two inventions were mentioned which I found extremely striking, worked out in detail, very appealing to a physicist. Both of them are, so to speak, structural innovations: the coming of specialized structural proteins and the first appearance of lignin. Both are elements making possible what is fashionably called a quantum jump in the behavior of two great kingdoms, plants and animals, when once they have acquired the ability to maintain large systems. Because the large living systems are showy, they are in the museums; they are like ourselves. The simple laws of scale suggest that large organisms, certainly on dry land and probably also in the ocean environment, must have stiff structures; otherwise they cannot survive easily. Mass goes up with the cube, and strength only with the square of linear dimension. You cannot survive without a pretty strong structure. Of course, we have an internal skeleton and our friends the bugs have something different, but that doesn't matter so much.

The notion that a few chemical inventions made this possible is striking. We begin to see a new power, shown by both of those papers today, to enter into the details, to try to suggest the biochemical pathways and various correlations. This should not end, it seems to me, until we have spelled out quite a number of such jumps. In population genetics and in evolutionary theories, as I see it, we have little predictive power as yet. We cannot predict rates. We cannot predict proportionalities. We have to take the evidence of change and work backward. It is unclear whether we shall ever have a profound predictive power. But at least our analysis should have a much richer texture than in the past. That is beginning to happen, and I am very encouraged to see it. I think we can expect results in terms of convergences, in terms of interacting biochemical inventions.

It is always fascinating to have the physiologist, the paleontologist, the general biologist working together, not only on the remarkable information-transfer story at the DNA-RNA level, but even at the much grosser level of the engineering inventions that enable living forms to work. I want to mention one invention that is especially striking to me. The internal organization of vertebrates, including ourselves and our kin, brings many implications that do not appear on the surface. Consider one behavioral invention, if you like. It has a quality reminding us of the proteins and the lignin, but it is not at all like them.

Let us look at a strategy: the strategy for catching food as a predator does. We are, after all, predators, both on berries and on bears; we belong to the hunting-foraging creatures. Suppose you were in the position of living on lobsters. That is a nice position to be in: it is achieved by some New Englanders and by all common octopi. Octopi are intelligent invertebrates — in some ways our analogs within the invertebrate kingdom.

Let us approach the octopus from the standpoint of a rational analysis of prey-seeking behavior. What is the situation? Lobsters are not as variable in behavior perhaps as some land animals, but still they are not all that uniform either. It is quite likely that any particular desirable game, like lobsters, appears in fluctuating numbers within the field of action of any carnivore. There is little likelihood of a steady flow of lobsters, one dropping down every hour. That is the Santa Claus or Big Rock Candy Mountain theory of life. Most of us hunters don't find it that easy. You've got to go out and scrabble around a little bit to get what you want.

When caribou are numerous you should of course hunt caribou. When the Eskimo or the Indian hunter, skillful person that he is, finds that caribou are unhappily in short supply, he will simply redouble his efforts, for he is hungry and back home the wife and kids are hungry. The whole situation is serious. The same tendency is found in every hunting-gathering mammalian predator activity, say for the leopard. We explain strange behavior on the part of mammals sometimes in that way: "Well, it was hungry. I forgot to feed my Siamese, so it tried to eat the curtain!"

On the other hand, an octopus has a much purer view. It behaves more like the theory of games. When lobsters become few, an octopus does not seek in a frenzy to find those few lobsters or put up with eating mere crayfish. Heaven forbid! Instead the octopus goes to sleep — a most intelligent thing to do. Every once in awhile it wakes up and looks out. "Any more lobsters around?" No. Back to sleep it goes again.

Such control over impatience, anxiety, and hunger is very hard for us to understand. Our thought is based on our design: namely, we have to generate 100 watts all the time. There is a basal metabolism, roughly 100 watts, that we expend. If we don't keep the machine fueled, we're in irreversible danger. But the octopus has no such base load demand. Cold-blooded, he is willing to relax to the ambient temperature of the warm sea environment in which he lives, provided only that every once in awhile he can scrape up a 100th of a watt, turn the ganglia on a little bit, open an eye. That isn't too hard to do. Once you look at it coolly, you realize that human behavior goes completely against the sound principles by which an organism would adaptively go hunting. Whenever it is hard to hunt, don't continue to hunt with greater frenzy over longer hours, as we all do. On the contrary, take it easy. If conditions are not good, there is not much use in hunting. We can't adopt that principle, though, because for a couple hundred million years conditions were generally good enough so that somehow or other it was worth paying to keep all our subtle electronics going in order to have an opportunity to hunt well. How different evolutionary structure can be! We need a special view of the lives of other creatures. If you now carry this logic over to some distant world, then it gives scope to the issues we are talking about.

The strategic discussions for SETI showed in circumstantial and detailed operational terms that we have already begun to make a clear plan. Certainly, a great deal of hope emerges. Indeed we have a remarkable new opportunity: we seem to be on the threshold of finding other planets. We are setting up apparatus dedicated to the purpose of finding planets, both by interferometry and by astrometry with modern techniques. I very much hope that the entire community will support and applaud this effort, because it seems to be one of the most important auxiliary searches that could be made. It is important even if we never have a chance of getting radio signals. It can give us something else to look at than just the single Sun-planet system to which we are so well adapted.

I will conclude with some remarks of a more philosophical nature. The first is straightforward; it is a distinction I have made before, but I think it bears repeating. We can characterize attitudes toward SETI by invoking the names of two philosophers. (Perhaps they should not have to bear this burden; they are not really responsible. But as often happens, it is convenient here to use the names of famous workers of the past as labels. I ask their forgiveness.) First is the Aristotelian view, which seems quite plain: Earth is the cosmic center; the heavens revolve around it, $1/R$ reaches infinity here, and here is the right place! On this view, of course, the whole outward-looking style is neither necessary nor desirable. Astronomers can hardly accept that view; at least they have not accepted it for several centuries now. They are not going to change, and I am pretty sure that most of the other sciences will follow in turn. (It was thus impressive to hear a well-known student of fossil man give us a detached and generous view of what other worlds might contain.)

The second point of view I like to attribute to Copernicus. Everyone knows what that implies, though I don't know if he actually said it anywhere: namely, that this green and blue Earth is not all that different from the planets and the Sun and all those other things that circle and shine in the sky. They are themselves earthy or gaseous or whatever, but they are physically real objects. Nowadays men have walked upon one such object and shown that it is not different in kind from the one we inhabit. Since we know Earth is also earthy, then it is clear to us that these are only relative categories. Circular, shining, and perpetual orbits was Aristotle's view; it was Copernicus who recognized that Earth was no less circular, shining, and perpetual. We take a very different view of the cosmos post-Copernicus. That has been the spirit of SETI. The radical Copernicanism of the very first efforts is still viable, though admittedly we have more judgment about where to look.

I would like to mention another point of view, which seems a little curious to me. It is less developed among the general public than in the scientific community, and it seems to be based on a very grand extrapolation indeed. I would like to call it the Malthusian view. Take a piece of semilog paper and plot the growth of more or less anything in human culture (people, telescopes, cities, motorcycles, whatever you want), extrapolate, and very soon it shoots right off the page. At some point the mass will exceed the mass of Earth, or the volume exceed the volume of the Universe, or the velocity of expansion exceed the speed of light. Therefore, there must be some catastrophe ahead. This is true of every exponential, all but independent of its rate. If you change the rate by a factor of 2, you just double the time: 100 years becomes 200 years. You are talking about the long-term future, so the rate doesn't make that much difference. Therefore, this is only a statement of the transient existence of exponential fits: certainly such a fit is a good thing to notice, but it is never totally realistic.

Ecology has brought us to understand that something always turns up to put a plateau above every exponential, some necessary resource. Californians will guess that it is gasoline that limits the Universe, but if that were not true, it would turn out to be something else.

What happens when we apply exponential arguments to space travel? We have heard several discussions this afternoon of the difficulty of interstellar travel. Several recent papers, however, have taken a longer range view. The most conspicuous of them is in a book by Freeman Dyson, a very able physicist indeed. He points out that we have not reckoned with self-multiplying systems of an artificial kind. (There are plenty of multiplying systems that are living; I am not so sure that artificiality makes a big difference, but I will accept the argument.) If you can make a self-multiplying system, put it in a spaceship, and give it an initial stock of capital resources, off it will go to find a suitable solar system. Eventually, it will be well enough ensconced that it will create replicas of itself, which will then be sent out to find other solar systems to be settled, and so on. If you make some calculations it turns out that you can cross the whole Galaxy with this scheme during a geological epoch, in a few hundred million years.

The argument goes next: since the Galaxy doesn't look as though it is densely occupied that way, since many more than 300 million years have gone by if our calculations are right, then nothing like this has ever happened. The most enthusiastic paper I have seen says that we ourselves will be able to start this process in 100 years. It is plain to that author that we must be the first ever to encounter such a possibility. No use, then, spending a lot of experimental effort on SETI; these calculations make it quite clear. Of course, they are not very robust calculations.

Nevertheless, I find it quite interesting that the mere examination of the sky should suggest that because the stars aren't arranged in tasteful letters, or free form Crescent or Cross, or anything else, that the Universe has not been modified by a gardener of any sort. This seems enough for some to say we should not make our patient search through the channels and through the spatial directions to find signals. I hold that this is a semiserious problem. Serious, that is, only to the extent that it develops within the scientific community an antagonism to what seems to me an already difficult but at least an empirically based scheme. It is the one we have been talking about for these two days. I hope we can come to some meeting of the mind with persons who represent this theoretical pessimism (or is it optimism?). Let me simply suggest a mixed strategy: instead of waiting 300 hundred million years to see if the theory is right, I might try waiting only 300 years to test it by some more active procedure.

At this conference we see a merging of disciplines, a coming together of many scientific points of view. We master technicalities that can hardly be put into the same box: radio astronomy and transfer RNA, paleontology and climatology. Yet they all bear sharply on one question. This is, of course, the strength of the matter, as Alastair Cameron said beautifully at the noonhour press conference. This strength, I think, will continue. It will support the enterprise and give point and wit to the enterprise as long as it goes on.

A curious situation has arisen under the power of intense modern instrumental specialization (in the broad sense, where *instrument* includes *method*). The instrumental specialization of our science has grown steadily since the 19th century. It is sharply reflected in the institutions of our universities, which are slow to change in the face of it. For example, many universities still have a Botany Department and a Zoology Department, and if you bring in any one of a number of microorganisms that Lynn Margulis will be very happy to show you on a slide, they don't know which department ought to study it! Perhaps it doesn't make any difference. But our research structure is inherited from institutional decisions in Scottish and German universities made 100 to 150 years ago. Sooner or later this will change. For all real large-scale engineering activities, such as NASA has carried out so successfully, we know that that is not the way to do it. Such activities require mission teams and a mix of specialties. The universities are going to have to learn. Perhaps through our mixture here between the academic and the operational, in this meeting at Ames, we have begun a style that will grow.

There is a narrowness of action, though not of intent, which characterizes university departments, and scientific publications and scientists in general: if it is too popular, it is somehow vulgar and wrong. You can't really speak to those people across the street. I live next to the chemists at MIT, but I never see them. I hardly know who they are, yet between physics and chemistry it is hard to know who should study what molecule. I myself am guilty. We form communities not based on the problems of science, but on quite other things. This is part of the general split between the intelligent informed member of the public and the scientist who speaks in narrow focus. But the great theoretical problems which I believe the world expects will somehow be solved by science, problems close to deep philosophical issues are the very problems that find the least expertise, the least degree of organization, the least institutional support in the scientific institutions of America or indeed of the world.

Two of these, of course, are the great questions, "Are we alone?" and "How did life begin?" These questions are treated now in the elementary textbooks, because of the vigor of a few people over the last 30 years, but they are hardly treated anywhere else. The further you go away from the freshman student, the less likely you are to find a colleague interested in it. This is beginning to change. Five or 10 years ago, the radio astronomers, just to name a group of people I know quite well, were pretty hard to talk to about SETI in any way. It wasn't so much that they would disagree, that's fine: they still do. But they laughed, and that was not very pleasant. Well, now at least they are only smiling; this is a kind of gain.

One cure for this ill, though a difficult one, is the pursuit of a scientific discourse on a more philosophical, more consciously aesthetic, better-illustrated style, one willing to grapple with large problems, even though only small solutions can at present be offered for them. This was demonstrated to us in two quite different tones: first by Lynn Margulis and second by William Schopf. I think that science requires this change. I expect to see an enlarging of the disciplines to form at last an interdisciplinary pool, aware of larger philosophical issues. We need not try to solve them or to prescribe their limits, but we must recognize their human importance, their intellectual existence as an increasing element within scientific thought. If that were the only positive result from the SETI investigation, I think it would still be judged by history to have proved extremely worthwhile.

Glossary

Glossary

abiotic — not involving or not produced by living organisms

absorption spectrum — array of lines and bands that results from the passage of radiant energy from a continuous source through a cooler, selectively absorbing medium

activation energy — amount of energy needed to initiate a reaction

adsorption — adhesion to a surface in an extremely thin (often monomolecular) layer

albedo — fraction of incident electromagnetic radiation reflected by a body such as a planet, star, or cloud

allotropy — existence of an element in more than one form

amino acid — acid containing the amino (NH_2) group; more than 80 amino acids are known, but only about 20 occur naturally in living organisms, serving as the building blocks of proteins

anaerobic — occurring in the absence of free oxygen

Ångstrom (Å) — unit of length convenient for measuring wavelengths of electromagnetic radiation: $1 \text{ Å} = 10^{-10}$ m

aperture — in the case of a telescope, an area open to the Universe; signifies either the maximum physical or the effective capture cross section of a telescope or radio antenna; often stated in terms of the diameter or an equivalent diameter if the aperture is noncircular

apoprotein — protein portion of a conjugated protein (i.e., a compound of a protein with a nonprotein)

arcsecond (arcsec) — measurement of angular separation: a 1-inch stick would subtend an angle of 1 arcsec at a viewer's eye at a distance of about 6.5 miles

Arecibo — short name for the National Astronomy and Ionospheric Center (NAIC) at Arecibo, Puerto Rico; often refers only to the NAIC 1000-ft (305-m) zenith ($\pm 20°$) antenna, the world's largest radio astonomy collector

437

artefact — material object modified purposefully for use by the person or animal that does the modifying

astrometry — branch of astronomy that focuses on measurements, especially those relating to positions and movements

astronomical unit (AU) — mean Earth-Sun distance: 1 AU = 1.496×10^{11} m = 8.31 light minutes; a convenient unit for measuring distances between planets and their stars

atmosphere — gaseous mass enveloping a planet or star

atom epoch — fourth epoch in the history of the Universe, lasting from about 100 sec to 10^6 yr, in which matter came to dominate radiation as the principal constituent of the Universe

atomic mass unit — convenient unit for measuring the mass of an atom or molecule: 1 atomic mass unit is defined as 1/12 the atomic mass of the most abundant carbon isotope, ^{12}C

AU — see astronomical unit

autotroph — organism whose metabolism requires only external sources of carbon dioxide and nitrogen

bacterium — any of an extremely flexible class of microscopic plants whose members have a variety of structures, come singly or in colonies, live just about anywhere, and derive energy from whatever source is available

Big Bang — hypothesized initial singularity that could have started the space and time of our Universe, now thought to have occurred 7 to 20 billion years ago

billion — one-thousand million, 10^9

bin — output signal channel in a multichannel spectrum analyzer or MCSA (q.v.)

binary system — two neighboring stars that revolve around their common center of gravity; the fainter of the two stars is called the companion

binwidth — nominal frequency bandwidth of a bin; see *resolution binwidth*

biosphere — general term for regions in which life can exist

blackbody — body capable of absorbing energy of all wavelengths falling on it; it is also capable of radiating all frequencies in a particular ratio to its absorbing properties. The value of the ratio depends only on the temperature of the body. The physical laws describing the properties of a blackbody are derived from the theory of thermodynamic equilibrium

black hole — gravitationally collapsed mass from which nothing — light, matter, or any other kind of signal — can escape; a theoretical concept so far

blue-green algae — any of a class of algae (a group of mainly aquatic, simple photosynthetic plants) whose chlorophyll is marked by bluish-green pigments

body plan — basic architectural arrangement of a biological structure

carbohydrate — organic compound consisting of a chain of carbon atoms to which hydrogen and oxygen, present in a 2:1 ratio, are attached

carrier signal — electromagnetic wave whose modulations are used as signals in television and other radio transmissions

cassegrain (telescope or focus) — some telescopes, particularly at microwave and shorter wavelengths, have a second reflector near the focus of the larger, primary mirror. This translates the focal point to a position near the apex of the primary where it is more accessible, and where practical antenna feeds are less responsive to radiations arriving from very wide angles relative to the nominal pointing direction

cation hydration — process of incorporating molecular water into a complex molecule with positively charged units of another species

celestial sphere — the visible, seemingly spherical surface that appears to surround Earth and to be centered at the observer

CETI — acronym for communication with extraterrestrial intelligence. Sometimes pronounced with a long e and short i, to distinguish it from SETI which is favored with two short vowels

chaos — hypothetical first epoch in the history of the Universe, lasting 10^{-23} sec — a period about which we cannot yet even speculate

chemically peculiar stars — stars manifesting anomalies in the relative abundances of elements, which may arise from mechanical rather than nuclear effects; so-called manganese stars, for example, show a great overabundance of manganese and gallium, usually accompanied by excess mercury

chemisorption — use of chemical forces to take up and hold onto something

chemoautotrophic — autotrophic and deriving energy from an inorganic compound (such as sulfur)

chemosynthesis — synthesis of organic compounds using energy derived from chemical reactions

chert — hard, flinty, siliceous rock, often arising through precipitation

chondrite — stony meteorite with small, rounded occlusions, usually of olivine or pyroxene, embedded in a matrix

chromosome — a gene-containing filamentous body found in cell nuclei

clade — group of organisms all descended from a single common ancestor

clay — earthy material, composed mainly of hydrous aluminum silicates and other minerals, which is plastic when moist but hard when fired

climate — long-term manifestations of short-term atmospheric variations

cluster (astronomical) — group of stars numbering from a few to hundreds of thousands of stars. Galactic clusters, sometimes called open clusters, contain up to a few hundred members and occur rather close to the plane of the Galaxy. Globular clusters contain tens of thousands of stars distributed about their center in a spherical manner and are found far from the plane of the Galaxy as well as in it toward the center of the Galaxy

collagen — insoluble fibrous protein used by vertebrates to hold themselves together (i.e., it is a chief constituent of connective tissue fibrils and occurs in bones)

complex output — multichannel spectrum analyzer (MCSA, q.v.) may be set to give complex, undetected signal amplitude streams from each output channel in the form of two orthogonal quantities $(a + jb)$. The more common format is to present detected signals by their amplitude $\sqrt{(a^2 + b^2)}$ or power $(a^2 + b^2)$, either with or without the associated angle $\theta = \arctan b/a$ (vide *detection*)

continuously habitable zone (CHZ) — region around a star in which a planet can maintain appropriate conditions for the existence of life (including the retention of a significant amount of liquid water) for a period sufficient to allow the emergence of life

core — central part of Earth, having a radius of about 2100 miles

cosmic rays — general term for the stream of atomic nuclei that constantly bombard Earth's atmosphere

crust — outer part of Earth, composed essentially of crystalline rocks

cryogenic — of or relating to the production of very low temperatures (on the Kelvin or absolute scale)

CW — continuous wave; a relative term of ancient lineage used to distinguish a simple wave with little or no modulation (e.g., a pure carrier (q.v.) or one simply keyed "on" and "off" as in manual radio telegraphy), vis-a-vis a highly complex, modulated wave (e.g., FM broadcasting)

dB — decibel; a unit of power ratio equal to 10 times the common logarithm of the ratio

decay constant — for an atom that undergoes radioactive decay, the decay constant is the proportionality factor between the time rate of decay and the total number of atoms present; it is the inverse of the mean lifetime of an atom

degassing — process whereby the atmosphere and ocean water have slowly accumulated through geologic time by emanating from Earth's interior in the form of volcanic gases

detection — in electromagnetics, an operation converting the vector electromagnetic wave to a scalar time series proportional to either the amplitude or the power of the wave, with or without an accompanying angular time series. A crucial aspect of detection is: the signal-to-noise ratio after detection is the square of the signal-to-noise ratio before detection (vide *complex output*)

differential thermal analysis — method of determining, for samples being heated at a controlled rate, the intensities of and the temperatures at which thermal effects occur

DNA — deoxyribonucleic acid; the molecule that is the basis of heredity in
  many organisms

Doppler shift — a change in frequency resulting from relative motion along
  the line between the transmitter and the receiver. If the source and the
  receiver are approaching each other, the frequency received is higher
  than the frequency transmitted by a factor, depending on the actual
  relative velocity. Knowledge of this shift is used to determine the relative velocity

drifting (signal) — refers to a signal with an apparent time rate of change in
  its typical frequency. All signals drift to some extent. In a SETI system,
  the dominant drift should be largely the result of only the time rate of
  change in the Doppler shift (q.v.)

ecology — study of the relationship between organisms and their environment

EIRP — equivalent isotropic radiated power; theoretical construct (all electromagnetic wave emitters are appreciably directive); product of real
  transmitted power and directive gain of the emitter (antenna) in the
  direction of the receiver (compared to the unit gain of the imaginary
  isotropic radiator)

electromagnetic radiation (wave, spectrum) — energy involving electric fields
  and magnetic fields oscillating in phase at right angles to each other,
  propagated in a direction at right angles to both fields with a velocity in
  free space equal to c (approximately 300,000 km/sec, or
  186,000 miles/sec), a universal constant

encephalization — process of phylogenetic concentration and increase in
  nervous tissue

entropy — tendency of systems to become more disordered (and thus more
  uniform) over time; also a measure of disorder

epitaxial — growing on a crystalline substrate in such a way as to mimic the
  orientation of the substrate

eras (geologic) — all of Earth's history since the appearance of the first life
  forms is divided roughly into four eras: Precambrian, from 3.5 billion
  to 570 million years ago; Paleozoic, from 570 to 225 million; Mesozoic,
  from 225 to 65 million; and Cenozoic, from 65 million to the present.
  The last two eras are broken down into the following periods: the
  Mesozoic into Triassic, Jurassic, and Cretaceous; the Cenozoic into
  Tertiary and Quaternary

erg — the unit for work in the cgs system of units: the work done by a force
  of 1 dyne moving through a distance of 1 cm

ETI — extraterrestrial intelligence; also used to signify extraterrestrial intelligent species

eucaryote — organism composed of one or more cells with clearly formed
  nuclei

eutectic — form of a compound that has the lowest possible melting point

evolution — change over time in the morphology and physiology of species of organisms; or of any object, such as stars; or of ideas; etc.

exobiology — extraterrestrial biology, that is, the study of life forms as they might occur outside the terrestrial environment

feed (antenna feed, line feed) — in a reflecting antenna system, the device that converts a guided (by wire, cable, or other wave guide) electromagnetic wave into an electromagnetic radiation field, and vice versa, when reciprocity theorem holds as it so often does. Commonly, feeds are some form of horn antenna, but they may be dipole arrays or their equivalent or linear devices like the Yagi antenna or the leaky multicavity line feed used with some spherical reflectors, as at Arecibo

Fischer-Tropsch-type (FTT) reaction — catalytic reaction in which hydrocarbons and their oxygen derivatives are produced through the reaction of hydrogen and carbon monoxide

flare (astronomical) — relatively rapid outburst of energy in a star

fossil record — remnants or traces of organisms of past geological ages embedded in Earth's crust

free radical — atom or group of atoms having at least one unpaired electron

freezing point shift — change in the freezing point of a solution compared to that of the pure solvent

fungus — plant of the Thallophyta subkingdom, which lacks chlorophyll and ranges in form from a single cell to massed bodies, including yeasts, molds, and mushrooms

Gaia hypothesis — hypothesis that the biosphere has an important modulatory effect on the surrounding atmosphere

gain stability — crudely defined by $\Delta G/G = g(t)$; the smaller this quantity over the relevant time interval, the less the gain instability or the greater the gain stability. Gain (amplification) of analog signaling systems always varies somewhat with time; $g(t)$ contains a variety of "noise" terms of zero mean and various secular terms; the latter dominate unless proper precautions are in force

galaxy — spelled with a lower-case g, galaxy means any of millions of stellar systems once called "island universes" or extragalactic nebulae. Depending on their form, galaxies may be called spirals, barred spirals, ellipticals, or irregulars. Spelled with a capital G, Galaxy refers to that particular stellar system which includes our Sun and all the stars visible to the naked eye. The Milky Way is our view of the Galaxy

galaxy epoch — fifth epoch in the history of the Universe, lasting on the order of 10^{10} yr, during which matter largely coagulated into galactic masses

gamma ray — high-energy electromagnetic particle or photon, especially as emitted by a nucleus in its transition from one energy level to another

gene pool — total amount of information in all the genes of all the reproductive members of a biological population at any given time

genetic drift — random fluctuations (or "walk") of gene frequencies from generation to generation that occur in small populations

genus — taxonomic category ranking below a family and above a species

giga — 10^9 (as in gigahertz, GHz); one billion (U.S.A.)

glycoprotein — conjugated protein in which the nonprotein group is a carbohydrate

gravitational wave — propagating field predicted by general relativity to occur as a result of any large-scale change in the distribution of matter (as in the collapse of a star)

greenhouse effect — general warming of the lower layers of a planetary atmosphere that tends to increase with increasing atmospheric carbon dioxide. As with Earth, this process occurs whenever an atmosphere is relatively transparent to visible-light radiation but opaque to the longer wavelength infrared radiation of the surface of Earth

hadron epoch — second epoch in the history of the Universe, lasting on the order of a second; named for the heavy elementary particles (protons, neutrons, mesons) that were the most abundant form of matter at the time

halloysite — porcelainlike clay with a composition similar to that of kaolin but with more water and a distinct structure

helium — atom consisting of two protons and two electrons

hertz (Hz) — measure of frequency; an oscillating system that completes a cycle a second has a frequency of 1 Hz

Hubble constant — constant of proportionality in the relation between distance of galaxies and their velocity of recession

hydrocarbon — organic compound containing only carbon and hydrogen

hydrogen — simplest atom, consisting only of one proton and one electron; the most abundant element in the Universe

hydrogenation — process of combining something (especially an unsaturated compound) with hydrogen

hydrosphere — aqueous envelope of Earth, including bodies of water and water vapor in the atmosphere

hydrothermal vent — opening in the sea floor produced by the hot magmatic emanations that are rich in water

Hz — see hertz

index of refraction — ratio of the speed of light in a vacuum to its speed in a given medium

infrared radiation — electromagnetic radiation with wavelengths longer than that of visible light and shorter than millimeter radio waves

interstellar clouds — masses of tenuous gas and dust between the stars

ionosphere — that region of Earth's upper atmosphere having an appreciable abundance of electrons and charged atoms (ions), ranging from 50 to 500 miles above the ground

isotope — one of two or more atoms whose nuclei have the same number of protons but different numbers of neutrons

jansky (jy) — convenient unit of incident spectral flux density used in radio astronomy; 1 jy = 10^{-26} W/m² Hz (named for Karl G. Jansky, initial discoverer of extraterrestrial radio radiations)

joule (J) — unit for work in the mks system of units: 1 J = 10^7 ergs

kaolinite — mineral (a hydrous silicate of aluminum) that constitutes a principal part of the fine clay, kaolin

Kardashev cultures — N. X. Kardashev has distinguished three types of technological societies according to the amount of power they can harness: Type I can engage the power available on a planet; Type II, the power output of a star; and Type III, the power output of a galaxy

kHz — kilohertz (see Hertz)

lepton epoch — third epoch in the history of the Universe, lasting about 100 sec, in which the lighter elementary particles such as electrons, neutrinos, and muons were the dominant form of matter

Life Era — era in the history of the Universe when life emerges as the dominant element

light year — distance traveled by light in a vacuum in one year: 1 light year = 9.46×10^{15} m

lignification — deposition of lignin, the chief noncarbohydrate constituent of wood, in cell walls

lithosphere — outer (rocky) layer of the solid Earth, usually taken to be about 50 miles in depth

luminosity — relative brightness

luminosity class — two stars may be classified as of the same spectral type yet differ, perhaps widely, in intrinsic brightness or luminosity (see *spectral type*)

M$_\odot$ — (see *solar mass*)

magma — molten matter under Earth's crust, from which igneous rock is formed

magnetosphere — region of the upper atmosphere, extending out thousands of miles, dominated by Earth's magnetic field, so that charged particles are trapped in it

magnitude (stellar) — relative measure of the brightness of celestial objects. A difference of 1 magnitude in brightness corresponds to a luminosity difference of $10^{0.4} \approx 2.51$; a difference of 5 magnitudes corresponds to a factor of $10^{5\,(0.4)} = 100$ in luminosity

Main Sequence — principal sequence of stars on the graph of luminosity versus effective temperature, encompassing more than 90% of observable stars. The lower mass limit for the Main Sequence is 0.085 $M_\odot$ and the upper limit is about 60 $M_\odot$

mantle — layer of Earth that lies below the lithosphere and above the core

Matter Era — collective name for the most recent three epochs in the history of the Universe (atom, galaxy, stellar), covering all of time after the Radiation Era

Maunder minimum — virtual disappearance of sunspots in the period 1645 to 1715

MCSA — short for MCSA/SD, a digital, energy-efficient, real-time, multi-channel spectrum analyzer and signal detector; hardware device that continuously accepts a significant portion of the electromagnetic spectrum and divides the spectral energy into a set of many contiguous frequency bins or output channels. Operational functions included are frequency analysis and signal detection, averaging, threshold testing, etc.

mega — 10^6 (as in megahertz, MHz); one million

metamorphic rocks — rocks formed by the action of great heat and/or pressure

metazoan — an animal whose body is composed of cells differentiated into tissues and organs and (usually) a digestive cavity

MHz — one million Hertz; megahertz (see *hertz*)

micron — 10^{-6} m

microwave — electromagnetic wave roughly in the range 0.01–1 m in wavelength (ordinary broadcasting utilizes waves in the 200–600 m range; the "short waves" used in long-distance communications are rarely shorter than 10 m)

midocean ridge — a 40,000-mile-long continuous median mountain range in the Arctic, North and South Atlantic, and Indian Oceans. The crest of the ridge is usually found at 500 to 1300 fathoms below the surface; its width is usually a few hundred miles, and it has a relief of 10,000–33,000 ft

mitochondrion — cytoplasmic organelle serving as a site of respiration

monomer — simple chemical compound of relatively low molecular weight which can undergo polymerization

montmorillonite — soft claylike mineral composed of hydrous aluminum silicate

multicellular — consisting of many cells

nebula — rarefied cloud of gas or dust observed in interstellar space

neoteny — attainment of sexual maturity during the larval stage

neutral point (in binaries) — point (barycenter) between a binary pair at which the gravitational attractions of the two stars cancel one another

neutron star — collapsed star whose core is composed primarily of neutrons at a density above 10^{14} gm/cm^3

noble gas — one of a group of rare but extremely stable gases with low reaction rates (helium, neon, argon, krypton, xenon, radon)

nucleic acids — long, chainlike molecules which, in the various combinations of constituent groups, embody the genetic code (DNA) and assist with its transmission (RNA)

nucleoside — compound consisting of a sugar and a purine of pyrimidine base

nucleosynthesis — production of the chemical elements from hydrogen nuclei or protons (such as occurs in thermonuclear reactions in stars)

nucleotide — compound composed of a nucleoside combined with phosphoric acid

oceanic basalts — rocks of the oceanic island volcanoes

organelle — specialized cellular part analogous to an organ

ozone layer — layer of Earth's atmosphere at about 20 to 30 miles, marked by a high ozone (O_3) content

paedomorphic transformation — phylogenetic change that involves retention of juvenile characteristics by the adult

panspermia — diffusion of spores or molecular precursors of life through space

parent compound — chemical compound that is the basis for one or more derivatives

parsec (pc) — parallax second, the distance at which 1 AU subtends an angle of 1 second of arc: 1 parsec = 3.086×10^{16} m = 3.26 light years

phase change — metamorphosis of a substance from one state to another, as from gas to liquid or from solid to gas

phase delay — signal travel time between two points in any signal path or circuit (measured in seconds or in degrees or radians of the signal frequency); to be distinguished from simultaneous phase difference between two points

photoautotrophic — autotrophic and deriving energy from light

photolysis — component process in photosynthesis in which water is dissociated and hydrogen is joined to a molecule of a substance called NADP under the indirect influence of solar energy

photon — quantum of electromagnetic energy, of value $h\nu$, where h is the Planck constant and ν is the frequency of the radiation

photosynthesis — process by which light energy and chlorophyll manufacture carbohydrates out of carbon dioxide and water

phylum — category of taxonomic classification just above class

plastid — any of various cytoplasmic organelles of photosynthetic cells that serve in many cases as centers for metabolic activity

plate tectonics — model of the structure of Earth in which the surface consists of a small number of semirigid plates floating on a viscous underlayer in the mantle; the clashing of plates concentrates most deformation, volcanism, and seismic activity along their peripheries

polarization — electromagnetic radiation that exhibits different properties in different directions at right angles to the line of energy propagation is said to be polarized

polymerization — process of forming long molecules (polymers) out of small units (monomers)

polynucleotide — polymeric chain of nucleotides

polypeptide — molecular chain of amino acids

prebiotic — relating to the chemical or environmental precursors of the origin of life

procaryote — cellular organism that does not have a distinct nucleus

Project Cyclops — a 10-week design study sponsored by NASA, Stanford University, and the American Institute for Engineering Education, of possible means for detecting extraterrestrial civilizations

prosthetic group — nonprotein group of a conjugated protein (i.e., a compound of a protein with a nonprotein)

protein — complex polymer built of amino acids that contains the elements carbon, hydrogen, nitrogen, oxygen, sometimes sulfur, and occasionally others such as phosphorus and iron. Proteins are essential constituents of all living cells; they are synthesized from raw materials by plants but assimilated as separate amino acids by animals

proteinoid coagulations — proteinlike linkages of amino acids produced under laboratory conditions from repeated heating and cooling of simple organic molecules

protist — any of a kingdom of living organisms (Protista) that includes algae, slime molds, protozoa, and fungi, usually characterized by unicellular reproductive structures, true nuclei, and chromosomes

proton-proton cycle — energy-releasing nuclear reaction chain believed to be important in energy production in hydrogen-rich stars

pulsar — rotating, magnetized neutron star, emitting very energetic radio pulses at a rapid and very regular rate

purine — nitrogen base that forms a component (with sugar and phosphate) of nucleotides and nucleic acids

pyrimidine — nitrogen base such as cytosine, thymine, or uracil that is a constituent of nucleotides and nucleic acids

quantum — a discrete small unit; in any given physical process, it is the minimum permissible unit of energy

quark — a possibly hypothetical particle that carries a fractional electric charge and is held to be a constituent of known elementary particles

quasar (quasi-stellar radio source) — object nearly stellar in appearance that emits visible light at a rate exceeding that of a large normal galaxy by a factor of about 100 and that emits radio waves at a rate comparable to the strongest radio galaxies

Radiation Era — collective name for the first three epochs of the history of the Universe (chaos, hadron epoch, lepton epoch), lasting overall about 100 sec and dominated by radiation rather than matter

radioactivity — spontaneous emission from the nuclei of atoms of certain elements, of charged particles, and/or strong radiation; the remaining nucleus is generally an isotope of another element

radionuclide — radioactive nuclear species or nuclide

rare earth elements — series of elements usually taken to include elements with atomic numbers 58 to 71, lanthanum, and sometimes yttrium and scandium

red giant — luminous red star that has exhausted the supply of hydrogen in its core and evolved off the Main Sequence; as the core contracts and becomes hotter, helium burning starts and an extensive envelope develops as the outer regions of the star expand

redox potential — voltage difference at an inert electrode immersed in a reversible oxidation-reduction system

reducing atmosphere — atmosphere comprised of substances that readily provide electrons

refraction — bending of a wave front when the wave encounters a medium with propagation properties different from the one in which it has been traveling

resolution binwidth — equivalent power bandwidth, $\Delta\nu$; if p_m is the maximum of the real bin frequency response, $p(\nu)$, then $\Delta\nu p_m = \int_0^\infty p(\nu)d\nu$

rift zone — elongated valley formed by the depression of a block of Earth's crust between two faults or groups of faults of approximately parallel strike

RNA (t, m) — ribonucleic acid; a nucleic acid generally associated with the control of chemical reactions. Transfer RNAs move particular amino acids to growing polypeptide chains in protein synthesis; messenger RNAs act as templates for the formation of proteins

roentgen — unit of X or gamma radiation dosage: The amount of such radiation sufficient to produce ions carrying 1 electrostatic unit of charge in 1 cm^3 of air

seeing — in astronomy, this is the quality, not the act of observation; it refers to the atmospheric turbulence degrading an image and is measured by the smallest detail of the image that can be readily distinguished

sessile — attached directly by the base (rather than being raised on a stalk)

SETI — acronym for search for extraterrestrial intelligence, primarily by means of microwave radio exploration (pronounced with short e and i)

shock wave — discontinuity in the flow of a fluid (including a gas or plasma) marked by an abrupt increase in pressure, temperature, and flow velocity at the shock front

sidereal time — time measured in relation to the fixed stars: the length of a sidereal day is 23 hr, 56 min, 4.09 sec of mean solar time

siderophile element — element with a weak affinity for oxygen and sulfur and readily soluble in molten iron (including iron, nickel, cobalt, platinum, gold, tin, and tantalum)

signal-to-noise ratio — ratio of the signal power in a signal channel to the (unwanted) noise power present

smectite — class of clay minerals characterized by distinct swelling properties and high cation-exchange capacities (including montmorillonite, nontronite, etc.)

solar constant — rate at which radiant solar energy is received normally per unit area at the outer layer of Earth's atmosphere; its value is about 1.94 gram calories/cm^2/min

solar mass (M$_\odot$) — mass of the Sun, 2×10^{30} kg, used commonly as a unit to measure the masses of stars

solar nebula — cloud of gas and dust out of which a star condenses

solar (stellar) wind — radial outflow of hot plasma from a star's corona, carrying both mass, angular momentum, and energy away from the star

speckle interferometry — method of using short-exposure photographs to recover information down to the diffraction limit of large optical telescopes

spectral analysis — study of the distribution by wavelength or frequency of the radiation emitted by an object of interest

spectral type (or class) — classification used to sort stars by photospheric temperature and intrinsic brightness. The seven spectral classes O-B-A-F-G-K-M, listed in order of decreasing temperature, include 99% of all known stars. Each spectral type is divided into a variable number of subtypes designated by Arabic numerals. Further, stars are sorted by intrinsic brightness into luminosity classes designated by the first five Roman numerals. In turn, these are subdivided into a small number of subclasses designated by the first few letters of the lower case English alphabet; for example, the Sun is a G2 V star (also sometimes denoted as a dwarf G2 star) and Betelgeuse (α Orionis) is classified as M2 Iab (i.e., intermediate between Ia and Ib)

spectroscopic binary — binary star that can be distinguished from a single star only through analysis of the Doppler shift of the spectral lines of one or both stars as they revolve about their common center of mass

spectroscopy — splitting of light into its constituent wavelengths

stellar epoch — sixth epoch in the history of the Universe, lasting perhaps 10^{10} yr from the galactic era to the present, dominated by the formation of stars

stromatolite — laminated sedimentary fossil formed from layers of blue-green algae

$^{87}Sr/^{86}Sr$ **ratio** — strontium isotope ratio is used as a corrective factor in rubidium-strontium dating for studying the ultimate age of origin of igneous rock

sublimation — process of passing from gas to solid state (or vice versa) without becoming a liquid

supernova — phenomenon in which a star, at the end of its nuclear burning life, increases its energy output several billionfold for a short time. Some supernovae become as bright as the whole galaxy in which they are observed. A large fraction of the star is exploded into interstellar space, leaving behind a core which may become a white dwarf, a neutron star, or a black hole

superphylum — taxonomic category lying between a kingdom and a phylum

symbiosis — intimate living together of two organisms (called symbionts) of different species, for mutual or one-sided benefit

synchrotron emission — radiation from electrons constantly accelerated in a magnetic field at a rate great enough for relativistic effects to be important. Predicted long ago, this radiation was first encountered in the particle accelerator called the synchrotron. Much of the radiation observed by radio astronomers originates in this fashion

synchrotron self-absorption — reabsorption of radiation from accelerated electrons by other nearby electrons; this is a possible source of low-frequency turnovers observed in the radio spectra of compact sources

systems analysis — analysis of the response to inputs of a set of interconnected units whose individual characteristics are known

taxon (pl. taxa) — group of organisms constituting one of the formal units in taxonomic classification (phylum, order, etc.) and characterized by common characteristics in varying degrees of distinction

tessera — unit of a mosaic

tool — material object used to make other objects or to facilitate activities such as resource extraction

transfer function — mathematical relationship between the output of a system and its input

trophic level — level within a food chain in which all members are equally far removed from the primary food producers

T-tauri stars — luminous variable stars associated with interstellar clouds and found in very young clusters; they are believed to be still in the process of gravitational contraction from their protostellar phase and have not yet arrived at the Main Sequence and begun to burn hydrogen

ultraviolet light — band of electromagnetic radiation from about 40 to 4000 Å

Universe — everything that came into existence at the moment of the Big Bang, and everything that evolved from that initial mass of energy; or everything we can, in principle, observe

valence state — state of electrical imbalance in an atom or molecule

visual binary — binary star system whose components can be identified with an optical telescope

volatile organic compounds — class of organics that is easily vaporizable at low temperatures and pressures

WARC — World Administrative Radio Conference, usually convened every 20 years (1959, 1979, etc.); sponsored by the International Telecommunications Union (ITU), an organ of the United Nations (UN). These conferences draft treaties allocating radio spectrum space to the various services, such as broadcasting, radar, etc. Individual nations make transmitter assignments within their own territories. Treaty enforcement relies entirely on the individual and mutual advantages of international cooperation. There is only one radio spectrum

wave guide — special transmission medium resembling a pipe and often having a rectangular cross section, inside of which radio waves may be propagated

whistler — electromagnetic ultralow frequency radiation observed in planetary magnetospheres; energized by lightning and other discharges

Wm^{-2} (or W/m^2) — watts per square meter of incident signal flux per whatever resolution bandwidth is in use. It is the total signal flux if the receiving bandwidth equals or is greater than the bandwidth of the signal

$Wm^{-2} Hz^{-1}$ (or $W/m^2 Hz$) — spectral flux density (see jansky)

x radiation — electromagnetic radiation in the range of approximately 0.05–100 Å

Index

9050